An introduction to the properties of condensed matter

An introduction to the properties of condensed matter

D.J. BARBER
B.Sc., Ph.D.
Professor of Physics
University of Essex, Colchester, Essex

and

R. LOUDON
M.A., D.Phil., F.R.S.
Professor of Theoretical Physics
University of Essex, Colchester, Essex

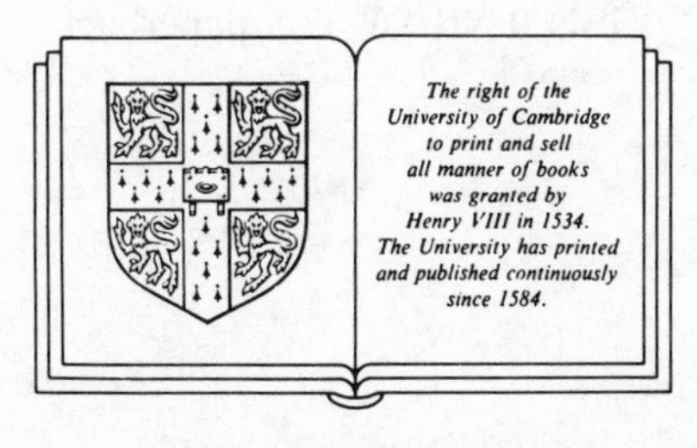

CAMBRIDGE UNIVERSITY PRESS
Cambridge
New York New Rochelle Melbourne Sydney

Published by the Press Syndicate of the University of Cambridge
The Pitt Building, Trumpington Street, Cambridge CB2 1RP
32 East 57th Street, New York NY 10022, USA
10 Stamford Road, Oakleigh, Melbourne 3166, Australia

First published 1989

Printed in Great Britain at the University Press, Cambridge

British Library cataloguing in publication data

Barber, D.J. (David J.)
An introduction to the properties of condensed matter
1. Condensed matter. Physical properties
I. Title II. Loudon, Rodney
530.4

Library of Congress cataloguing in publication data

Barber, D.J. (David J.)
An introduction to the properties of condensed matter/
D.J. Barber and R. Loudon.
p. cm.
Based on lectures given at the University of Essex.
Includes bibliographies and index.
ISBN 0 521 26277 1. ISBN 0 521 26907 5 (pbk.)
1. Condensed matter. I. Loudon, Rodney. II. Title. III. Title:
Properties of condensed matter.
QC173.4.C65B37-1989
530.4′1-dc19 88-25909 CIP

ISBN 0 521 26277 1 hard covers
ISBN 0 521 26907 5 paperback

APL

Contents

Glossary of symbols and list of physical constants

A_0	Activation energy
A	Amplitude, area, anisotropy factor, or a constant
a	External radius, dimension, or a constant
a	Capillary constant
B	Amplitude, or a constant
b	Internal radius, dimension, or a constant
$\mathbf{b}$	Burgers vector
C, C_i	Constants
C_0, C_v	Concentrations
c_{ij}	Stiffness coefficients
D, D', D_0	Diffusion coefficients
D_i	Deflections
d	Diameter, or differential operator
E, E_s	Energy, surface energy
e	Mathematical constant, 2.7183
F, $F(r)$	Force
f	Shape factor, volume fraction, or function
G	Rigidity modulus
g	Acceleration due to gravity
h	Height or vertical dimension
I	Second moment of area
J	Polar second moment of area
K	Bulk modulus, or a constant of proportionality
K_c	Thermal diffusivity
k	Torsional constant, or order parameter
L	Linear dimension (length)
L_0, L_t	Initial length, length at time t
l_r	Aspect ratio

M	Moment
m	Mass
N	Force of reaction, or a normal
n	Exponent or refractive index
$\mathbf{n}$	Director, or arbitrary vector
P, p	Pressures
q, $\mathbf{q}$	Wavevector
R	Radius
r, $\mathbf{r}$	Position
S	Area
s	Length of arc
s_{ij}	Coefficients of elastic compliance
T	Temperature, or a tensile force
t	Time, or thickness
U	Internal energy
u, u_i, $\mathbf{u}$	Position, or an amplitude
V	Volume
v, v_i, $\mathbf{v}$	Velocity
W	Load, or work
w	Load
x, y, z	Coordinates
α	Angle, or exponent
β	Angle, or exponent
Γ	Torque
γ	Surface tension, or exponent
Δ, δ	Increments
∂	Partial differential
ξ	Emissivity, or energy of a pair of atoms
ε	Strain, or energy
$\dot{\varepsilon}$	Strain rate
η	Viscosity coefficient (dynamic)
Θ	Dilatation
θ	Angle, or contact angle
$\dot{\theta}$	Angular velocity
$\ddot{\theta}$	Angular acceleration
κ	Instrumental constant
λ	Wavelength
ν	Poisson's ratio, viscosity coefficient (kinematic) or jump probability
Ξ	Microstructural parameter
π	Mathematical constant, 3.1416

ϱ	Density or polar radius	
Σ	Summation	
σ	Stress	
τ	Relaxation time	
υ	Atomic volume	
ϕ	Angle	
χ	Susceptibility	
Ψ	Angle of twist per unit length	
ψ	Angle	
ω	Angular frequency, or probability	
∇	Vector operator	
G	Gibbs free energy	
$\mathscr{E}$	Film elasticity	
$\mathscr{F}$	Fluidity	
$\mathscr{G}$	Constant	
$\mathscr{I}$	Moment of inertia	
e	Elementary charge	1.602×10^{-19} C
m_e	Mass of electron	9.109×10^{-31} kg
k_B	Boltzmann's constant	1.381×10^{-23} J K^{-1}
N	Avogadro's number	6.022×10^{23} mol^{-1}
R	Universal gas constant	8.314 J mol^{-1} K^{-1}
ε_0	Permittivity constant	8.854×10^{-12} F m^{-1}
h	Planck's constant	6.626×10^{-34} J Hz^{-1}
G	Gravitational constant	6.672×10^{-11} N m^{-2} kg^{-2}
c	Speed of light	2.998×10^{8} m s^{-1}
eV	Electron volt	1.602×10^{-19} J

Preface

The term 'properties of matter' is associated in the minds of many older physicists with rather boring, classical, and sometimes unsatisfactory treatments of the behaviour of liquids and the 'mechanical' properties of solids. For this and other reasons, the subject largely disappeared from undergraduate courses for some years, and the textbooks dealing with most of the topics went out of print. However, there is interesting and valuable physics in much of the subject matter that was covered by the old heading. In addition, it links up with more modern and 'fashionable' topics, which together have considerable importance to many quarters of industry. In our treatment of the properties of condensed matter (i.e. solids and liquids) we attempt to meld some of the old with the new, and we touch upon ways in which the subject matter is relevant to exciting new materials and current techniques. We also aimed to write an introductory, modest-priced textbook with a lower 'undergraduate boredom factor' than would now be accorded the older, more traditional textbooks. In these days of performance indicators, value for money, and accountability in universities, we trust this is a laudable aim and that we have not fallen too far short of our intentions.

The book has its origins in a second year course of 30 lectures given at the University of Essex, but the topics have been expanded somewhat and new material added. The book would never have been written but for the initial impetus from Simon Capelin of Cambridge University Press and his subsequent patience with its seemingly overcommitted authors. Its eventual completion owes much to the forbearance and encouragement of our wives, Jill and Mary, and particularly the former's diligence with the word-processor, which we gratefully acknowledge.

David Barber and Rodney Loudon
February 1988

Preface

1

Condensed matter

The basic constituents of matter, as we normally experience them on Earth, are particles known as atoms. This 'particulate' view of matter, which was first formulated by the Greek philosophers, need not be modified when explaining what we shall call the *macroscopic* properties of matter. This remains true despite recently-acquired knowledge about sub-atomic particles – leptons, hadrons and quarks, etc. If human beings were not inhabitants of the Solar System, but instead were better acquainted with neutron stars, pulsars and the like, we would need to embrace a wider view of the ways in which matter can be condensed than that which follows.

For many purposes we are able to treat atoms as approximately spherical lumps of matter with diameters of the order of 10^{-10} m. Although we know that an atom consists of a minute, dense, positively-charged nucleus surrounded by a cloud of negatively-charged electrons, we can often assume that it is indivisible. Of course, the chemical properties of matter are manifestations of the electronic structures of the various species of constituent atoms and many of the physical properties that concern physicists also reflect the behaviour of the electrons in solids. In this book we concentrate on treatments of the macroscopic properties of solids and liquids that are successful even though they largely ignore the details of atomic and sub-atomic structures. While taking this continuum approach to materials and emphasizing its many successes, it is nevertheless important that we indicate its limitations and how our treatments sometimes connect with theories at the microscopic level. In this chapter we therefore give brief introductions to some of the ideas and theories of the microscopic that are fundamental to understanding real (as opposed to ideal) materials and treating aspects of their properties that are not well explained by the macroscopic approach. We then call upon these ideas in later chapters.

1.1 Atoms and interatomic forces

Before proceeding further, we need to define the term *macroscopic*. This can be done by contrasting it with the term *microscopic*, which can be related to the diameter of an atom. With the advent of electron microscopes with sufficient resolution to 'see' the arrangements of atoms in crystals, as illustrated in fig. 1.1, the terms microscopic and atomic have become overlapping, but microscopic is usually taken to embrace phenomena relating to objects which require more than the unaided eye for study. Correspondingly, the macroscopic phenomena which we shall discuss largely arise from assemblies of atoms that, in one or more dimensions, are more than about 10^6 atoms (or 0.1 mm) in thickness.

The electron micrographs reproduced in fig. 1.1 remind us that many materials are not only tightly-packed collections of atoms but that they are often regimented or *ordered* into rows, planes and closely-packed, three-dimensional arrangements that minimize unoccupied space. The two micrographs effectively show the positions of columns of atoms aligned parallel to the electron beam, but comparing them also demonstrates that the arrangements of atoms in crystals are not necessarily 'frozen', a point which will be discussed in the next section.

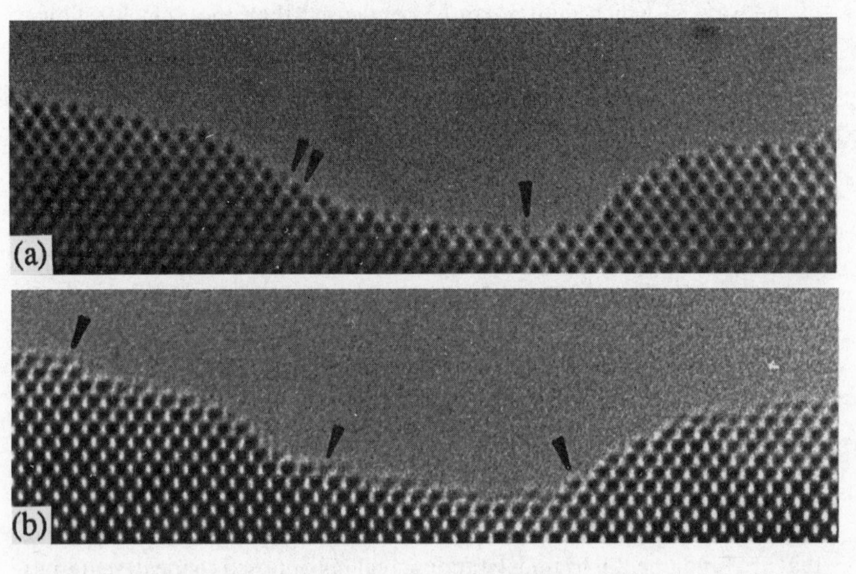

Fig. 1.1 High resolution transmission electron micrographs, recorded sequentially, illustrating the regular arrangement of atoms in a crystal of uranium oxide and the mobility of atoms at the surface of the crystal. The electron micrographs are effectively projections of the crystal structure and show the positions of atomic columns and not individual atoms. (Courtesy of D.J. Smith, Arizona State University.)

We must now consider why atoms *condense* into minimum volumes under some conditions, while expanding to fill every corner of a large vessel under slightly different conditions. These differences in behaviour are a consequence of the nature of the forces that act between atoms, which are of two types – attractive and repulsive. Both types are short range, but the repulsive forces vary more rapidly with interatomic separation and they dominate at very small distances. In one model which gives a satisfactory description of the interactions between certain types of simple atoms, the force of mutual repulsion is approximately proportional to r^{-13} while the force of attraction is proportional to r^{-7}. The dependences of the total force, and its components, on the separation r of two atoms are illustrated in fig. 1.2(a). Whether a material will exist in a condensed state or become widely dispersed (i.e. as a vapour or gas) depends on the potential energy that would be stored by bringing all of the atoms together and allowing their mutual forces of interaction to achieve some average equilibrium interatomic separation. The energy $\mathrm{d}U$ required to displace one atom a distance $\mathrm{d}r$ against the total force $F(r)$ is equal to the work done, so that

$$\mathrm{d}U = -F(r)\,\mathrm{d}r. \tag{1.1}$$

Thus the potential energy stored by bringing a pair of atoms, initially at infinite separation, to a separation r is

$$U = -\int_{\infty}^{r} F(r)\,\mathrm{d}r. \tag{1.2}$$

Figure 1.2(b) shows the dependence of U upon r corresponding to the force law illustrated in fig. 1.2(a). (This is a plot of what is known as the *Lennard-Jones potential*, and it is used to discuss elasticity in §2.4). The minimum in the potential energy curve corresponds to the equilibrium separation of the atoms. Increasing the number of atoms to simulate the situation in a crystalline solid does not qualitatively change the form of the curves in fig. 1.2.

1.2 States of matter

When the internal energy of a substance is large (which occurs when its temperature is high), each of the atoms or molecules possesses a high kinetic energy. The particles travel in straight lines until they make collisions with the walls of the containing vessel or with other atoms or molecules. Only during these brief encounters do the interatomic forces come into play. In this situation we say that the state of matter is *gaseous*. Lowering the temperature of such a gas will reduce the average kinetic

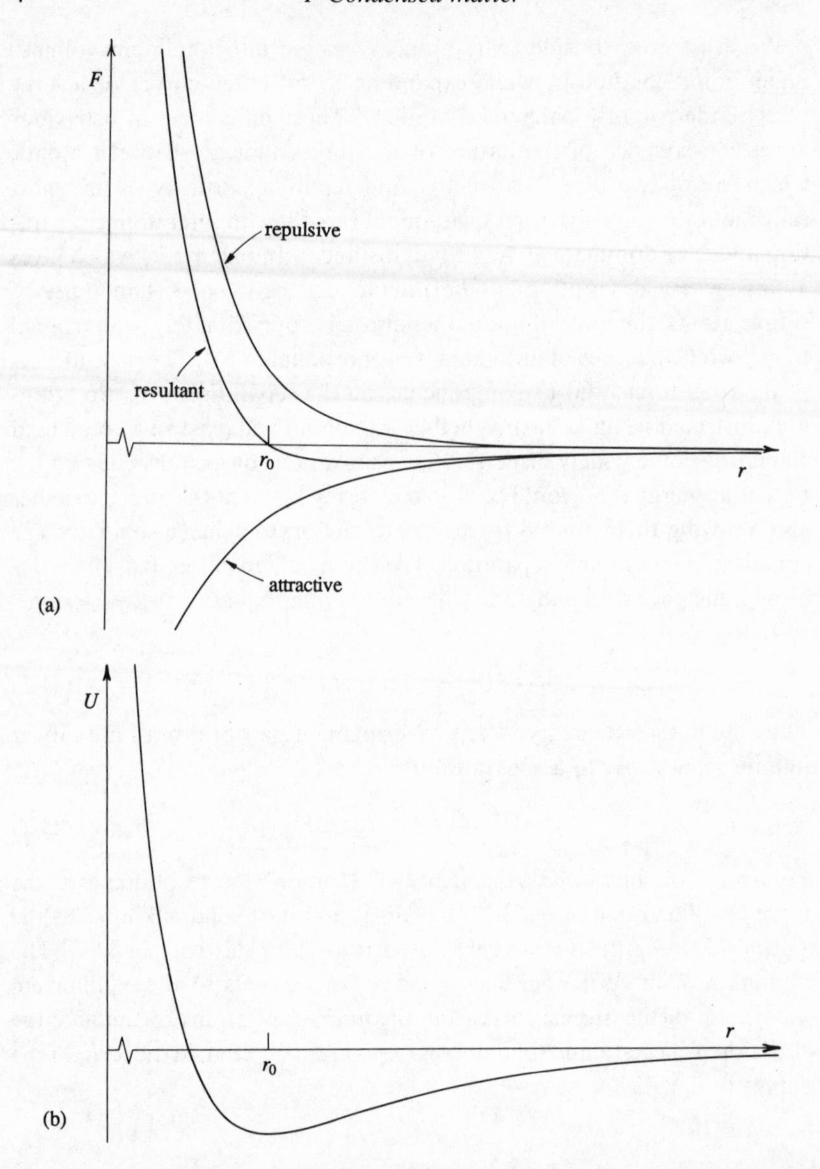

Fig. 1.2 (a) Curves to represent the forms of the attractive, repulsive and resultant forces between two atoms as a function of their separation, according to the Lennard-Jones potential; (b) Plot of the Lennard-Jones potential energy between two atoms as a function of their separation, corresponding to fig. 1.2(a).

energy of the molecules or atoms, thus reducing their velocities of rebound after collisions and thereby increasing the time during which interatomic forces can exert influence. Further lowering of the temperature will eventually cause at least some of the atoms to bind together after collisions.

This condensation process will create initially small volumes of material in which the atoms have separations of the same order as their diameters. The precise nature of their internal arrangements will depend mainly on the properties of the atoms and the interatomic forces. Since the atoms will still have some internal energy they may continue to move about, keeping a roughly constant distance from their nearest neighbours, or perhaps merely vibrating about fixed positions with respect to all the other atoms. The former situation corresponds to the *liquid* state while the latter arrangement is to be found in the *solid* state and essentially describes a frozen liquid.

We can anticipate that both liquids and solids will have characteristic energies, since they have both kinetic and potential energy components that depend upon the nature and the strength of the interatomic forces. Because in the condensed states of matter (liquids and solids) these forces hold the atoms together at a well-defined separation, we commonly call them binding forces or simply *bonds*. Bonds can arise from several mechanisms but they all reflect the electronic structures of the atoms involved. (The reader should consult a textbook on solid state physics for a discussion of bonding.) It is sufficient to point out here that for example, the distributions of electronic charge around many atoms are not spherical and that some atoms form directional, *covalent* (electron-sharing) bonds. When covalently-bonded materials solidify, their atoms tend to take up very specific relative positions to form a regularly-repeating three-dimensional array. This stable, ordered, pattern of atoms or molecules is what we call a *crystal*. Even before the effects of such arrangements were demonstrated by means of X-ray diffraction, the symmetries of many natural mineral crystals and the properties of cleavage shown by those such as halite, NaCl, and Iceland spar (calcite), $CaCO_3$, had hinted at some internal geometrical pattern of atoms.

Many solids consist not of one, but of many, small crystals – *crystallites* or *grains*. We then say that they are *polycrystalline* or, less specifically, crystalline. By means of special laboratory methods we are able to grow large, individual crystals of predetermined composition. These solitary synthetic crystals, which have similarities to the gemstones found in nature, are called *single crystals*. They are very useful for research into the fundamental properties of materials and in recent years they have become important for applications in modern electronic and optical devices.

If a crystalline solid is heated to a temperature close to, but still well below its melting point, the amplitudes of the atomic vibrations will induce some of the bonds to break, thus making it possible for atoms to exchange positions and to wander randomly through the solid. This process of atomic migration is called *diffusion* and is discussed in §6.1. Diffusion occurs more readily at surfaces than in the interior of a crystal, which is largely the reason why it is possible to demonstrate the rearrangement of atoms at the surface of a specimen within an electron microscope as in fig. 1.1. Diffusion has many uses and effects; for example it may play a part in processes whereby materials change shape under the influence of external forces (see §6.9).

At the melting point of a solid, the atomic vibrations cause bonds to break in sufficient numbers and at a sufficiently rapid rate to prevent the material from retaining its solid form, but the attractive interatomic forces indicated in fig. 1.2(a) still dictate the existence of the condensed state that we recognize as a liquid. Not all solids are crystalline. Some naturally solidify with a glass-like structure in which the long-range, ordered arrangement of atoms that we associate with crystals is lacking; others can be induced to retain a similar disordered atomic arrangement by cooling them very rapidly. Such solids are called *amorphous* and many of them are increasingly of commercial importance. The atoms in amorphous materials still have characteristic separations and small clusters of atoms may be positioned relative to each other to give local arrangements which loosely resemble those in crystals. However, such arrangements are not continued over distances of more than a few interatomic separations, so that long-range order is lacking and the structure is more like that of a liquid. Such solids and most liquids thus possess what we call *short-range order*. The existence of amorphous solids serves to emphasize that the major distinction between liquids and solids is not their internal structure but their rigidity. (This is discussed further in §§1.6 and 3.5). The transition to the gaseous state occurs when the atomic motions become so vigorous that atoms acquire enough energy to break the bonds with their neighbours and exceed the range of the attractive forces exerted by atoms in the liquid surface. The details of behaviour during changes of state require thermodynamical and statistical treatments which are not appropriate here (see suggestions for further reading in §1.13).

The fact that the same substance can exist in three states, solid, liquid and gaseous, each of which has a different distribution of atoms, is not the end of the story. Within both the liquid and the solid state further complexity arises from the possibility of more than one atomic arrangement. We refer to these various arrangements of atoms or molecules as *phases*; some are stable under easily obtainable laboratory conditions, others are not. Some can only be produced under extreme conditions of temperature and pressure, while others remain a matter for conjecture – for example, the possibility of a cubic-structured silicate phase deep in the Earth's mantle. However, we can say with certainty that every substance will in principle have three phases, one of which will be solid, one liquid and one gaseous.

1.3 Solids

1.3.1 The nature of solids

The regular arrangement of atoms in a crystalline solid mentioned in § 1.2 can easily be explained as the only stable form which allows *every* atom to experience a null resultant force and therefore to be completely at rest at 0 K. Until now we have only considered the nature of the bonding forces between two atoms. The variety of crystal symmetries that are observed are a result of nature's need to achieve this zero net force condition (which corresponds to a minimum total energy) for the atoms of whole crystal assemblies, notwithstanding the differing types of bonding that exist between atoms. The various atomic structures and crystal symmetries that can occur are most easily described in terms of repeating *unit cells*, each of which contains all the atoms necessary to build up the complete structure, in positions which are defined in terms of a set of unit *translation vectors* for the cell.

The variety and relative strengths of the bonds that can exist between atoms and molecules govern both the breadth of physical properties and the structure of solids. Structure and bonding are, of course, intimately connected but it is possible to create an artificial division and to discuss crystal structures somewhat in isolation, in terms of symmetry. Crystals are also frequently classified in terms of their bonding and many conveniently fit well under the broad headings; *ionic*, *covalent*, *metallic*, *molecular* and *macro-molecular*. The names of these categories are almost self-explanatory. However, for the sake of completeness, we consider each category very briefly in turn, after a short introduction to some very basic crystallography.

Figure 1.3 illustrates the case of a simple solid, caesium chloride, which can be represented in terms of a *space lattice* generated by a set of three orthogonal unit translation vectors of equal length. The chlorine atoms are situated at the *lattice points* (the intersections of the lattice lines) and the caesium atoms are positioned at the centres of every cell. If we call the lattice points A-sites and the centres of each basic lattice cell B-sites, we see that in a perfect crystal of this compound the two species of atoms are placed on the A and B sites in an orderly way. This causes the caesium atoms always to have chlorine atoms as their closest neighbours arranged around them in a particular octahedral geometry, and vice versa.

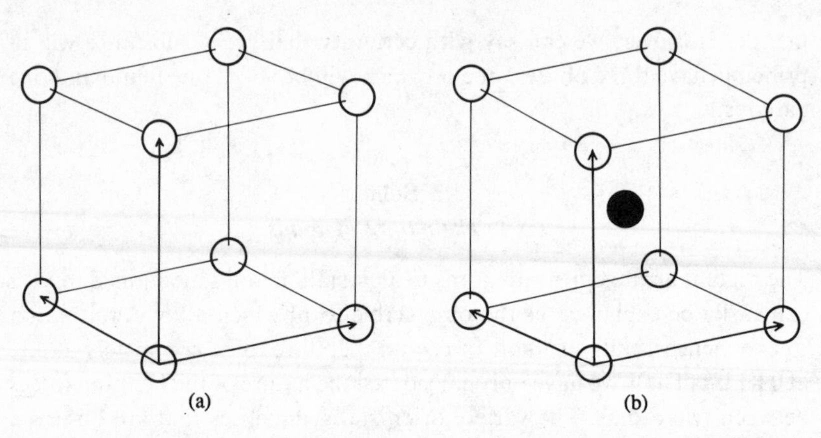

Fig. 1.3 (a) Unit cell of a crystal with the hypothetical simple cubic structure, showing basic translation vectors; (b) unit cell of the caesium chloride structure with its basic translation vectors.

In the caesium chloride structure the A and B sites are actually identical in that they have a similar number and spatial disposition of nearest-neighbour atoms – this can be seen easily by displacing the origin of the lattice to the centre of a cell. The structure of caesium chloride can be generated by placing the atoms on the lattice points of two interpenetrating *simple cubic* lattices. Figure 1.3(a) shows the unit cell of a simple cubic crystal structure that is used to build up the CsCl structure, and the three basic translation vectors that define the lattice which is produced by repeating the cell in three dimensions. Although this simple cubic cell is often used in theoretical models of the solid state, no real material has this structure. Figure 1.3(b) shows the unit cell of caesium chloride. The origin of one of its simple cubic *sublattices* is at the corner of the cell while the origin of the other sublattice is at the atom located at the centre of the cell defined by the first sublattice. With this structure it is not important which unit cell is chosen, i.e. whether it is a Cs atom or a Cl atom at the centre of the cell. The significant point is that adjacent A- and B-sites do not hold atoms of the same species, so that the line joining the centres of two adjacent A and B atoms is not a translation vector of the CsCl structure.

Having introduced the simple cubic structure, we now illustrate two important unit cells that form the structural basis of many real crystals. These are shown in fig. 1.4(a) and 1.4(b) and are known as the *face-centred cubic* (f.c.c) and *body-centred cubic* (b.c.c.) unit cells, respectively. These both give rise to *close-packed structures*, which means that, in reality, the atoms 'touch' their nearest neighbours. In the f.c.c. structure, the distance between nearest neighbours is half the length of a face diagonal, whereas in the b.c.c. structure it is half the body diagonal, these being the directions

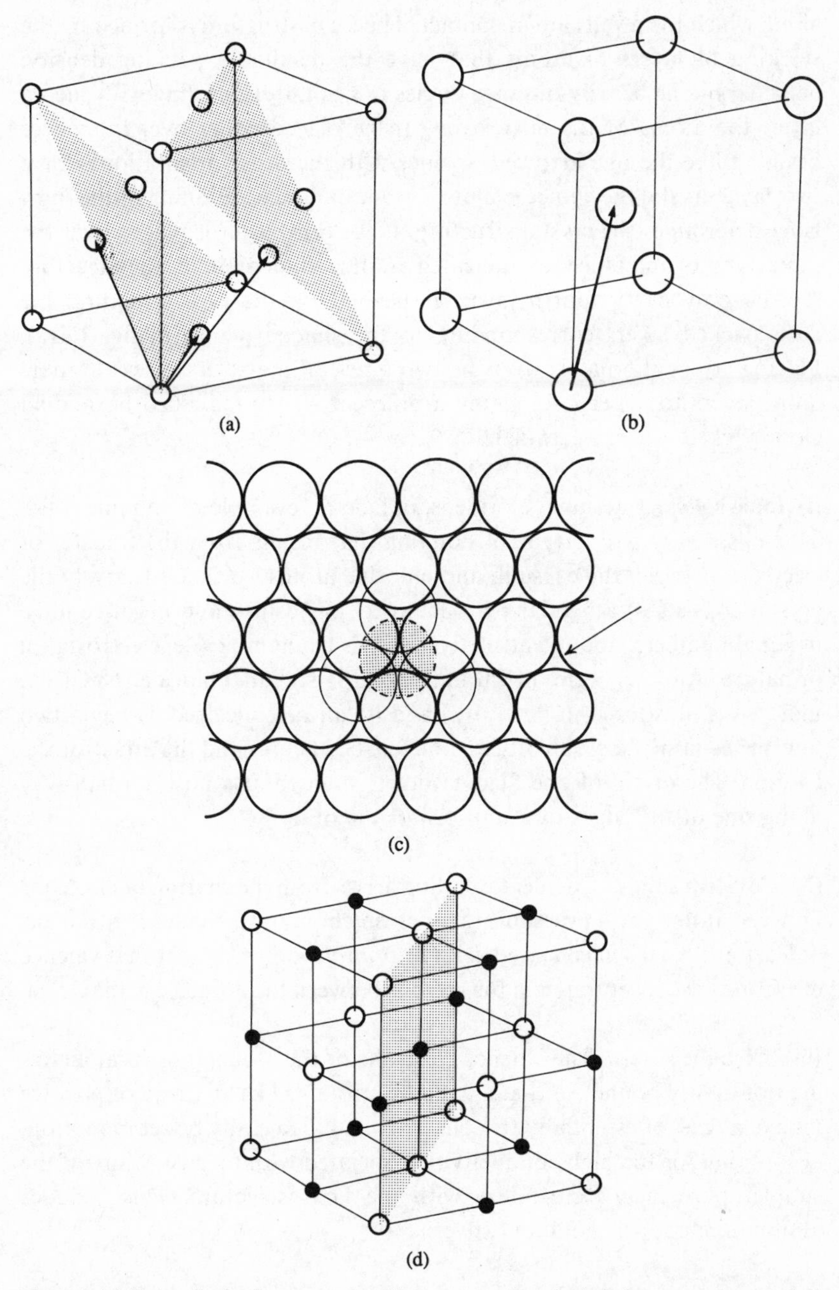

Fig. 1.4 (a) Unit cell of a face-centred cubic structure. The shaded planes are close-packed planes and three of the close-packed directions are indicated by vectors that join nearest neighbour atoms; (b) unit cell of a body-centred cubic structure with vector between two nearest neighbour atoms; (c) illustrating the possible ways of stacking close-packed layers of atoms; (d) unit cell of the sodium chloride structure.

along which the atoms are in contact. The f.c.c. structure is formed by the stacking of layers of atoms that have the maximum packing density, depicted by the heavily-outlined circles in fig. 1.4(c). Starting with such a layer, the atoms of the next layer can be placed either over interstices oriented like the one arrowed or ones with the other orientation. For a two-layer crystal the choice is unimportant but the positioning of the third layer determines the crystal structure. If the third layer is placed over the same type of interstice, as indicated by the shaded circle, the structure repeats only every fourth layer. It is, in fact, the f.c.c structure, the close-packed layers corresponding to the shaded planes in fig. 1.4(a). Placing layers alternately over the two types of interstice causes alternate atom layers to superpose, giving a different structure, called hexagonal close-packed (see fig. 1.15(a)).

(i) Ionic solids. Caesium chloride is, in fact, an example of an ionic solid. This has a very strong type of bonding that results from the transfer of electrons between the caesium and chlorine atoms, so that effectively the crystal consists of a regular periodic array of positive and negative ions, in equal numbers, and the attractive force is Coulombic (i.e. electrostatic) in nature. Another well known ionic solid is sodium chloride, NaCl. Its unit cell is illustrated in fig. 1.4(d) and it can be conceived as being two interpenetrating f.c.c. sublattices, one for the Na ions and the other for the Cl ions. The origin of one f.c.c. lattice is situated at a position half-way along one of the edges of a unit cell of the other.

(ii) Covalent solids. Covalent bonding arises from the sharing of electrons between atoms and a prerequisite for strong bonding is that each atom has at least one partly-filled *energy level* or *orbital*, with the result that valence electrons are concentrated in the regions between the atoms, e.g. diamond.

(iii) Metallic solids. The valence electrons of the atoms that form metals are not tightly bound, so that a metallic crystal is like an array of positive ions in a 'gas' of essentially free electrons that can easily be set in motion, accounting for the high conductivity associated with metals. Many of the simpler pure metals crystallize with the f.c.c. structure. This includes aluminium, copper, gold and silver.

(iv) Molecular solids. In molecular solids the crystal contains identifiable molecular subunits that are generally held in place by weak *van der Waals* (fluctuating dipole) forces of attraction between them. A typical example is provided by crystalline iodine, where the forces binding the atomic components of the I_2 molecules together are strong (the I–I bond is covalent) while the forces between the molecules are weak. Van der Waals forces are also responsible for the atomic binding in the crystalline forms of the inert gases, e.g. argon.

(v) Macromolecular solids. The term, macromolecular solids, refers to substances in which the molecules or atoms are joined into chains or rings, which in turn may be loosely linked or built into more complex large-scale network structures. The atoms forming the chains are usually linked covalently, but the subsidiary links are typically van der Waals bonds. Clearly this category is just an extension of (iv) but since it embraces many important polymers whose physical properties differ greatly from those of simple molecular solids, they merit separate consideration. We consider a particular example, polyethylene, and we enlarge upon the nature of polymers in § 1.3.3.

1.3.2 Order and disorder

As a consequence of the ionic nature of caesium chloride, mistakes in the regular ordering of atoms on the A- and B-sites never happen (the energy of repulsive interaction between similarly-charged ions is too high). But in other compounds, the difference in total energy between an ordered arrangement of atoms and a random, or *disordered* arrangement may be quite small, so that both may exist. Indeed, both arrangements may occur within the same piece of material, because achievement of the fully-ordered condition will be consequent upon the atoms having sufficient time to select the 'right' site (i.e. the lowest energy site) at the growing surface of the crystal or, alternatively, upon diffusion subsequently enabling internal rearrangement after the crystal formed. The equations governing the diffusion of matter are considered in §§ 6.2 and 6.3 and further considerations of order are to be found in § 1.7.

An example of a simple solid that can be either ordered or disordered is an alloy, called β-brass, with a composition of 50 at.% copper and 50 at.% zinc. Figures 1.5(a) and 1.5(b) show its structures in the ordered and disordered states respectively. When disordered there is a 50 per cent chance that any atomic site will be occupied by a copper or zinc atom. However, on ordering, the corner and body-centre sites of the body-centred cubic structure of the disordered alloy are no longer equivalent and the symmetry is reduced or *broken*. As pointed out earlier, the two types of atoms in the resulting ordered, caesium chloride-like structure may be

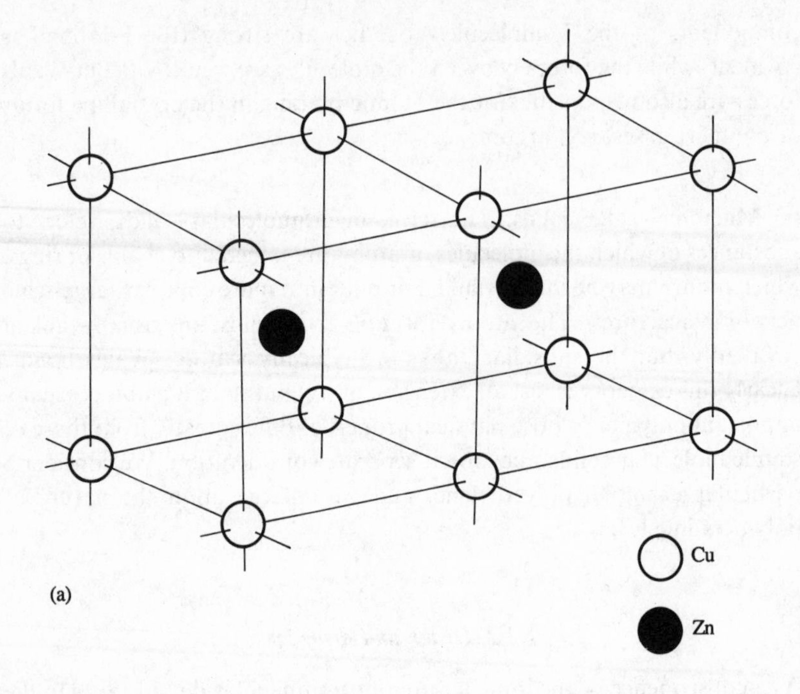

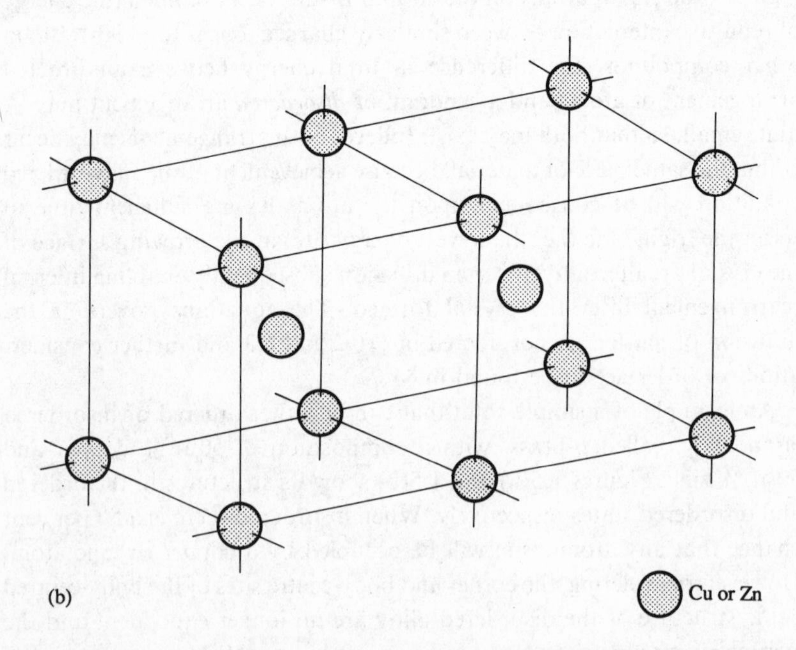

Fig. 1.5 Representations of the crystal structures of an alloy of 50 : 50 at.% CuZn (β-brass) in the states (a) fully ordered, and (b) completely disordered.

pictured as being positioned on two interpenetrating simple cubic sublattices, constituting a *superlattice*. Such ordered materials are therefore often said to have a *superstructure*.

The degree of order of a material is measured in terms of an order parameter k, which is defined so that it vanishes for the completely disordered condition and has finite values for other degrees of order. For β-brass, the long-range order parameter can be defined as

$$k = \frac{|\omega_{Cu} - \omega_{Zn}|}{\omega_{Cu} + \omega_{Zn}}, \quad (1.3)$$

where ω_{Cu} and ω_{Zn} are respectively the probabilities of copper and zinc atoms being at a particular lattice site.

We have stated above that regular, periodic arrangements of atoms and ions are preferred in nature because they have the minimum possible energy and are therefore the stable equilibrium form of the solid phase. It therefore follows that amorphous materials – substances like glasses and plastics, with non-periodic structures, are unstable and should ultimately transform and crystallize. In practice this can only take place if the atoms in their positions of quasi-stable equilibrium receive sufficient opportunities (i.e. thermal activation) occasionally to jump to more stable, lower energy positions, thereby eventually nucleating embryo regions of crystalline material that grow in the solid state. Glassy or *vitreous* materials after many, many years at normal temperatures do devitrify (i.e. start to crystallize), often with catastrophic results. Glass vessels made by peoples of ancient civilizations, retrieved from the Mediterranean Sea, are often devitrified. New glass held just below its softening temperature will start to crystallize immediately. Modern industry makes use of our knowledge about glassy–crystalline transitions, e.g. in the manufacture of rapidly-solidified glassy metal alloys with special magnetic properties and in the controlled devitrification of glasses to produce uniformly fine-grained glass-ceramics. The underlying causes of phase transitions are introduced in § 1.7.

Research into the condensed state continues to produce new insights into the complexities and variety of nature. For example, recent studies have shown that some aluminium–manganese alloy compositions can be crystallized in structures with *icosahedral* (five-fold) symmetries (Schechtman *et al.*, 1984). Since the five-fold symmetry lies outside the conventional formalism of crystallography, such solids are currently referred to as *quasicrystals*.

1.3.3 Polymers

Plastics or, more correctly, polymeric materials are formed from *high polymers*, which are enormous molecules formed as a chain of smaller basic groups of atoms. The latter, known as *monomers*, consist mainly of carbon and hydrogen atoms that are linked covalently, in the manner described in § 1.3.1. The term 'high' is appended to polymers to distinguish the long chain molecules, with molecular weights around 10^5 or 10^7 from polymers consisting of a few monomers. Figure 1.6(a) shows the monomer ethylene which is the basic unit of the polymeric chain of one of the simplest high polymers, polyethylene, a well-known 'plastic' in which a piece of one molecule would look like fig. 1.6(b). The chains may be terminated by different types of chemical groups, but since the chains are very long such terminations do not necessarily have much effect on the properties of the polymer.

Polyethylene is called a *linear polymer* because the monomers are joined end-to-end. Replacement of one of the hydrogen atoms in the ethylene monomer with a different type of atom or an atomic group greatly modifies the properties, creating such well-known plastics as polyvinyl chloride (PVC), represented by fig. 1.6(c), and polypropylene, shown in fig. 1.6(d). The binding between the molecular chains is by means of the very weak van der Waals forces that occur between hydrogen atoms. Thus the simple linear polymers soften rapidly as the temperature is raised and melt at relatively low temperatures. They form the basis of materials known as *thermoplastics* which can be shaped and moulded at between 300 and 400 °C. These polymers will reversibly soften and regain their rigidity on changing temperature between 20 °C and a temperature in the range just quoted.

Other types of polymer have branching chains or take the form of three-dimensional giant networks, involving strong covalent bonding. Such materials have very different properties from linear polymers. Many do not soften appreciably before melting and some are the basis for *thermosetting* plastics. These are formulations which, once polymerized, cannot be softened or worked by raising the temperature without structural or chemical decomposition occurring.

Even for linear polymers we have not yet described how the large molecules relate to one another in the solid phase, apart from the question of their generally-weak interchain bonding, which is called *cross-linking*. By very carefully cooling a solution of a low-to-medium molecular weight linear polymer one can grow relatively perfect lamellar single crystals. In these lamellae the chains are well-aligned with each other and generally fold back and forth within the thickness of the crystal, as shown schematically in fig. 1.7(a). In contrast, polymers cooled from their molten states, especially polymers of high molecular weight, are either completely amorphous (i.e. the chains are random) or at best only contain small

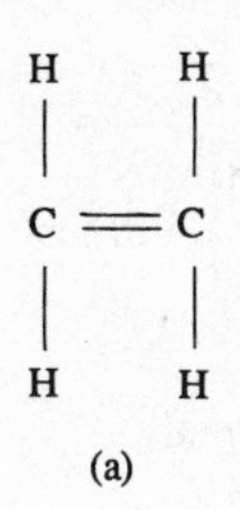

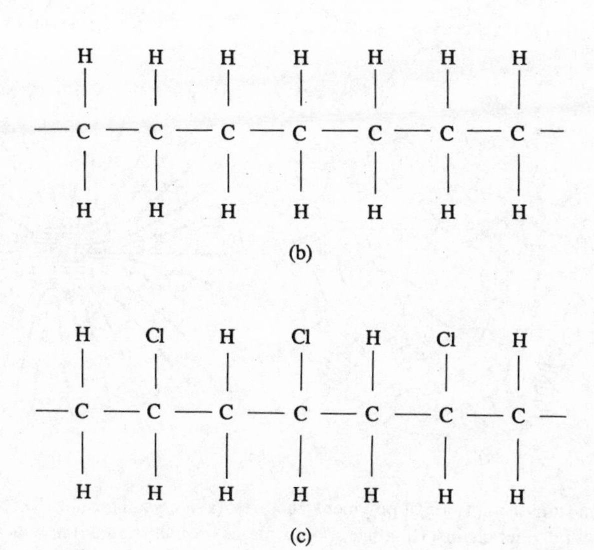

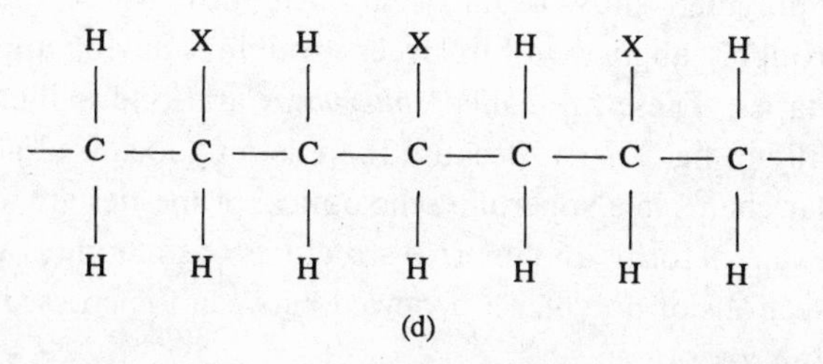

Fig. 1.6 The chemical structures of some simple polymers: (a) the ethylene monomer, (b) part of a polymeric chain of ethylene, (c) the form of the polymeric chain of polyvinyl chloride, (d) the form of the polymeric chain of polypropylene, where $X = CH_3$.

regions where the chains are folded or aligned to give ordered volumes. Figures 1.7(b) and 1.7(c) show schematically fully amorphous and partially crystalline polymers. In bulk crystalline polymers the ordered regions are called crystallites because of their small size, typically 0.01 to 0.1 μm.

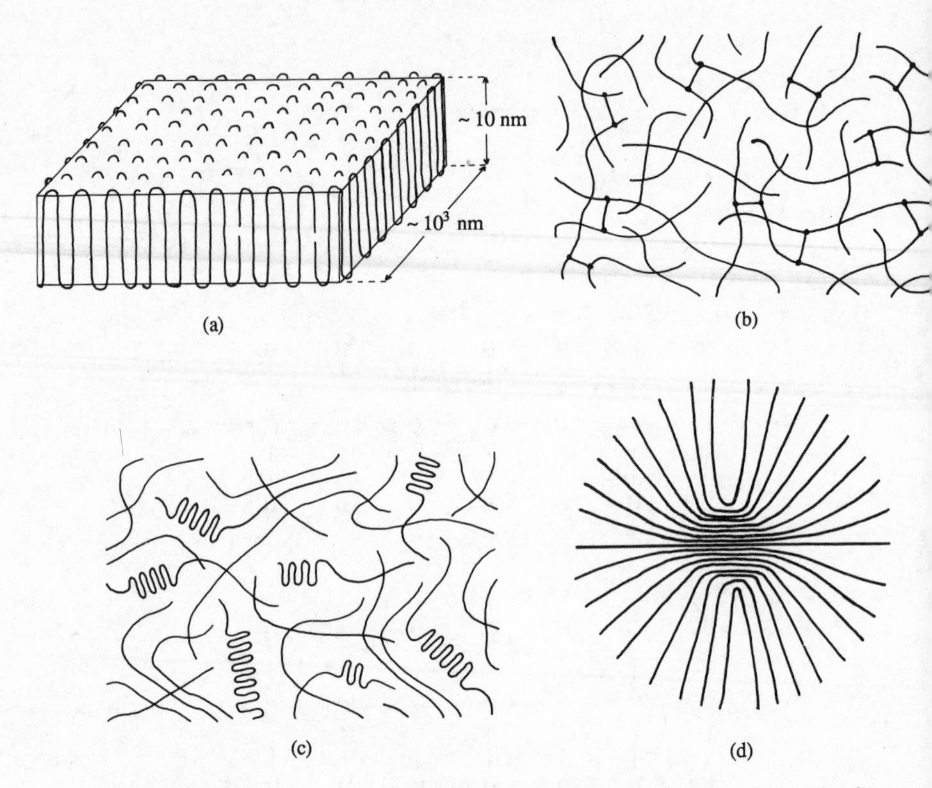

Fig. 1.7 Some important types of polymer structures: (a) a crystalline lamella resulting from the folding of polymer chains, (b) a linear amorphous (non-crystalline) polymer, (c) weakly cross-linked chains in an amorphous polymer, (d) a single spherulite.

Crystalline polymers most often occur in a form where the molecular chains are roughly aligned within larger groupings having approximately spherical shapes. These are called *spherulites* and one is illustrated diagrammatically in fig. 1.7(d). Because there is only localized alignment of the molecular chains in a spherulite, the optical properties are not uniform, so that the external forms and internal structures of spherulites are apparent when thin sections of polymer are viewed between the crossed polars of a polarizing microscope.

1.4 Liquids

When an atomic solid melts, its atoms become positionally disordered but the average distance between the nearest neighbour atoms does not change greatly, even though close packing arrangements are destroyed and the characteristic symmetry of a crystalline structure disappears. Bonding between atoms becomes transitory. The atoms move with mean velocities determined by the temperature and there is generally a finite possibility of near-surface atoms escaping to give a gaseous phase. Each of the main types of solid listed in the previous section produces its own type of liquid, reflecting the fact that bonding between the atoms still occurs, albeit of a more transient nature. It will be obvious from the description of amorphous solids in § 1.2 that the solid–liquid phase transition in such cases is not so dramatic as in the case of the crystals of simple atomic solids. For example, there is no well-defined melting point for the majority of glasses.

As a simple solid approaches its melting point an increasing proportion of the lattice sites are vacated by atoms which receive sufficient thermal excitation to break free of their bonds, a subject discussed more fully in §§ 1.11 and 6.1. Because adjacent atoms can then fill the vacant lattice site, often called a *vacancy* for brevity, the vacant site is able to move through the crystalline material, i.e. to diffuse. When the solid melts, the atoms have only slightly fewer nearest neighbour atoms, on average, than they did when bonded to them in the solid phase. According to one model of simple liquids, described more fully in § 4.3, there are again many 'holes' in the structure, so that the transition at the melting point is not very marked if we only focus on the volume fractions of vacancies and holes. However there is, of course, a more abrupt change from long-range to short-range order and the concepts of lattices or lattice sites are inappropriate for liquids.

The melting of a polymer crystal corresponds to the orientational disordering of the long molecular chains and only subsequently will other changes occur in the nature of the chains and the molecular bonding. Orientational disorder arises when many molecular crystals and crystals containing charged polyatomic groups approach their melting points, again with the result that the precise point at which the solid becomes a liquid may be difficult to define. For completeness, and at the risk of possible confusion, it should also be pointed out that orientational disorder of atomic groups within crystals can commence well below the melting point and it takes the forms of both recognizable solid state phase transformations and *librations* (freely rotating groups, e.g. the nitrate group in $NaNO_3$ above 275 °C).

A liquid does not necessarily have the same structure over the whole range of temperature for which it is stable. Many liquids have *pseudo-* or *quasi-crystalline* properties near to their freezing points and there is a gradual change with increasing temperature, so that the liquids exhibit *quasi-gaseous* behaviour close to their critical temperatures (meaning the temperature at which a liquid becomes gas at standard pressure). This distinction in liquid behaviour is sometimes alluded to by calling a liquid,

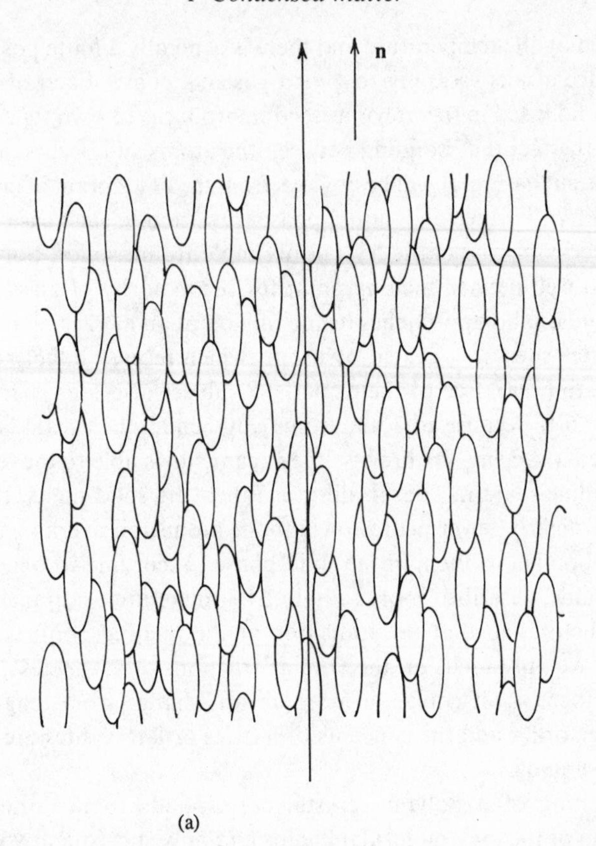

(a)

Fig. 1.8 Schematic representations of the arrangements of the molecules in the main types of liquid crystals: (a) nematic, (b) cholesteric or chiral–nematic, in which the director traces out a helical path, (c) smectic A, and (d) smectic C.

above but close to its freezing point, a *melt*. The quasi-crystals with five-fold symmetries mentioned in § 1.3.2 are, significantly, formed by quenching melts (i.e. cooling them very rapidly).

1.5 Liquid crystals

Nowadays everyone's frequent encounters with 'liquid crystal displays' (LCDs) require that an explanation be given of how liquid crystals relate to the foregoing introductory descriptions of solids, crystals and liquids. They occur within a range of organic substances that, when heated, do not melt abruptly to form liquids. Instead they make transitions through a series of new phases having structures with a degree of symmetry, so that they can be pictured as intermediate between crystalline solids and simple atomic liquids, hence the name *liquid crystals*. They form from organic

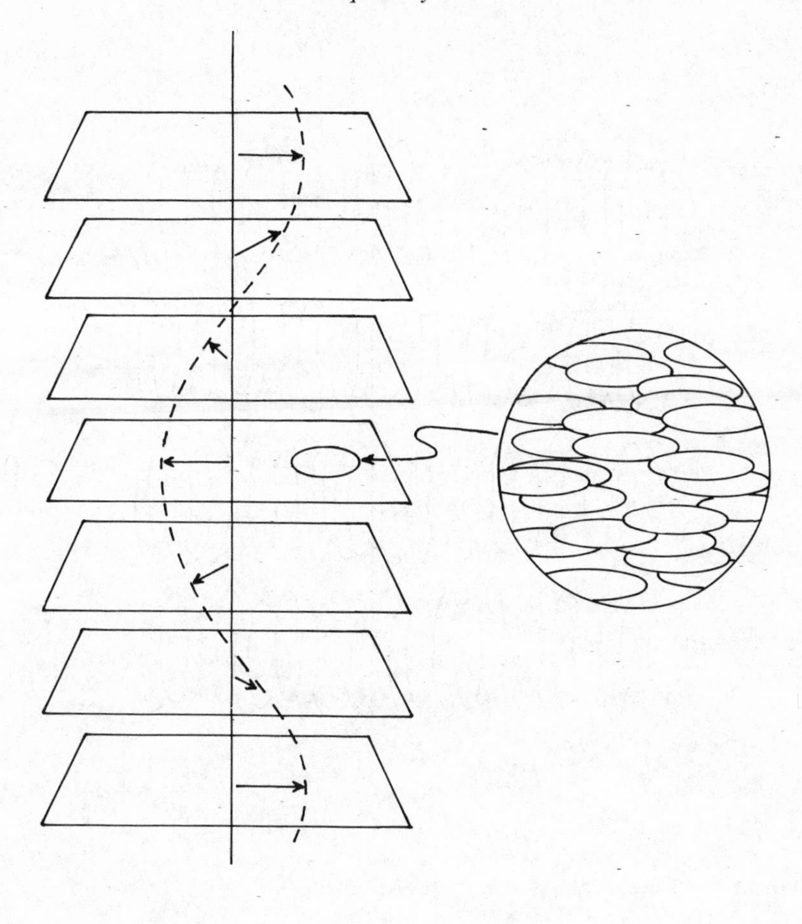

Fig. 1.8 Continued.

compounds whose molecules are very non-spherical, indeed the molecules are commonly rod-shaped.

Three basic types of liquid crystal, or *mesomorphic phase* are known – they are the *nematic*, *cholesteric* (or *chiral-nematic*) and *smectic* types. (Mesomorphic means having an intermediate form, and phases with such forms are often simply called *mesophases*). Figure 1.8 illustrates the main characteristics of the three types. The longest axes of the molecules in a nematic liquid crystal, fig. 1.8(a), are roughly aligned with one another (there being some statistical scatter) within small, local volumes of the mesophase, but positional order is lacking. Changes in the orientation of the local molecular alignment from place to place within nematic liquid

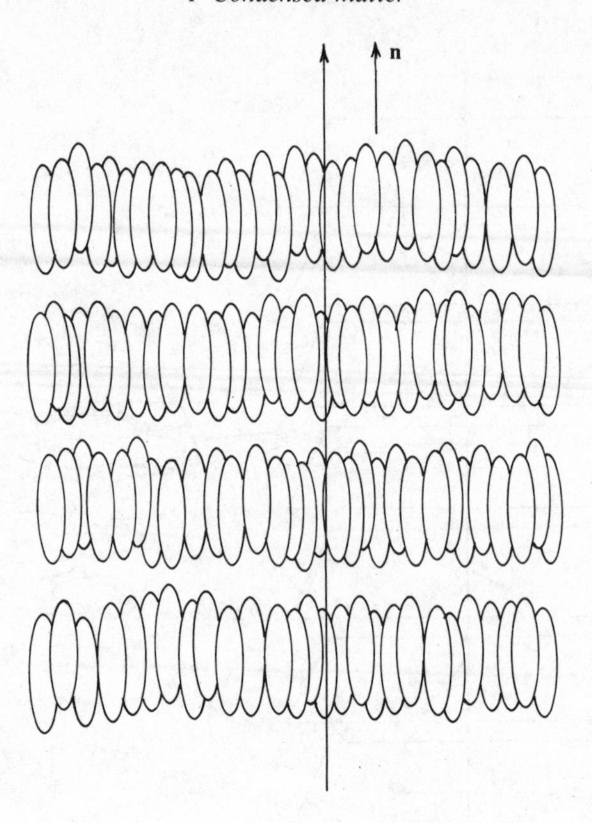

(c)

Fig. 1.8 Continued.

crystals normally give them a turbid appearance, but the arbitrary orientations can be influenced by external forces, such as the effect of the container walls or an electric field.

The cholesteric, or chiral–nematic, mesophase is essentially a special case of the nematic liquid crystal that occurs with optically-active substances. If the crystal is constrained between two parallel plates, the molecules become oriented in sheets, as sketched in fig. 1.8(b). Within each sheet the long axes of the molecules are roughly aligned with a direction lying in the sheet, which we can denote by a vector **n**, known as the *director*. Unlike the general nematic case, the direction of **n** varies from sheet to sheet in a helical fashion around the z-axis (the normal to the plates) – this is the origin of the word 'chiral', which means twisted. The pitch of the helix is usually very sensitive to temperature. Because light of a wavelength equal to the pitch is selectively reflected by the molecular sheets, such cholesteric liquid crystals now form the basis of coloured-

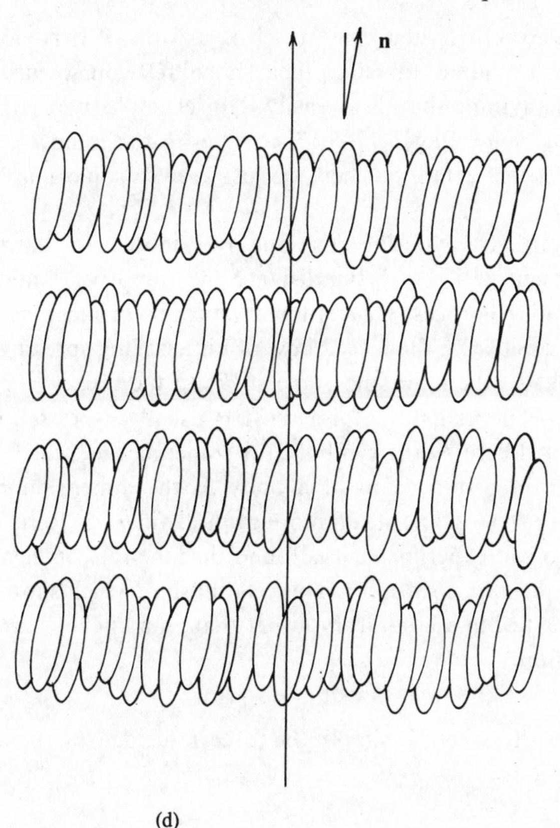

Fig. 1.8 Continued.

display temperature-indicating instruments. In the smectic phases there is, in addition to the orientational order of the nematic phase, an ordering of the molecules into layers, within which there is some degree of freedom for motion and the possibility of both tilted and orthogonal molecular arrangements. These factors explain the occurrence of three smectic subclasses, two of which are illustrated in figs. 1.8(c) and 1.8(d). The transitions of a substance through a series of mesophases mentioned at the start of this section may, and often do involve more than one of the smectic phases.

1.6 The distinction between solids and liquids

It should be clear by now that there can be considerable difficulties in distinguishing between some solids and some liquids on structural grounds alone. It is therefore often more convenient and simpler to make the distinction in terms of the response of substances to the application of a small stress, whether they *flow*, which is a form of behaviour that we associate with a liquid, or retain their shape. (Of course, gases also flow and we call anything that flows easily – for example, under the influence of gravity, a *fluid*). In §§ 5.12–5.17 we investigate the flow properties of materials in some detail, but here we just keep to some simple facts and observations.

When layers of a material move one over another so that the amount of displacement varies with the distance measured perpendicular to the layers, we call the process *shear*. Flow is a process of shear, as can be seen readily by considering the forced flow of butter being spread with a knife – some sticks to the knife and some sticks to the bread, in between the butter flows. The magnitude of a shear is usually expressed in terms of *strain*, denoted here by the symbol ε (or sometimes θ), which is the ratio of a change in a dimension of a body to the dimension before flow occurred. We shall define strain more carefully in § 1.8. Returning to our butter and bread experiment, it is found that the rate of shear $\mathrm{d}\varepsilon/\mathrm{d}t$ is a function of the applied *stress* σ. This relationship, which should probably be put to the test in a laboratory rather than in a kitchen, can be written as an equation:

$$\mathrm{d}\varepsilon/\mathrm{d}t = f(\sigma). \tag{1.4}$$

The applied stress here is simply the force F divided by the area A over which it acts,

$$\sigma = F/A. \tag{1.5}$$

We can assume that $\mathrm{d}\varepsilon/\mathrm{d}t = 0$ when $\sigma = 0$ and for small stresses we can then expand $f(\sigma)$ as a power series, i.e.

$$f(\sigma) = \mathscr{F}\sigma + \mathscr{G}\sigma^2 + \text{higher terms}. \tag{1.6}$$

We thus obtain in the limit of very small stresses:

$$\mathrm{d}\varepsilon/\mathrm{d}t = \dot{\varepsilon} = \mathscr{F}\sigma. \tag{1.7}$$

This is one expression of Newton's law of viscous flow, where $\mathscr{F}$ is stress independent. If higher-order terms in eqn (1.6) are significant the flow law is non-linear. $\mathscr{F}$ is the *fluidity*, which is the reciprocal of the *viscosity*, η (see eqn (5.121)). Rewriting eqn (1.7) in terms of η we obtain

$$\sigma = \eta\dot{\varepsilon}. \tag{1.8}$$

Viscosity is measured in Pa s or *poise* (10 poise $\doteq$ 1 Pa s) and its values for different materials cover many orders of magnitude. Water at room temperature has a viscosity of about 10^{-2} poise whereas the viscosity of the rocks in the Earth's mantle is about 10^{22} poise. Simply in terms of viscosity, the distinction between liquids and solids is taken somewhat arbitrarily at a (low stress) viscosity of 10^{15} poise. An uncontained liquid will eventually flow under gravity to an equilibrium state in which it has an essentially flat surface, except for the effects of surface tension (Chapter 4) at the periphery of the liquid. This is not quite the whole picture, however, because the time over which the applied stress is allowed to act on the substance before its possible effects are assessed is also very important. The microscopic flow processes in many materials may have a range of *relaxation times*, but these are usually dominated by one characteristic relaxation time which we can denote by τ. If we observe a given material over a time longer than τ it may appear to flow like a viscous fluid, while it will behave like a rigid solid on shorter time scales. The significance of this behaviour led Reiner (1969) to introduce the Deborah number†, D, the ratio of the time of relaxation to the time of observation t_0:

$$D = \tau/t_0. \tag{1.9}$$

A material behaves as a fluid for very small D and as a solid for very large D!

These matters are further discussed in §§ 1.9–1.12, where we indicate the microscopic mechanisms whereby strong solids, such as rocks, can flow and become folded into mountain belts over geological timescales (when D is very small) and yet can behave in a brittle manner when, for example, struck with a hammer (i.e. when D is large). The branch of physics which deals with the flow and deformation of matter is known as *rheology* (Greek – *rheos*, a stream). It provides a phenomenological description of the mechanical properties of materials; it is indeed a field close to classical mechanics. Rheology embraces the viscous, elastic and plastic behaviour of matter.

1.7 Structural phase transitions and the melting of solids

We have so far postulated that solid phases have uniform structures and compositions. Less has been said about liquids, but in Chapter 4 it is shown that they are macroscopically homogeneous and uniform, and have sharp interfaces with their vapours. We shall therefore assume here that the interfaces between the three phases of matter for a particular substance are sharp, and that at these boundaries discontinuous changes in structure and/or composition occur. A *phase transition* occurs when a particular phase becomes unstable under a given set of thermodynamic conditions. Classical thermodynamics offers a general and sound framework for

† So named after the prophetess Deborah, who prophesied that the mountains would flow before the Lord (Judges 5.5).

understanding phase transitions in condensed matter. In order to explain the relationships between different states of the same substance (and also the possibility of different structures within the liquid and the solid phases), we must show how phases can be characterized thermodynamically.

In principle, various thermodynamic quantities can be used, but current practice is to base discussion upon free energy, in particular the *Gibbs free energy*, G. The free energy can be determined for any substance in equilibrium, whatever state or phase of matter it occupies. It is a function of both pressure, P and temperature, T, and we can relate it to a particular phase state by subscript, e.g., for a solid

$$G_S = f_S(P, T). \tag{1.10}$$

It can be shown that, when two states of matter are in thermodynamic equilibrium, the free energies, per unit mass of each state, are equal. The free energy of the system is a continuous function of P and T during phase transitions, but other thermodynamic quantities such as internal energy, U, entropy, S, volume, V and heat capacity, C, undergo discontinuous changes.

The Gibbs free energy is given by

$$G = U + PV - TS \tag{1.11}$$

so that

$$\mathrm{d}G = \mathrm{d}U + P\mathrm{d}V + V\mathrm{d}P - T\mathrm{d}S - S\mathrm{d}T, \tag{1.12}$$

but

$$\mathrm{d}U = T\mathrm{d}S - P\mathrm{d}V, \tag{1.13}$$

hence

$$\mathrm{d}G = V\mathrm{d}P - S\mathrm{d}T. \tag{1.14}$$

The first and second derivatives of the free energy are then expressible as

$$(\partial G/\partial P)_T = V, \quad (\partial G/\partial T)_P = -S, \tag{1.15}$$

$$(\partial^2 G/\partial P^2)_T = (\partial V/\partial P)_T = -V\beta, \tag{1.16}$$

$$[\partial/\partial T(\partial G/\partial P)_T]_P = (\partial V/\partial T)_P = V\alpha, \tag{1.17}$$

and

$$(\partial^2 G/\partial T^2)_P = -(\partial S/\partial T)_P = -C_P/T, \tag{1.18}$$

where α, β and C_P are respectively the volume thermal expansivity, the compressibility and the heat capacity at constant pressure.

From these equations we see that a phase transition in which there is a discontinuous change in volume and entropy must also exhibit a discontinuous change in the first derivative of the Gibbs free energy. A change in entropy implies latent heat of transformation. Such phase changes are called *first order* transitions, the terminology indicating that the first derivatives $\partial G/\partial T$ and $\partial G/\partial P$ of the Gibbs free energy are discontinuous. *Second order* transitions correspond to discontinuous changes in thermal expansivity, compressibility and heat capacity, i.e. the second derivatives of the Gibbs free energy are discontinuous. We shall not concern ourselves with higher order transitions.

Many substances undergo one or more *structural phase transformations* when in the solid state. In these transformations, the internal arrangement of atoms alters, so that there is a change in *crystal symmetry* – the mathematical measure of the arrangement. Such transitions are often accompanied by a change in the density, elastic properties, etc. The reduction or *breaking* of the symmetry operations that characterize a phase of higher symmetry is the essential feature of a phase transition to a structure of lower symmetry. The breaking of symmetry occurs because atoms are displaced from the positions that they occupy in the higher symmetry phase, so that the lower symmetry phase may be envisaged as a distorted version of the other. It is usually found that in the distorted phase the displacements are identical (both in magnitude and direction) for given atoms in a number of adjacent unit cells. In other words, there occur domains within which the common displacements create long-range order of the atoms. This type of *displacive* transformation can be described in terms of an order parameter. Order–disorder transitions, which involve the indiscriminate mixing of two or more types of atom between lattice sites, mentioned in § 1.3.2, may also be specified in terms of an order parameter, as defined in eqn (1.3). The order parameter for a structural phase transition measures the size of the displacements of the atoms from their sites in the higher symmetry phase. If, as the temperature is raised, the order parameter goes discontinuously from a finite value to zero, then the phase transition is first order. If the degree of order changes continuously, then the transition is one of second order.

Widening the discussion to transitions between solids and liquids, the *Gibbs phase rule* of thermodynamics tells us that a simple one-component substance can exist simultaneously in the solid, liquid and vapour phases at only one combination of pressure and temperature, the *triple point*. Many solids have low vapour pressures at their triple points, which makes the triple point temperature almost identical with the melting point at atmospheric pressure. For example, for water the temperature and pressure

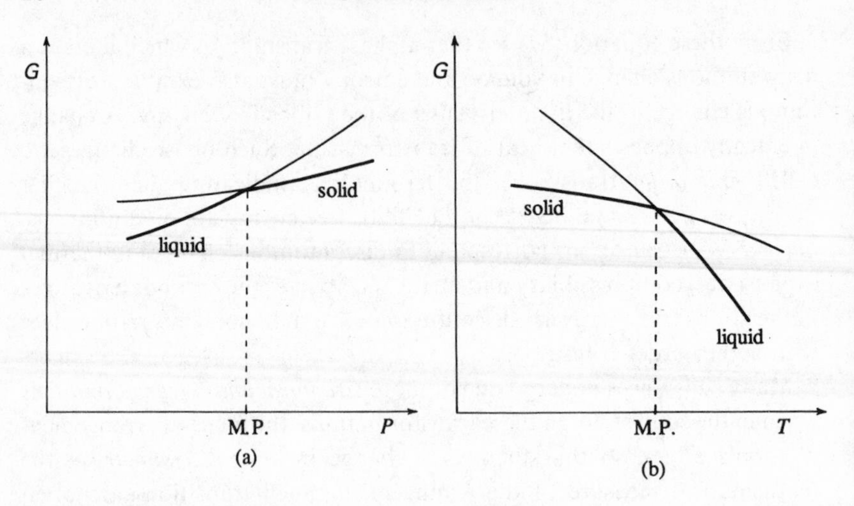

Fig. 1.9 Variation of the Gibbs free energy of a simple atomic substance near its melting point as a function of (a) pressure, and (b) temperature.

at the triple point are 0.0098 °C and 4.579 torr, respectively. When considering melting, the thermodynamic quantities that are of practical importance are the change in volume and the change in entropy. In plots of the dependence of the free energy on pressure and temperature, which typically look like figs. 1.9(a) and 1.9(b), the intersections of the tangents to the curves at the melting point, from eqns (1.15), give

$$(\partial G_L/\partial P)_T - (\partial G_S/\partial P)_T = \Delta V_{M.P.}, \quad (1.19)$$

$$(\partial G_L/\partial T)_P - (\partial G_S/\partial T)_P = -\Delta S_{M.P.}. \quad (1.20)$$

These equations imply that there are abrupt changes in the gradients of the tangents to the free energy surfaces for a substance at its melting point and at the pressure where it condenses to the solid phase. Moreover, theory says that the G, P and G, T surfaces are independent of each other. The latter, in turn, implies that by suppressing the transition and preventing a new phase from appearing, the free energy of the original phase can be studied experimentally for values of P and T for which it is not stable. It is relatively easy to do this by the *supercooling* of a melt, but much more difficult to *superheat* a solid (i.e. retain its solid form above the melting point).

In practice, the entropy of fusion, $\Delta S_{M.P.}$ can often be calculated without undue difficulty from the absorption of heat, determined calorimetrically, when a substance changes from solid to liquid. The volume change is less easy to measure and so often it is derived from measured values of the entropy change at the melting point, using the *Clausius–Clapeyron equation*:

$$\Delta S_{\mathrm{M.P.}}/\Delta V_{\mathrm{M.P.}} = \mathrm{d}P/\mathrm{d}T. \quad (1.21)$$

The result, commonly expressed as a fractional change with respect to the solid volume, $\Delta V_{\mathrm{M.P.}}/V_{\mathrm{S}}$, is potentially a clue to how the structure of a solid becomes modified as the transition to the melt occurs, and therefore of value to the understanding of the liquid state.

Almost all substances expand on melting, ice being one of the very few exceptions. The behaviour of ice is generally attributed to its possessing a very 'open' network structure, which collapses at the melting point, allowing the atoms to adopt smaller average separations. Melting is a complex phenomenon that is still not well understood and is a subject of active research.

The initiation of melting in simple crystalline materials is thought to start at places where the solid is already partially-disordered, i.e. at sites which can be envisaged as concentrations of crystal defects, such as grain boundaries, discussed in § 1.11. In more complex solids, a variety of other factors come into play. For example, in high polymer solids the weakest links may be literally the cross-links between chains, so that the solid may change its rigidity as these links break, without completely melting in the conventional sense.

1.8 The elasticity of homogeneous bodies

We have concluded that a liquid differs from a solid in having no rigidity. Viscosity, a topic covered in §§ 5.12 and 5.13, acts like a kind of rigidity but, strictly speaking, rigidity is a solid-state elastic property, as we shall see. The elastic behaviour of solids and liquids determines their response to the application of force or impulse. Before we can consider elasticity and rigidity we must first examine the kinds of forces (stresses) that can be applied to a piece of material and the kinds of deformation (strains) that result. We must therefore define stress and strain more carefully than hitherto.

1.8.1 Stress

Although the physical agent for the deformation of a body of material may be a force, quantitative treatments of deformation mostly work in terms of stress, given by eqn (1.5), $\sigma = F/A$. Not only does this procedure take the surface area of the body into account, but it also more easily represents the local conditions pertaining to deformation existing at various points within the body.

In mechanics two definitions of stress are commonly encountered, and it is probably helpful to consider these in relation to the simple case of the deformation of a cylinder or wire subjected to an axial stress, fig. 1.10. One definition, that of what we may call conventional stress, gives stress σ as

$$\sigma = F/A_0, \tag{1.22}$$

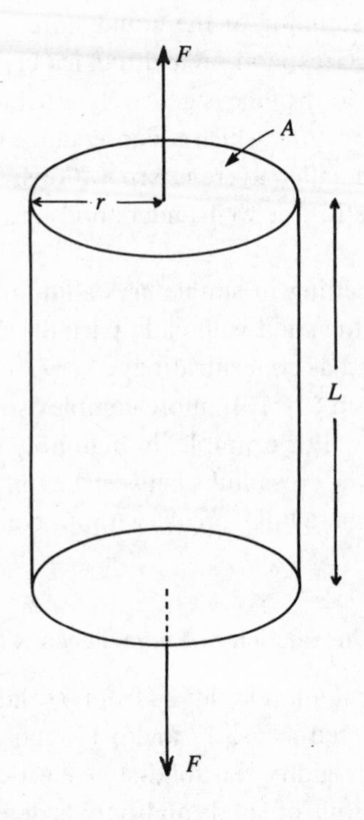

Fig. 1.10 A cylindrical rod or section of a wire subjected to axial tensile forces.

where A_0 is the initial cross-sectional area of the cylinder. A better definition is one which recognizes that the cross-section of the body changes during deformation. This defines the *true stress* as

$$\sigma_t = F/A \tag{1.23}$$

so that the true stress at any time is the force applied at that time, divided by the area A over which it acts at that time. The SI unit of stress is the Pascal ($1\,\mathrm{N\,m^{-2}}$).

If a force is normal to the plane over which it acts, we say that it produces a *normal stress*. A *shear stress* is produced if the force acts parallel to the plane. Figures 1.11(a) and (b) show normal and shear stresses applied to the face of a cube while fig. 1.11(c) indicates that any

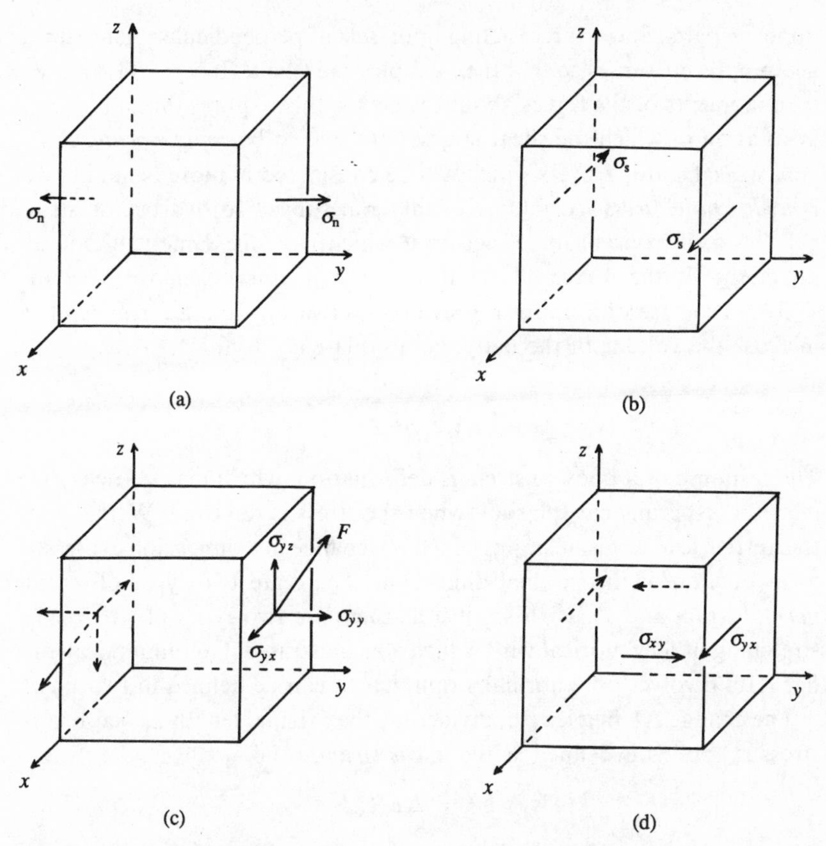

Fig. 1.11 A cube of solid subjected to various systems of stresses: (a) balanced normal stresses applied to the faces perpendicular to the *y*-axis ('*y*-faces'), (b) balanced shear stresses applied to the *y*-faces, (c) balanced arbitrary stresses acting on the *y*-faces, and the components into which they may be resolved, written in terms of the double suffix notation, and (d) complementary shear forces applied to the *x*- and *y*-faces.

force F may be resolved into normal and shear components to give the corresponding normal and shear stresses. One frequently-used convention is to adopt different symbols σ_n and σ_s for normal and shear stresses, respectively. (Often the Greek symbol τ is used for shear stress but we shall continue to use τ to denote characteristic times.) By means of fig. 1.11(c) we also introduce another notation, that of the double suffix. It can be seen that the first suffix indicates the plane on which the stress acts (strictly-speaking the direction of the normal to the plane) while the second defines the direction in which it acts. Thus a repeated suffix denotes a normal stress and a mixed suffix denotes a shear stress. Since we are concerned with the deformation of solid bodies and not with their integral displacements or rotations, forces must be applied to parallel planes as equal and opposite pairs. Shear forces acting upon sets of perpendicular planes must act in opposite senses so that their couples cancel out. In fig. 1.11(d), if we take moments of the forces about the z-axis, this requires that $\sigma_{xy} = \sigma_{yx}$, a situation in which the shear stresses are said to be *complementary*.

A special state of stress which will be considered in more detail in § 2.1 is *hydrostatic stress*. A volume of material subject to hydrostatic stress behaves as if external forces act on it which have the same values in all directions. If the forces all act towards each other, then we have the well-known case of a uniform pressure. Outwardly-acting forces tend to increase the volume of the body, i.e. to dilate it.

1.8.2 Strain

The response of a body to stress is deformation, which may be reversible (elastic) or permanent (plastic), when the stress is removed. We measure this response in terms of strain, which is a change in a dimension expressed as a function of the original dimension. There are two types of strain, *normal strain* and *shear strain*, just as there are two types of stress. The stretching of a cylindrical wire, which was illustrated diagrammatically in fig. 1.10, involves only normal strain, which can be defined in two ways:

The change ΔL in a length, divided by the original length L_0 before any stress is applied is defined as the *linear strain*, ε, i.e.

$$\varepsilon = \Delta L / L_0. \tag{1.24}$$

Strain is clearly a dimensionless quantity. Materials scientists are sometimes more concerned with the instantaneous value of the strain referred to the actual length L_t of the specimen at time t. This can be obtained by integrating the infinitesimal strain defined as

$$\mathrm{d}\varepsilon = \mathrm{d}L/L \tag{1.25}$$

to obtain the time-dependent *logarithmic strain* or *true strain* as

$$\varepsilon_t = \int_{L_0}^{L_t} \mathrm{d}L/L. \tag{1.26}$$

Hence

$$\varepsilon_t = \ln (L_t/L_0) = \ln [1 + (\Delta L_t/L_0)], \tag{1.27}$$

where ΔL_t is the change in length at time t. The difference between the linear strain ε and the logarithmic strain ε_t is negligible for strains up to a few per cent (which is always the case for elastic deformations). For larger strains (corresponding to significant plastic deformation) however, the magnitude of the linear strain is larger than that of the logarithmic strain in the case of tension (for the same extension) and *vice versa* for compression.

The fractional change in the volume of a body that is elastically deformed in tension or compression by forces applied along a single axis is generally smaller than the linear strain parallel to the axis. This implies that the cross-section of a wire elongated in tension, or a cylindrical specimen shortened by compression must change unless specially constrained to prevent changes in lateral dimensions (which would clearly necessitate additional forces). Provided that the material is elastically isotropic and homogeneous the relative changes in dimensions are independent of the direction of measurement in a plane transverse to the stress axis. The *transverse strain* can then be defined by analogy with the linear strain, using the specimen shown in fig. 1.10 for example, as

$$\varepsilon_{tr} = (r - r_0)/r_0. \qquad (1.28)$$

The ratio ν of the transverse strain ε_{tr}, in a direction perpendicular to the applied stress, to the normal strain ε

$$\nu = \varepsilon_{tr}/\varepsilon \qquad (1.29)$$

is called *Poisson's ratio*. It is found experimentally to be a constant for a given material.

The definition of shear strain can be illustrated by means of the simple shear deformation of a rectangular block into a parallelepiped, as shown in fig. 1.12. Balanced shearing forces are applied to one pair of faces so that the faces perpendicular to the x-axis become parallelograms and those initially perpendicular to the y-axis become inclined and elongated. Consider two initially-orthogonal lines inscribed on one of the faces that is perpendicular to the x-axis. On shearing the body, the lines are no longer orthogonal. The traditional definition of shear strain, commonly used in engineering applications, is that it is equal to the tangent of the change in the angle between the two lines. For very small strains (which do not change the volume of the body) the shear strain is

$$\varepsilon_s = \theta = d/h. \qquad (1.30)$$

In the previous section, the condition of hydrostatic stress was introduced. When applied to an isotropic solid, the volume of the solid will change, but not its shape. If the volume changes from an initial value V_0 to V, then the *dilatation* or *volume strain* is defined as

$$\Theta = \Delta V/V_0 = (V - V_0)/V_0. \qquad (1.31)$$

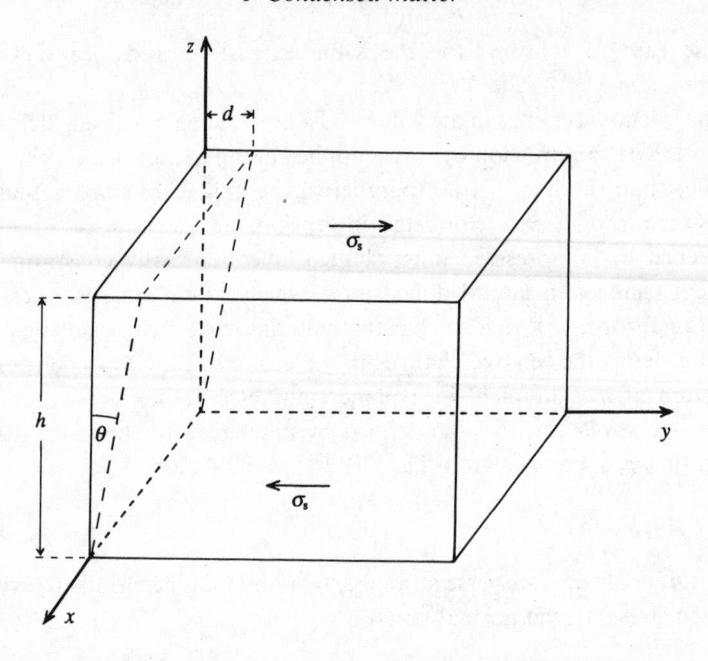

Fig. 1.12 One definition of shear strain ('engineering' shear strain) illustrated by the elastic deformation of a rectangular parallelepiped.

1.9 Elastic deformation

Elastic behaviour is unique amongst the rheological properties because it is the only one that is thermodynamically reversible. A *perfectly elastic* body to which stresses are applied acquires strains instantaneously. When the stresses are removed, the strains vanish instantaneously. Up to a small value of strain, which is characteristic for a given material, the application of a force to a body results in a strain that is proportional to the stress. This is a statement of Hooke's law (1676) which was originally published as an anagram of its Latin form 'Ut tensio sic vis' – 'As the tension, so the force'. (This concise Latin form gives us the shortest law in physics, but it may also sound confusing because the Romans apparently muddled the concepts of tension and extension!) For an interesting account of the early work of Hooke and others, see Timoshenko (1953).

Expressing Hooke's law as an equation we obtain

$$\varepsilon = s\sigma, \tag{1.32}$$

where the constant of proportionality s is the *elastic compliance*. Alternatively, the relationship can be written the other way round, i.e.

$$\sigma = c\varepsilon, \quad c = 1/s, \tag{1.33}$$

where the constant of proportionality c is the *elastic stiffness*. The dimensions of stiffness, c are $\mathrm{N\,m^{-2}}$, i.e. Pa.

In practice, every type of solid has at least two different stiffnesses – one corresponding to deformation by normal stresses and the other to shear deformation, denoted by c_{11} and c_{12} respectively. In other words, there are linear relationships between stress and strain for various modes of deformation, but the constants of proportionality are different (although of similar order). The full significance of the double suffixes will become apparent in §2.3, which deals with the elasticity of anisotropic materials, for which more than two coefficients of stiffness need to be specified. A substance with a small stiffness is compliant in the sense that it undergoes a large distortion for a small applied stress σ. A material of large stiffness is rigid in the sense that even to induce a small strain ε requires a large stress σ. The reader may wonder why the letter c should be used for stiffness and the letter s for compliance, but that is the universal convention and its total illogicality makes it easy to remember. The reciprocal of the quantity s_{11} is better known as *Young's modulus*, E. If we determine the constant of proportionality between stress and strain by the method used by Hooke himself (a weight suspended on a wire), it is E that we obtain. This modulus is named after Young (1773–1829), a philosopher, mathematician, and physicist who was also an Egyptologist, and was apparently the first person to use it in lectures. Young's modulus ranges from about 3.5 Pa for polymers to about 10^3 GPa for ceramic materials, with values for metals lying somewhere between them. The other moduli which may be familiar to readers, the *rigidity modulus* (or *shear modulus*) and the *bulk modulus* are functions of E and Poisson's ratio ν, as we show in §2.3. For the purposes of the rest of this chapter, we shall use these moduli, defined in the following ways:

(*a*) We take the case of a block of isotropic material sheared by balanced stresses acting tangentially on two opposite faces, as shown in Fig. 1.12. If Hooke's law is obeyed, then the shear stress, σ_s $(=F/A)$ is related to the strain by

$$\sigma_s = G\varepsilon_s = G\theta, \tag{1.34}$$

where G is the rigidity modulus. For a given material the value of G is usually about one-third to one-half that of Young's modulus.

(*b*) When Hooke's law is obeyed for a block of material subjected to hydrostatic pressure, the dilatation Θ and the pressure p $(=F/A)$ are related by the equation

$$p = K|\Theta|, \tag{1.35}$$

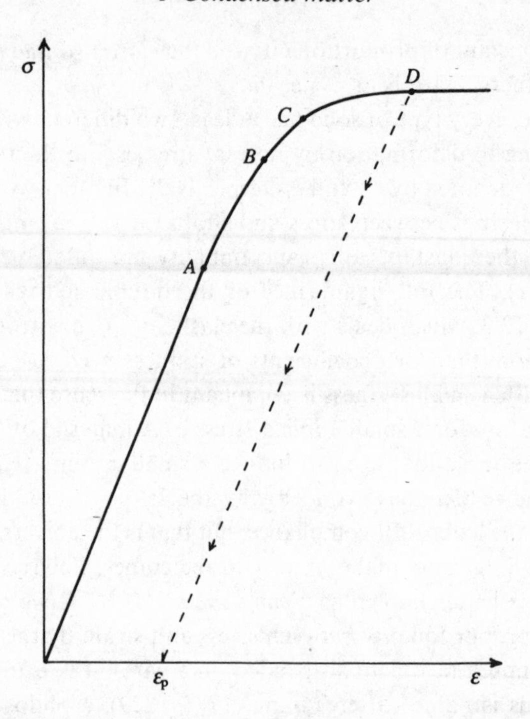

Fig. 1.13 Typical form of the relationship between stress and strain for a simple metal at low strains: the deformation is elastic and completely reversible for stresses equal to or less than that corresponding to point A; see text for an explanation of the significance of the points labelled B to D.

where K is the isothermal bulk modulus, a positive quantity since pressure acts so as to decrease the volume, the work also being performed at constant temperature. K is the inverse of the compressibility defined by eqn (1.16).

The regime of elastic deformation for which Hooke's law holds is easily seen if a plot is made of the stress versus strain for a material. This will typically look as shown in fig. 1.13. Here the solid is Hookean up to a point A called the *limit of proportionality*. The point B, the *elastic limit* indicates where *plastic* (i.e. non-reversible) *strain* begins. Up to point B all the strain vanishes if the applied stress is removed. Beyond B, at higher stress there is often a sudden change in the slope of the stress–strain curve; this is the *yield stress*, point C in fig. 1.13. For some materials these points may scarcely be separate or distinguishable, but with other solids, small but important deviations from elastic behaviour occur (usually covered by the 'umbrella' term *microplasticity*) before any elastic non-linearity or macroscopic plasticity is apparent from their stress–strain curves. Such effects may be evident from the examination of the carefully prepared surfaces of a testpiece under a microscope, for reasons which will become apparent in § 1.10. If the test is discontinued at the point D, the unloading path for the specimen will be the dashed curve, so that on removal of all stress there remains an irreversible plastic strain, ε_p.

1.10 Plastic deformation and the strength of crystals

Since elastic deformation involves the stretching of the bonds between the atoms in a solid, it might be anticipated that the termination of the elastic regime, corresponding to bond breaking, would result in the solid rupturing in a somewhat catastrophic manner. Some solids indeed behave like this, and are termed *brittle*, but the majority do not and we mostly confine our considerations to this larger group. Essentially, plastic deformation or *plastic flow* occurs because solids do not, in general, have the ideal and homogeneous structures that we discussed in earlier sections. A typical solid has *microstructure*, one form of which is polycrystallinity. Other forms include secondary phases, small voids, mistakes in the ideal stacking sequence of atoms, etc. Such defects in the perfect structure cause a material which has been stressed beyond the elastic limit to yield inhomogeneously, so that it does not generally fail in a brittle manner and 'fly apart'. Instead it first shows some degree of plastic or *ductile* behaviour. Thus the *yield stress*, which is a crude but convenient way of determining when a solid begins to deform irreversibly, is a *structure sensitive* property.

The yield stress will depend on the history of the material – whether it was cooled slowly or chilled rapidly when made, was rolled or forged into its present shape, etc. We have already indicated that high polymers, for example, may easily have varying proportions of crystalline and amorphous phase. It will be no surprise that the plastic flow behaviours of the two such phases of the same polymer differ and therefore the yield stress of a polymer will be very structure-sensitive and dependent on its thermal history. Because most fairly pure high polymers have long molecular chains which are either folded regularly or else intimately interwoven and weakly cross-linked, they yield at low stresses and may flow plastically up to large strains without rupturing, as the chains straighten.

The plastic flow of metals is somewhat different and also depends on whether we are talking about a single crystal or a polycrystalline specimen (see § 1.2). Figure 1.14 shows the typical forms of the curves relating applied force (or load) to extension (or displacement) for the tensile deformation of a pure, soft metal with a close-packed atomic structure (e.g. aluminium) in the two cases. After yielding, the single crystal flows

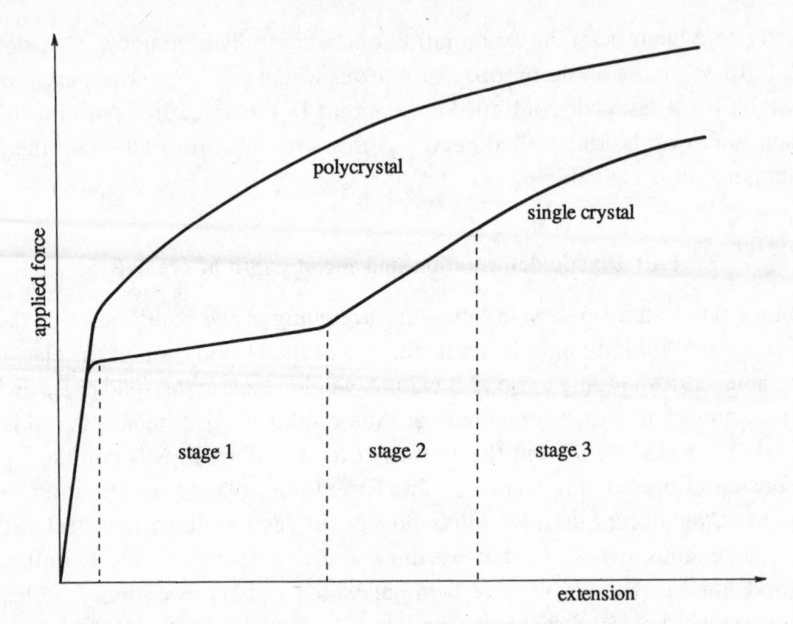

Fig. 1.14 The general shapes of the applied force–versus–extension curves for a simple metal plastically deforming in tension for both single crystal and polycrystalline forms.

for some time at a roughly constant stress level (a regime often called stage 1), after which there is a markedly greater resistance to flow, as witnessed by an increase in the force needed to continue the deformation. This behaviour is known as *strain hardening*, or *work hardening*, and it can be sub-divided into two further stages, 2 and 3. During stage 2 the relationship between stress and strain is approximately linear, while during stage 3, which is sometimes called the parabolic stage, the strain varies approximately as the square of the stress. In contrast, the force–extension curve of the polycrystalline specimen scarcely shows any evidence for plastic deformation without strain hardening seen in stage 1 with a single crystal. The behaviour of atomic solids can only be understood by a more detailed examination of the types and forms of crystal defects that can occur in them. The differing flow laws demonstrated by the force–displacement curves in fig. 1.14 are a result of the changing numbers of crystal defects within the specimens as deformation proceeds and the increasing effects of the interactions between their elastic strain fields.

As an introduction to crystal defects, we look at the reasoning that led Taylor (1934) to conclude that the observed strength of single crystals of pure, simple solids could best be explained by invoking the concept of a defect in the atomic stacking, called a dislocation and described in the following section.

When a single crystal specimen of a metal, like aluminium or copper, is stressed beyond its elastic limit, the resulting plastic flow takes place by means of the sliding of planes of atoms, one over another, throughout the crystal. In solids with simple structures, like the metals already mentioned (they have face-centred cubic unit cells), the atoms are in an arrangement of maximum close packing. In these solids, at least, the sliding action, which is called *slip*, is observed to take place on the planes that have the closest packing. (We show why this happens later in this section, but it may be intuitively understood as giving the smoothest motion of two atomically-corrugated planes moving one over the other.) Moreover, it is also found that slip occurs in a direction that is parallel to the closest-packed rows of atoms in the close-packed planes. Figure 1.4(a) indicated the close-packed planes in the f.c.c. structure, which have Miller indices $\{111\}$. Only one set of $\{111\}$ planes is shown in fig. 1.4(a); symmetry of the cubic structure requires that there are three more equivalent sets. There are three close-packed directions (six, if one counts both positive and negative directions), of type $\langle 110 \rangle$ at 60° to one another in any $\{111\}$ plane. (Readers who are unfamiliar with the indexing of crystallographic planes and directions should consult a basic textbook on crystallography, e.g. Kelly and Groves, 1970; Barrett and Massalski, 1980). It should be mentioned here that, in ionic solids, where electrostatic interactions between ions are very strong and important, the simple arguments about close-packing are not necessarily paramount. For example, sodium chloride crystals slip on the diagonal $\{110\}$ planes. A plane of this type is the plane shown shaded in fig. 1.4(d). (Such planes are commonly called dodecahedral planes, because the body bounded by the twelve possible planes is a dodecahedron.) Slip in NaCl still occurs in the close-packed $\langle 110 \rangle$ directions, of which there is one in each $\{110\}$ plane (again not counting both positive and negative directions). This combination of slip plane and slip direction avoids bringing ions of like charge into too close proximity during slip.

The metal, zinc, does not have the f.c.c. structure, but the hexagonal close-packed structure mentioned in § 1.3. In this structure the crystallographic planes that correspond to one of the atomic layers illustrated by fig. 1.4(c) are closer packed than any other type of plane. The structure is shown in Fig. 1.15(a) and the close-packed layers are parallel to the base of the cell. Slip occurs on these basal planes, again in a close-packed direction, of which there are three in the plane. The surface of a long cylindrical zinc crystal that has been compressed plastically, when examined under a microscope, is found to exhibit many fine parallel sets of *slip-lines* which correspond to the traces of this particular plane in the surface of the

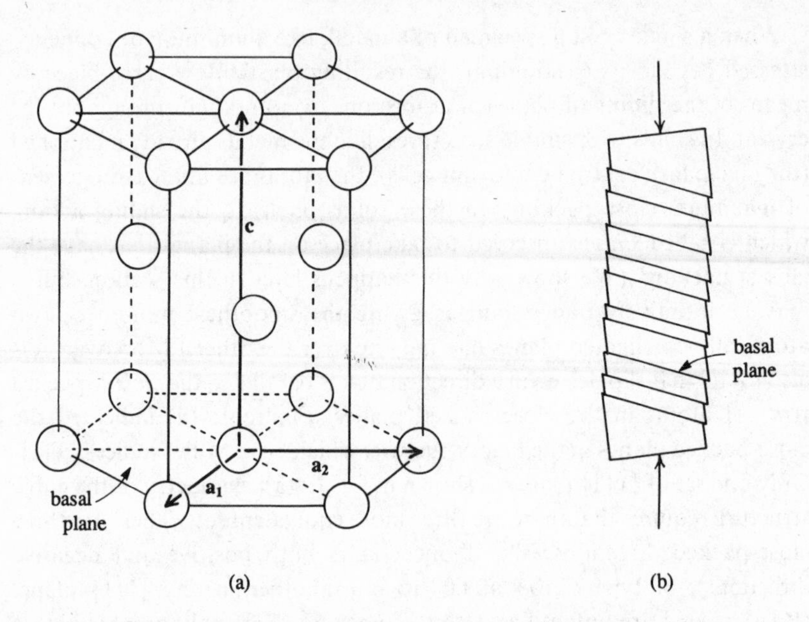

Fig. 1.15 (a) Close-packed hexagonal structure of zinc crystal, indicating the basic translation vectors and the basal plane, upon which slip occurs; (b) sketch to show the effect of slip on a single crystal rod of zinc that has undergone plastic deformation.

cylinder, as indicated in the sketch, fig. 1.15(b). The slip-lines show where the bonds between the atoms in two adjacent close-packed planes have been subjected to a sufficiently large shear stress that they have broken, allowing a microscopic translation to take place (one plane with respect to the other) before new bonds form.

Before considering slip in a little more detail, it should be mentioned here for completeness that some crystals do have other means of changing their external shapes without a change in symmetry of the basic crystal structure. For example, one mechanism known as *deformation twinning* (although often simply referred to as *twinning*) results from essentially instantaneous, cooperative displacements of the atoms in a small volume of the solid in response to applied stress. The effect of these atomic translations is illustrated in fig. 1.16. The structure of the twinned volume is seen to have 'flipped' into an orientation that causes its lattice to be a mirror image, in some simple crystallographic plane, of the lattice of the untwinned surroundings. We shall not give any further consideration either to this or other processes that are usually unable to generate large plastic strains unless slip also occurs.

In practice the plastic deformation, by slip, of either a single crystal, or a grain within a polycrystalline solid, is initiated at a level of shear stress

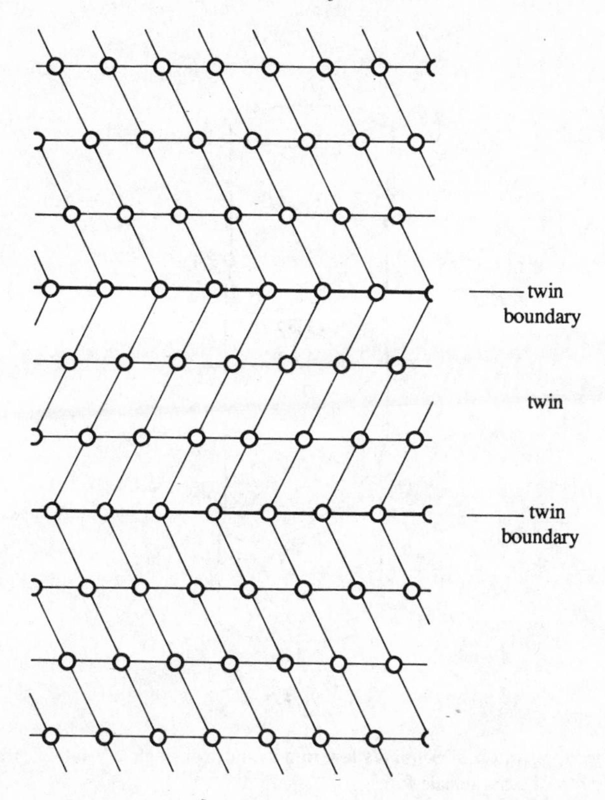

Fig. 1.16 Two-dimensional representation of a twinned region in a crystal (a rhombic form of unit cell is used for clarity).

which is characteristic of the material, the temperature, and the slip plane that is activated. The reason why the stress at which a crystal starts to flow plastically should be an intrinsic property becomes apparent later in this section. The critical stress needed to initiate slip on a particular plane of atoms is determined by means of *Schmid's law* which we illustrate by taking the simple case of a cylindrical single crystal in tension, shown in fig. 1.17. Consider the shear stress on a plane whose normal N makes an angle ψ with the axis of the cylinder as a result of tensile forces F applied perpendicular to its ends. Let the direction along which slip is possible within the plane be OS, making an angle λ with the stress axis. Then the component of F acting in the possible slip direction is $F \cos \lambda$ and this force acts over a plane of area $A/\cos \psi$, where A is the cross-sectional area of the crystal. The resolved shear stress on the plane is therefore

$$\sigma_s = (F/A) \cos \psi \cos \lambda. \tag{1.36}$$

Schmid's law says that slip occurs when σ_s attains a value $\sigma_{s\,crit}$ known as

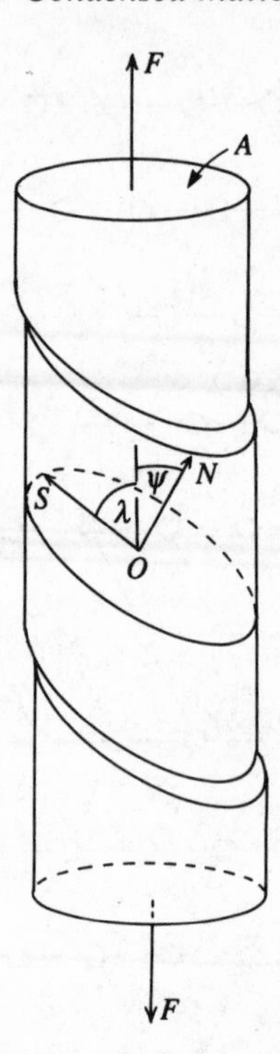

Fig. 1.17 The application of Schmid's law to a cylindrical single crystal deforming by slip under the action of axial tensile forces.

the *critical resolved shear stress* (c.r.s.s.) of the material. The product of the two cosine terms is sometimes called the *Schmid factor*.

In the majority of simple solids, slip is possible on more than one type of crystallographic plane. As we saw in § 1.3, different planes have different atomic arrangements and different packing densities. As already mentioned, in solids of high symmetry such as f.c.c. metals, the most closely-packed plane is the usual slip plane, since there are no other important considerations as in ionic crystals where electrostatic interactions play a role. Deformation of a solid by slip is specified in terms of one or more *slip systems*, $(hkl)[uvw]$, the convention being to write the Miller indices (hkl) of the slip plane first and the direction indices $[uvw]$ second. Solids of high crystal symmetry may have several equivalent slip systems, e.g. all the possible $\{111\}\langle 1\bar{1}0\rangle$ combinations in an f.c.c. metal. For a given direction of applied stress, slip occurs first on the slip system for which the Schmid factor is largest. However, the different densities of atomic packing in different crystallographic planes result in specific values of the c.r.s.s. for every possible type of slip system. Because different values of c.r.s.s. are

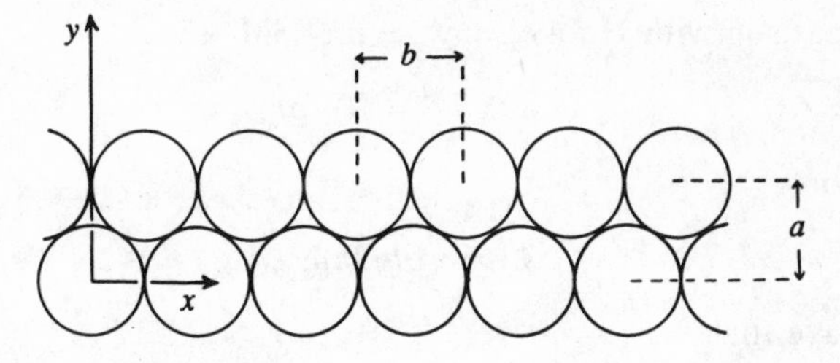

Fig. 1.18 The method for calculating the ideal critical resolved shear stress for a defect-free crystalline solid.

associated with the possible slip systems in simple solids of high symmetry, it follows that slip will be initiated on one slip system, but that other types may come into play as the level of internal stress rises.

It is not difficult to estimate what the strength of a simple crystalline solid should be, on the assumption that all the bonds between two adjacent blocks of crystal which slip over each other are broken simultaneously. This was first done by Frenkel in 1926 and we follow his method. We consider a crystal at the absolute zero of temperature and neglect the zero point energy of the atoms. We let the spacing between the planes on which the slip is to occur be a, and the spacing between the atoms in the closest-packed rows in the planes be b, which is therefore the spacing in the slip direction. We take coordinate axes x, y with x parallel to the slip direction and y perpendicular to the slip plane, as shown in fig. 1.18, with the origin at the centre of an undisplaced atom. If we take σ_x as the shear stress required to displace the upper block a distance x with respect to the lower one, σ_x will be zero at positions $x = 0, b, 2b$, etc., since the atoms are then restored to equilibrium positions. For values of $x = b/2, 3b/2, 5b/2$, etc., σ_x will again be zero, by virtue of symmetry arguments, since the displaced plane of atoms just above the slip plane will be in metastable equilibrium. We thus deduce that the shear stress σ_x is a smoothly varying periodic function of x with a period b. Frenkel assumed that the variation in σ_x would be reasonably well represented by a sine wave

$$\sigma_x = A \sin(2\pi x/b), \tag{1.37}$$

where A is a constant. When x is small

$$\sigma_x \simeq A2\pi x/b, \tag{1.38}$$

and the shear strain ε_{xy} can be expressed simply as

$$\varepsilon_{xy} = x/a. \tag{1.39}$$

Then comparison with Hooke's law, eqn (1.34), gives

$$A 2\pi x/b = Gx/a \tag{1.40}$$

and therefore

$$A = Gb/2\pi a. \tag{1.41}$$

We thus have that

$$\sigma_x = (Gb/2\pi a) \sin (2\pi x/b). \tag{1.42}$$

The maximum value of σ_x, which is clearly $Gb/2\pi a$, is equal to the critical resolved shear stress, since this is the stress needed to generate a displacement greater than $b/2$, i.e. a permanent offset of the upper block. Equation (1.42) supports the assertion, made at the beginning of this section and based on experimental observations, that slip occurs in the direction of closest atomic packing. It also suggests that slip should occur most readily on planes that have the greatest separation, which will be planes of closely-packed atoms, thereby confirming our other assertion, which was also deduced from observations.

Since $a \simeq b$ for simple cubic-structured crystals, eqn (1.42) also predicts that the c.r.s.s. should be about $G/10$. A less crude model than the one used above modifies the result somewhat, to about $G/45$. For most materials G is about 5×10^{10} Pa, i.e. 50 GPa, which means that the theoretical shear stress to cause plastic deformation by slip is about 1 GPa. This is some four orders of magnitude greater than experimentally-observed values for pure f.c.c. metals (e.g. aluminium). Only very thin, specially-prepared forms of single crystals, called *whiskers*, exhibit near-theoretical strengths and these were not available when the huge discrepancy between theory and experiment was first recognized. It was this discrepancy that caused three researchers, Taylor, Orowan and Polanyi, to put forward, more or less simultaneously in 1934, models for the slip of crystals which dispose of the assumption that all the bonds transverse to the slip plane are broken simultaneously. Taylor's model has been the most influential. It invokes the concept of a line defect, a *dislocation*, which delineates the boundary between the part of an atomic plane over which slip has occurred and the part where it has not.

1.11 Dislocations and other defects in crystals

In the Taylor model, slip propagates by the advance of the dislocation into the unslipped area of the plane on which the dislocation rests, causing fresh atomic bonds to be broken, while broken bonds are reformed in its

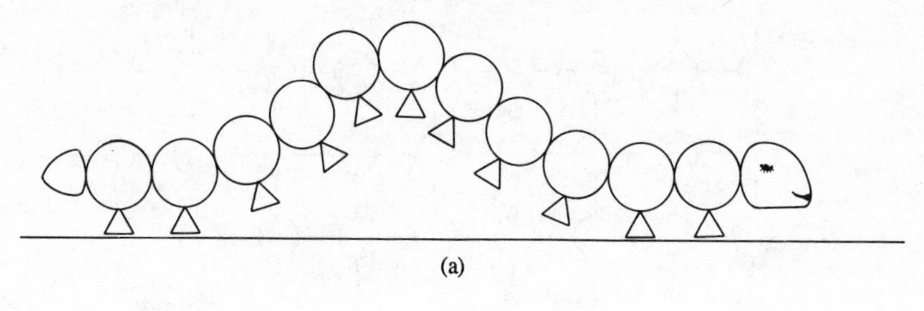
(a)

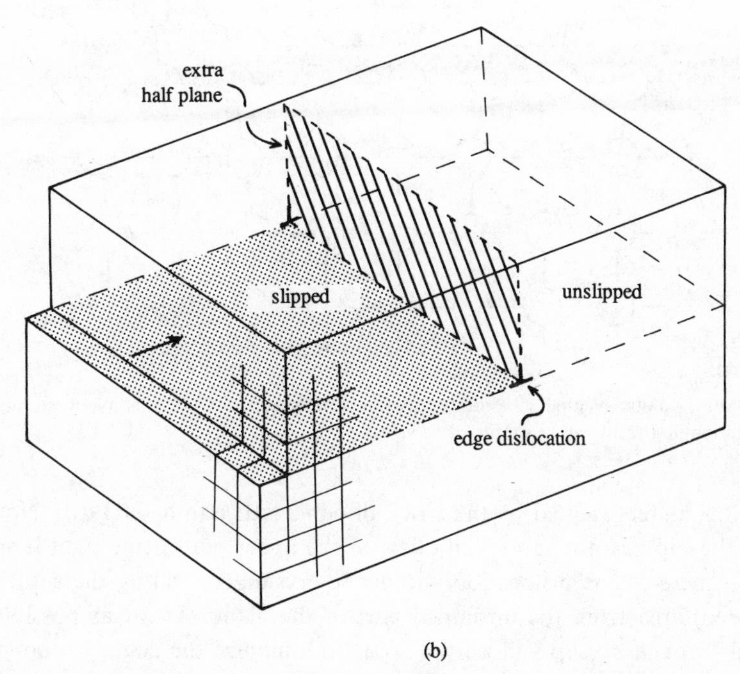

(b)

Fig. 1.19 (a) 'Caterpillar analogy' to illustrate why a dislocation facilitates the slip of a crystal; (b) partial slip of the upper part of a crystalline block over the lower part, accompanied by the creation of an edge dislocation.

wake. A dislocation, which does not have to be straight, marks the boundary between the slipped and unslipped areas on a slip plane of the crystal. The motion of the dislocation in the slip plane is called *glide*. The effect of a dislocation in facilitating slip can be envisaged by comparing the glide of a dislocation with the motion of a caterpillar, fig. 1.19(a), which advances by propagating a hump along its body and only has a few of its pedicles (viz. bonds!) off the ground at any instant. The atomic arrangement very close to the dislocation can easily be pictured if we take a rectangular parallelepiped of initially perfect crystal and force the left hand side of the upper part to slip over the lower by just one interatomic

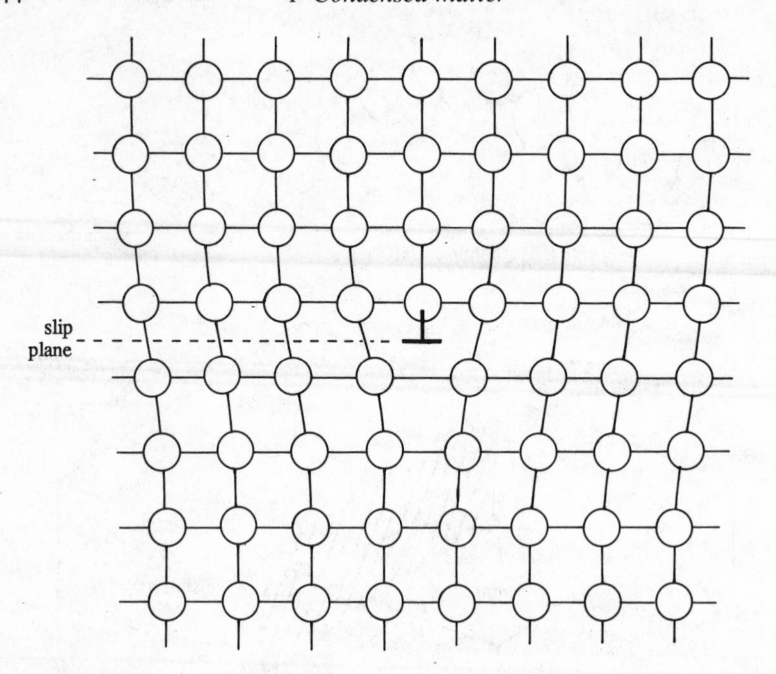

Fig. 1.20 Two-dimensional view of the atomic arrangement in the vicinity of an edge dislocation in a solid with a simple cubic structure.

spacing, as represented by the block of cubic lattice in fig. 1.19(b). Note that the slip has not caused an offset of the upper part at the right hand end – there is one dislocation within the crystal, separating the slipped (shaded) area from the unslipped part of the plane. As far as possible, bonds broken by slip will reform so as to minimize the degree of bond-stretching and departures of atoms from their preferred coordination. As a result the atomic arrangement at the core of the dislocation may be expected to look rather like fig. 1.20. The method that we have chosen to create this defect is not general and indeed we have thus depicted one particular sort of dislocation, called an *edge dislocation*, because there is always an incomplete plane of atoms terminating at it. This is the hatched plane in fig. 1.19(b).

Other sorts of dislocation also occur, and fig. 1.21 illustrates the curved dislocation line which results from slip that occurs non-uniformly across a block of crystal; the shaded area again indicates the extent of slip. This should be compared with fig. 1.19(b). We see that where the dislocation emerges on the face of the block at B, it is an edge dislocation. However, at position A it is a *screw dislocation*, so named because of the helical nature of the lattice in its vicinity. There is often no reason why a dislocation

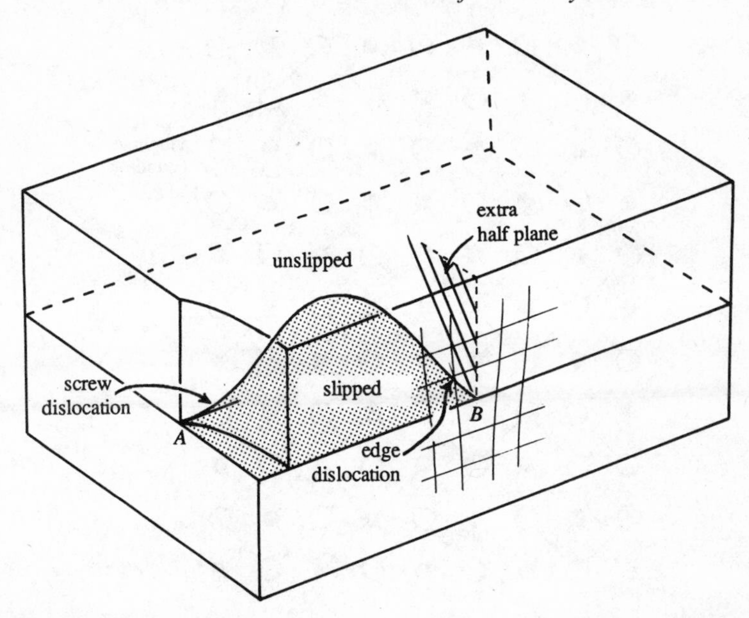

Fig. 1.21 How a curved dislocation varies in character along its length, possessing edge character where its direction is perpendicular to the slip direction and screw character where its direction is parallel to the slip direction.

should be either pure edge or pure screw in character, so that in real materials such dislocations of 'mixed' character do commonly occur. As a dislocation moves, the lattice distortion associated with it is not dissipated or changed in any way. This property is therefore employed as a means of characterizing a dislocation. The characteristic parameter of a dislocation is its *Burgers vector*, usually denoted **b**, which specifies the magnitude and direction of the local lattice displacement that the motion of a dislocation propagates. It should be clear from fig. 1.19(b) that the Burgers vector of an edge dislocation is perpendicular to the direction of the dislocation, whereas the Burgers vector is parallel to a screw dislocation.

When one dislocation encounters another, there are several possible results of the interaction, one of which is the formation of kinked dislocations. A *kink* is an offset in a previously straight dislocation, the offset being equal to the modulus of the atomic translation vector characteristic of the other dislocation that 'inflicted the damage'. Kinks are essentially one-dimensional dislocations and some types of kink are able to glide through the crystal, like dislocations that are involved in slip processes. The glide of dislocation kinks has close parallels with the propagation of *solitons*, discussed in § 5.10 (see, for example, Weertman, 1985).

Dislocations are line defects in the perfect stacking that is to be expected

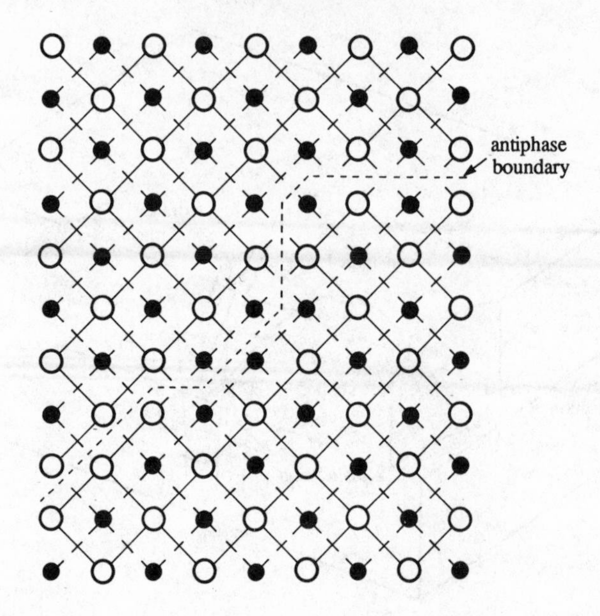

Fig. 1.22 Two-dimensional representation of an antiphase boundary between two ordered domains in a crystal of a binary compound.

of an ideal crystal. Various other types of crystal defects also exist, but which we shall not discuss in any detail. They include:

(i) *point defects* – this category includes vacant lattice sites, or vacancies, where atoms are missing (previously mentioned in § 1.4). These are often the most important type of point defect, partly because they must occur in all crystalline atomic solids (explained at the end of this list). Foreign (impurity) atoms occurring at normal lattice sites or in the interstices of the main atomic array also constitute point defects.

(ii) *planar defects* – mistakes in the regular atomic packing which may occur over comparatively large areas; this heading includes the *twin boundaries* that separate the parts of crystals which have undergone twinning (mentioned in § 1.10) from the rest of the crystal. The atoms on the two sides of a twin boundary occupy 'mirror' lattice positions (see fig. 1.16). The term planar defects also embraces *antiphase boundaries*. These are interfaces across which, for example, the regular A B A B A B . . . sequence of atoms in an ordered binary compound are out of phase, as illustrated schematically by the two-dimensional crystal in fig. 1.22. This may result from the nucleation of ordered *domains* at different places within a crystal, followed by their subsequent growth and impingement. An antiphase boundary could thus be called a domain wall and, indeed, is somewhat analogous to a magnetic domain wall. By adopting the term antiphase boundary, we avoid confusion in the case of solids where both types of defect occur. An antiphase boundary can also be created in an ordered compound by the glide of a type of dislocation that does not displace atoms by a proper lattice translation vector. It is apparent from fig. 1.22 that the adjective 'planar' is misleading because such interfaces are not necessarily required to be plane on an atomic scale, unless created by slip.

Thermodynamic arguments concerning the energy of formation of a vacancy, E_v and the minimization of the free energy of a solid show that there is an equilibrium concentration of vacancies in a crystalline solid which depends on the temperature, T. The result, which is derived for the simplest case in most solid state physics textbooks, gives the ratio of the number of vacancies, n, per unit volume to the number of atoms per unit volume, N, in terms of a Boltzmann factor

$$n/N = \exp(-E_v/k_B T). \tag{1.43}$$

There are long-range elastic stresses associated with dislocations, so that they not only interact with each other, but they may also attract point defects, particularly impurity atoms, to form relatively stable assemblies. The dislocations can then glide only if they can be pulled away from the cloud of point defects and they are said to be *pinned*, thus affecting both ease of plastic deformation and the removal of dislocations from crystals where their presence is undesirable (e.g. in semiconductors, in which they influence the lifetimes of charge carriers and affect the structure of the energy levels). The behaviour of dislocations and their role in relation to the bulk properties of materials has been widely studied during the last decades, particularly using direct methods of observation such as transmission electron microscopy.

In practice, many materials are used in polycrystalline form and often they intentionally contain more than one solid phase. For example, metal alloys rely for their strength on fine internal dispersions of solid phases which differ from the bulk in chemical composition, and often also in structure. In addition to point, line and planar defects, therefore, we must also recognize the existence and effects of grain boundaries, interphase boundaries, microcracks, external surfaces, etc. on the properties of solids. Indeed, surfaces have very important effects in relation to strength, since from the viewpoint of continuum theories of solids, surfaces are major defects. Any cracks, steps or other discontinuities on surfaces are also important, since they act as concentrators of stress and can initiate slip or brittle fracture at stress levels well below those for the initiation of comparable mechanisms in the interior. The particular distributions and varieties of defects occurring in a given crystalline material are often referred to by the general term microstructure.

The crucial role of crystal defects in relation to the onset of plastic deformation has been confirmed by the properties of some whiskerlike single crystals, already mentioned in §1.10. After the ideas of Taylor, Orowan and Polanyi had gained acceptance, it was found that single crystal whiskers of many substances could be grown under rather special conditions. The important feature is that one dimension is very much larger than the other two, and whiskers are typically a few millimetres long and have transverse dimensions of the order of 10 μm. The thinnest and most perfect of these whiskers contain no dislocations and have surfaces free from defects other than atomic-scale steps. As stated in §1.10, these perfect whiskers can have strengths close to the values to be expected on the basis of the refined Frenkel-type strength calculation. Their small dimensions and physical perfection play major roles in giving them their strength and their behaviour shows that defects set a limit to the strength of crystalline solids.

Some of the characteristics of grain boundaries are easily visualized by means of the two-dimensional physical modelling of atoms as hard spheres, for example with bubble rafts (first demonstrated by Bragg & Nye, 1947) or single layers of ball-bearings. Figure 1.23(a) is a photograph of part of a 'polycrystalline' raft of ball-bearings and the disordered grain boundaries are clearly visible. Even with this somewhat limited model it is clear that they are quite small in width, comparable with the dimensions of the 'atoms'. In practice, grain boundaries in real materials often form in a manner and with an orientation vis-a-vis the adjacent grains that achieves optimum packing and minimal extent of disorder and width of boundary. Figure 1.23(b) illustrates a special type of boundary, composed of edge dislocations, which achieves a high degree of registry between the lattice sites of the two grains on either side, thereby minimizing the local elastic strains. Such boundaries correspond to only small misorientations ($\simeq 1°$) between the two lattices and are called *sub-grain boundaries*. It is not the intention of this book to examine such phenomena and the many facets of the defect solid state in any detail. The existence of crystal defects should however always be borne in mind in regard to physical properties that are structure-sensitive, even though our treatment in the following chapters is mainly concerned with continuum theories of the macroscopic properties of solids and liquids.

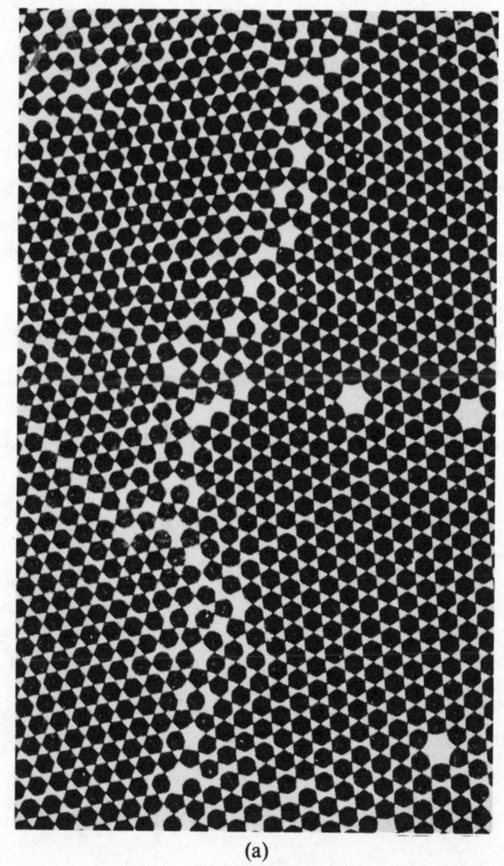

(a)

Fig. 1.23 (a) A 'grain boundary' in a two-dimensional crystalline solid represented by a close-packed layer of ball bearings; (b) a sketch to show the idealized structure of a symmetrical tilt sub-grain boundary in a simple cubic solid.

1.12 Time-dependent strain – anelasticity, viscoelastic behaviour and creep

In §§ 1.9 and 2.3 we assume that elastic strain is a single-valued function only of stress. This is not always true, since there are materials for which the attainment of maximum elastic strain lags behind the application of the maximum stress responsible for the strain. The stress- and time-dependence of elastic strain is known as *anelasticity*. It will be apparent that when structures made from an anelastic material are loaded and unloaded cyclically, the stress–strain curve will exhibit *elastic hysteresis* (Greek: husteros – coming after). Energy is lost as heat on each cycle and this produces a form of damping.

An ideal elastic material stores all the energy expended as work by

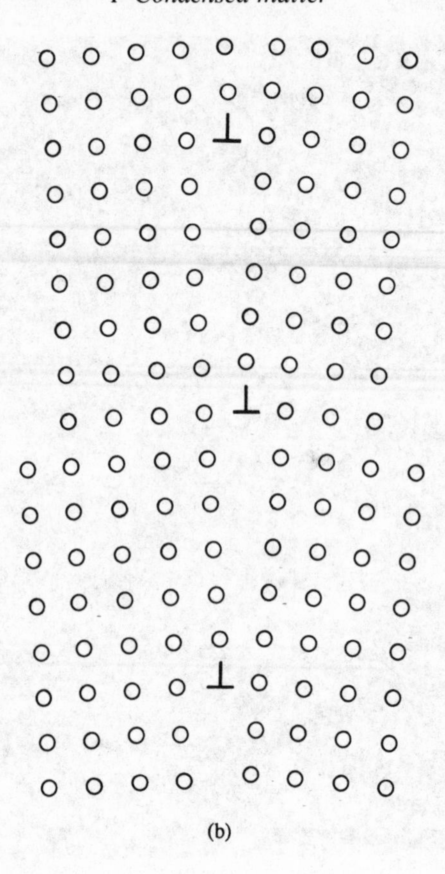

Fig. 1.23 Continued.

external forces in producing a change in shape, and this energy is then available to return the body to its original shape when the constraining forces are removed. A viscous fluid, however, has no predetermined boundaries and it flows irreversibly under the action of external forces. Real materials normally have mechanical properties that are intermediate between those of an elastic solid and a viscous fluid and they are therefore said to be *viscoelastic*. Both Hooke's law of elasticity and Newton's law of viscous flow are linear relations. A simple combination of these laws is sometimes used to model the behaviour of viscoelastic solids, on the assumption that the shear stresses relating to strain and strain rate are additive, thus

$$\sigma = G\varepsilon + \eta\, \mathrm{d}\varepsilon/\mathrm{d}t. \tag{1.44}$$

This equation is taken to describe the *standard linear viscoelastic solid* and, although this is an idealized concept, it gives a tolerably good description of the behaviour of some amorphous polymers.

In a simple tensile test where a heavy weight, sufficient to cause the elastic limit to be surpassed, is hung on a wire, it is found that the variation in length is strongly dependent on time. In other words, the rate of plastic flow, which we can express as the strain rate, $d\varepsilon/dt$, is not constant. In practice, arrangements are made to adjust the load, either continuously or periodically, to compensate for the change in cross-section as the sample elongates at constant volume. The instantaneous change in length is then expressed as a longitudinal strain (or stretch), and it is the variation of this strain as a function of time that is of interest. Usually, the strain rate is high after applying the load and later it may, for some time, reduce to a constant value. This phenomenon, in which plastic flow occurs at constant stress, is known as *creep*. It has much practical importance, for example, in static structures such as cranes and bridges where unplanned stress concentrations may occur, and in equipment embodying moving parts, small clearances and high temperatures, as in jet turbine engines.

Creep may occur as a result of several different microscopic processes, some involving dislocations, one caused by the diffusion of point defects to grain boundaries (discussed in §6.9) and one by the sliding of crystals past each other along their grain boundaries. We are less concerned with the details here than the fact that creep experiments provide another way of examining the rheological properties of a solid. The curves illustrated in figs. 1.13 and 1.14 correspond essentially to constant strain rate. A creep curve, shown in fig. 1.24 for deformation in tension, is a plot of strain versus time, at constant stress. The portion AB is the elastic regime, with the specimen yielding plastically at B. Creep is not greatly concerned with loading-induced transient effects, such as behaviour near the yield point, but with the way in which a material flows when allowed to deform 'at its own pace'. The slope of a creep curve is the *creep rate*. Transient effects are usually assimilated under the heading *primary creep*. This is the region BC in fig. 1.24, a regime which was found by Andrade (1910) to be well described by an equation of form

$$L_t = L_0(1 + at^{1/3})e^{bt}, \tag{1.45}$$

where L_t and L_0 are respectively the initial length of the sample and its length at t, while a and b are constants. In Andrade's time it was common to plot L_t on the vertical axis, rather than ε_t. The equation of the creep curve in its modern form can be obtained from eqns (1.27) and (1.45) and it is

$$\varepsilon_t = \varepsilon_0 + \ln(1 + at^{1/3}) + bt, \tag{1.46}$$

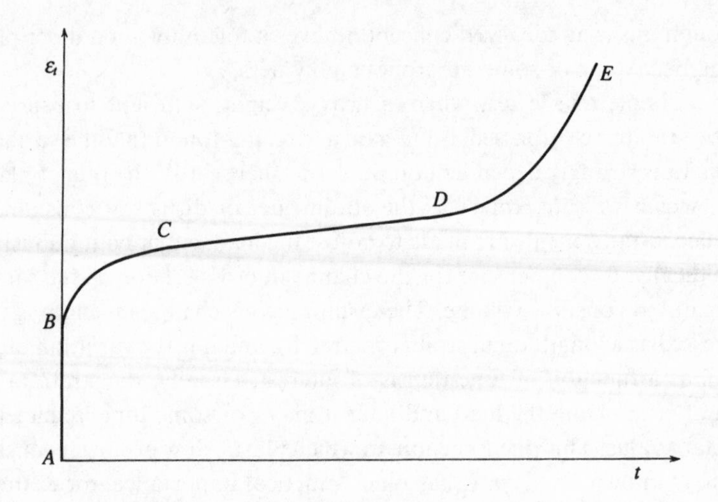

Fig. 1.24 The general form of a creep curve, giving true strain as a function of time; see text for an explanation of the significance of the points labelled A to E.

where ε_0 embraces the elastic strain in the sample. For small values of t, and provided that $a \ll b$, eqn (1.46) becomes

$$\varepsilon_t = \varepsilon_0 + at^{1/3} + bt. \tag{1.47}$$

As fig. 1.24 indicates, the creep of most materials achieves a quasi-*steady state*, i.e. at least one part of the curve can reasonably well be represented by a straight line, the region CD. From eqn (1.47) we see that the total creep strain is the superposition of primary creep, dependent on $t^{1/3}$, on a steady state viscous flow. From the linear, steady state region the viscosity of the material is given directly since, using eqns (1.8) and (1.47)

$$\eta = \sigma\dot{\varepsilon}^{-1} = \sigma/b. \tag{1.48}$$

Often a large strain is achieved, and then the deformation accelerates as some instability sets in (e.g. a crack, localized 'necking down' of the cross-section, etc.), rapidly followed by rupture (point E in fig. 1.24). For a clear introduction to creep, the reader is recommended to consult Poirier (1985).

Another way of looking at creep is to examine the creep rate (i.e., the strain rate) as a function of stress. Quite generally, the strain rate will be a function of several variables, expressed by a *constitutive equation*:

$$\dot{\varepsilon} = f(\sigma, T, P, \Xi) \tag{1.49}$$

where Ξ is a parameter representing the internal state of the material. Thus Ξ will embody the previous thermal and deformational history of the material, as recorded in terms of its microstructure. Perhaps surprisingly, it is sometimes possible to keep Ξ essentially unchanged by starting with a microstructure which causes internal crystal defects to be created and to annihilate each other at approximately equal rates. In this situation the last three variables in eqn (1.49) can be kept constant during the test, so that $\dot{\varepsilon}$ is only a function of stress. This is often approximately represented by means of a power law

$$\dot{\varepsilon} = A\sigma^n, \tag{1.50}$$

where the constant A and the power n are characteristic of particular creep mechanisms (one example involving vacancies is discussed in § 6.9).

Creep is mainly a low strain rate phenomenon (typically 10^{-6} to $10^{-12}\,\mathrm{s}^{-1}$). A material that will creep at a low strain rate may behave quite differently if an attempt is made to achieve higher strain rates by increasing the load. This can be recognized in the behaviour of the rocks of the crust and upper mantle of the Earth. We can only touch upon the reasons why most non-metallic, inorganic solids show brittle behaviour when intense stresses are applied over short time intervals. The explanation often hinges on the fact that crystals of low symmetry have few slip systems (unlike cubic crystals such as NaCl, discussed in § 1.10) and also one or more of these slip systems may have a very high c.r.s.s. (critical resolved shear stress) at normal temperatures. A law, known as the *von Mises condition*, requires that a crystal must possess five independent slip systems to undergo a general and arbitrary change of shape by means of dislocation glide. This applies to each grain within a polycrystalline body if the material is to deform plastically without internal pores forming at the grain boundaries. Moreover, for a given individual grain within a polycrystalline solid of low crystal symmetry, there will not, in general, be a slip system favourably oriented with respect to the stresses acting on the grain. It cannot therefore easily respond to the stresses by changing shape and it will even then only be able to deform to a significant strain if it breaks contact with some of the neighbouring grains. Thus cracking occurs within any polycrystalline aggregate of grains of low symmetry and such cracks tend to propagate catastrophically if the aggregate is subjected to tensile stresses. Rock-forming minerals are typically crystals of low symmetry and thus rocks do not flow plastically over short timescales, as do most metals. However, when under conditions of elevated temperature and hydrostatic pressure (as in the Earth's crust), which suppress cracking (see § 2.1), such minerals are more plastic and their parent rocks can therefore flow when sheared over geological timescales. Mountain building (*orogeny*) and *continental drift* (*plate tectonics*) are evidence of the low strain-rate, plastic behaviour of rocks under conditions of constant stress or near-constant stress. Loosely speaking, earthquakes are evidence of a more brittle, catastrophic failure that occurs when stress levels become too high because they have not been adequately alleviated by one or more flow processes. However, the details of the processes contributing to the generation of earthquakes are still not fully understood.

Rate-dependent plasticity is a subject that is of growing interest (the name embraces time-dependent and strain rate-dependent processes). Normally, the numbers and distribution of defects in a crystalline solid change as it deforms, so that its resistance to deformation (measured in terms of strength or hardness) also changes. This evolution of the microstructure is at the root of the strain hardening of metals exhibited in their stress–strain curves and mentioned in § 1.9. Materials have been developed with microstructures which, at a particular strain rate, do not change much as deformation proceeds to large strains and so the materials do not strain-harden appreciably. Such behaviour has much commercial merit, because fewer stages are needed in the shaping of such materials and also less energy is consumed. For this reason, materials that show a high strain rate behaviour known as *superplasticity* have become important in recent years. Superplasticity is a quasi-viscous flow, macroscopically speaking, so to that extent it has some similarity to creep. The deformation of at least one type of superplastic material appears to depend upon quasi-stable grain-boundary structures. But in detail the processes at work are not similar and the two phenomena are just different examples of rate-dependent plasticity.

1.13 Suggestions for further reading

Brittle behaviour

Cottrell, A.H. (1964). *The Mechanical Properties of Matter*, New York: Wiley.

Tabor, D. (1979). *Gases, Liquids and Solids*, 2nd edn. Cambridge: Cambridge University Press.

Creep

Poirier, J.P. (1985). *Creep of Crystals*. Cambridge: Cambridge University Press.

Crystallography

Barrett, C.S. & Massalski, T.B. (1980). *The Structure of Metals*, 3rd (revised) edn. New York: McGraw–Hill.

Kelly, A. & Groves, G.W. (1970). *Crystallography and Crystal Defects*. London: Longman.

Dislocations

Hull, D. (1965). *Introduction to Dislocations*. Oxford: Pergamon Press.
Humphreys, F.J. & Goodhew, P. (1985). *Dislocations*: Illustrative programs for the Acorn micro-computer, model BBC B. Engineering Software Series on disc. London: The Institute of Metals.
Kelly, A. & Groves, G.W. (1970) (see above).

Elasticity

Timoshenko, S.P. & Goodier, J.N. (1970). *Theory of Elasticity*, 3rd edn. New York: McGraw–Hill.

Phase Transitions

Adkins, C.J. (1983). *Equilibrium Thermodynamics*, 3rd edn. London: McGraw–Hill, chapter 10.
Zemansky, M.W. & Dittmann, R.H. (1981). *Heat and Thermodynamics*. Singapore: McGraw–Hill, chapters 10 and 13.
Walton, A.J. (1983). *Three Phases of Matter*, 2nd edn. Oxford: Clarendon Press.

Plasticity of crystals

Cottrell, A.H. (1964). *The Mechanical Properties of Matter*. London: John Wiley.
Feltham, P. (1966). *Deformation and Strength of Materials*. London: Butterworth.
Schmid, E. & Boas, W. (1950). *Plasticity of Crystals*, English edn. London: F.A. Hughes.

1.14 References

Andrade, E.N. Da C. (1910). *Proceedings of the Royal Society of London*, A**84**, 1–12.
Bragg, W.L. & Nye, J.F. (1947). *Proceedings of the Royal Society of London*, A**190**, 474–81.
Reiner, M. (1969). *Deformation, Strain and Flow*. London: H.K. Lewis.
Schechtman, D., Blech, I., Gratias, D. & Cahn, J.W. (1984). *Physical Review Letters*, **53**, 1951–3.
Taylor, G.I. (1934). *Proceedings of the Royal Society of London*, A**145**, 362–404.
Timoshenko, S. (1953). *History of Strength of Materials* ('With a Brief Account of the History of Theory of Elasticity and Theory of Structure'). New York: McGraw–Hill.
Weertman, J. (1985). *Metallurgical Transactions*, **16**A, 2231–6.

Problems

1. Use dimensional analysis to find the dimensions of stress, stiffness, compliance and viscosity. Although it is recommended that stress should be quoted in SI units, it is not unusual still to find stress given in terms of bars, atmospheres and pounds

per square inch (p.s.i.). Work out the factors necessary for converting between these units and the SI units of stress. State whether there are units for the measurement of strain and strain rate and, if so, what they are.

2. A copper rod must be capable of carrying a load of 500 kg while suffering an elastic strain of no more than 0.01 per cent. What is the minimum permissible cross-sectional area if Young's modulus for copper is $1.1 \times 10^{11}\,\mathrm{N\,m^{-2}}$?

3. Iron crystallizes from a melt with a face-centred cubic structure, transforming to a body-centred cubic form at lower temperatures. Assuming that both forms represent close packings of spherical atoms, calculate: (i) the dilatation that occurs when iron transforms from f.c.c. to b.c.c.; (ii) the void fraction in both structures.

4. The separation of sodium and chlorine ions in crystalline sodium chloride is 0.281 nm. The atomic weights of Na and Cl are 22.99 and 35.45 respectively. Calculate the density of NaCl.

5. Welded steel rails are laid in place on a hot summer's day (25° C) and anchored so that shrinkage cannot subsequently occur. If the coefficient of linear expansion of steel is $1.13 \times 10^{-5}\,^{\circ}\mathrm{C}^{-1}$ and Young's modulus is $2.1 \times 10^{11}\,\mathrm{N\,m^{-2}}$, calculate the tensile stress in the rails on a winter's day (– 5° C).

6. A single crystal rod of pure aluminium with a radius of 1 mm is subjected to axial forces of extension. The axis of the rod is accurately parallel to a $\langle 100 \rangle$ crystallographic direction. From the stress–strain curve for the rod, plastic deformation is seen to have started when the applied forces reached a value of 6.50 N. Find the critical resolved shear stress for plastic flow in aluminium.

7. A single crystal cube of sodium chloride, with $\{100\}$ faces and with edges 10 mm long, is compressed in a direction perpendicular to a pair of faces. The cube begins to yield plastically when the applied force reaches 0.2 kN. Find (*a*) the Schmid factors for the various $\{110\}\langle 1\bar{1}0 \rangle$ slip planes; (*b*) the critical resolved shear stress for $\{110\}\langle 1\bar{1}0 \rangle$ slip.

8. A rectangular block of crystal that has balanced shear stresses σ applied to one pair of faces deforms plastically while maintaining ϱ_d dislocations per unit area of the cross-section containing the stresses. If the dislocations have Burgers vector **b** and move with velocity v, show that the block deforms with strain rate, $\dot{\varepsilon} = b\varrho_\mathrm{d}v$.

9. The energy for the formation of a vacancy is 1.5 eV for a simple f.c.c. metal in which the edge of the unit cell is 0.4 nm. The metal has a melting point of 800° C. Find the vacancy concentration and the average distance between vacancies at (i) room temperature; (ii) just below the melting point.

10. A bolt of high tensile steel holds two flat rigid plates together. The bolt is tightened until the stress in it reaches 70 MPa and the assembly is then placed in a high temperature environment. Under these conditions the bolt steel is known to creep at a strain rate which is proportional to the cube of the applied stress, one data point being that the creep rate is $1.5 \times 10^{-12}\,\mathrm{s^{-1}}$ for a stress of 30 MPa. What is the stress remaining in the bolt after six months, given that its Young's modulus is 230 GPa? State any assumptions that you make.
[Hint: note that, when there is stress relaxation by creep, the initial elastic strain is progressively replaced by plastic (creep) strain.]

2

Properties of elastic solids

In Chapter 1 we introduced some of the basic flow properties of solids and showed how, somewhat arbitrarily, solids are distinguished from liquids. In particular, we explained what we mean by stress and strain, and discussed how the two quantities are related in the elastic and plastic regimes of deformation. In this chapter we examine the elastic behaviour of solids in more detail and for this purpose it will be necessary to expand our treatments of stress and strain to cover three-dimensional bodies in the case of *elastically isotropic* solids, i.e. solids whose elastic response to an applied stress does not depend upon the direction in which it acts. In this chapter and in the following one we shall avoid the full rigour of the theory of elasticity and, for example, we shall not prove the basic theorems upon which the theory is constructed nor introduce those that are not essential (see end of chapter for titles of comprehensive texts on elasticity). We shall make three assumptions, namely: (i) the existence of perfect elasticity, which implies complete recovery of form after the removal of stress; (ii) that energy is conserved during elastic distortion and on recovery from it; (iii) that Hooke's law is obeyed.

2.1 Generalized stress

In Chapter 1 we gave a working formula for stress (eqn (1.5)) and we now require a definition of stress that is satisfactory when applied to infinitesimally small elements of a solid, which we need to examine in the theory of elasticity. Consider a small, planar elementary area, δA at a point P within a stressed body and take the resultant of all the forces acting on one side of the element (because of the material which is to that side) to be a force, δF. If we took away all the material on the other side we would have

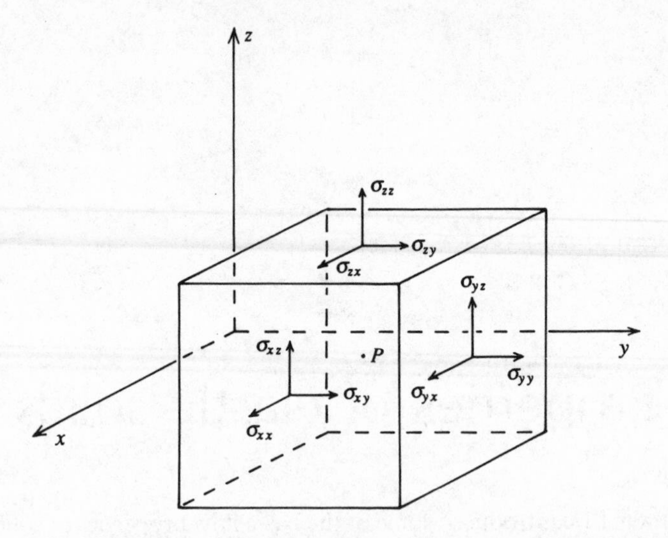

Fig. 2.1 A small cube centred on a point P within a stressed perfectly elastic body, showing the resolution of the stresses acting on three faces of the cube into the normal and shear components.

to apply a force δF to the newly-exposed side of the element δA in order to keep it from being displaced. Thus, previously equal and opposing forces δF must have been acting on the two sides of the element. The stress which acts at the point P in the original body is defined as

$$\lim_{\delta A \to 0} \delta F / \delta A$$

with the convention that the stress is positive when it is in the direction of a normal to the element drawn from P to the positive side of the element, which is usually determined with respect to a set of Cartesian axes.

We have not said that the force and the planar element are at right angles so that, calling upon the ideas in § 1.8, we see that the force δF may be resolved into normal and tangential components which, in turn, define the normal and shear stresses at the point P. Figure 2.1 shows a small cube of material about a point P within a body which is assumed to be acted upon by external forces. The cube has its faces perpendicular to a set of Cartesian coordinate axes, labelled x, y and z. Since the body as a whole is in a state of stress, a force $\mathbf{F}_x$ is exerted on the cube across the face of area A_x, by the adjoining material. This force can be resolved into components F_{xx}, F_{xy} and F_{xz} along the three coordinate axes (note that the first suffix refers to the face on which the force acts and the second to its direction). Thus the stress components at face A_x are

$$\sigma_{xx} = F_{xx}/A_x, \quad \sigma_{xy} = F_{xy}/A_x, \quad \sigma_{xz} = F_{xz}/A_x. \qquad (2.1)$$

A similar analysis of the faces A_y and A_z shows that there is a total of nine stress components σ_{ij}, where i and j can be any of the coordinates x, y and z. However when the cube is in static equilibrium, free of any torques tending to rotate it, we have complementary stresses with

$$\sigma_{xy} = \sigma_{yx}, \quad \sigma_{yz} = \sigma_{zy}, \quad \sigma_{zx} = \sigma_{xz}, \tag{2.2}$$

so that only six independent components are then required to describe the state of stress at a point (three normal stresses and three shear stresses).

The components of the force acting on a unit area of a plane passing through P and normal to an arbitrary vector $\mathbf{n}$ are given by

$$F_i = \sigma_{ix} n_x + \sigma_{iy} n_y + \sigma_{iz} n_z \quad (i = x, y, z)$$

i.e.

$$F_i = \sum_j \sigma_{ij} n_j \qquad (i, j = x, y, z). \tag{2.3}$$

The nine components of stress (six independent) are conveniently written in a 3×3 matrix array

$$\begin{bmatrix} \sigma_{xx} & \sigma_{xy} & \sigma_{xz} \\ \sigma_{yx} & \sigma_{yy} & \sigma_{yz} \\ \sigma_{zx} & \sigma_{zy} & \sigma_{zz} \end{bmatrix}. \tag{2.4}$$

It is clear from their definition that stresses may be positive or negative. If the orientation of the cube's axes are changed, the components of the stress will change as a result of the transformation (2.3) of the forces. It can be shown that there is always an orientation of the axes for which the shear components become zero. The double suffixes of the normal stress components can then be reduced to single suffixes, so that the matrix array becomes

$$\begin{bmatrix} \sigma_x & 0 & 0 \\ 0 & \sigma_y & 0 \\ 0 & 0 & \sigma_z \end{bmatrix}. \tag{2.5}$$

The resultant stresses are now perpendicular to the planes on which they act. These stresses are called the *principal stresses* at the point P, their directions are known as *principal axes* and the planes on which they act are *principal planes*. There are, of course, still six variables defining the stress system: three for the principal stresses and three to determine the orientations of the principal axes. In an *anisotropic* material the orientation of the principal axes in relation to the crystallographic axes is important in all cases except when

$$\sigma_x = \sigma_y = \sigma_z = \sigma = -p, \tag{2.6}$$

where p is the hydrostatic pressure on the body. The system is then said to be under *hydrostatic stress.*

The principal stress matrix (and therefore any stress system) may be separated into two parts as follows:

$$\begin{bmatrix} \sigma_x & 0 & 0 \\ 0 & \sigma_y & 0 \\ 0 & 0 & \sigma_z \end{bmatrix} = \begin{bmatrix} \frac{1}{3}(\sigma_x + \sigma_y + \sigma_z) & 0 & 0 \\ 0 & \frac{1}{3}(\sigma_x + \sigma_y + \sigma_z) & 0 \\ 0 & 0 & \frac{1}{3}(\sigma_x + \sigma_y + \sigma_z) \end{bmatrix}$$

$$+ \begin{bmatrix} \sigma_x - \frac{1}{3}(\sigma_x + \sigma_y + \sigma_z) & 0 & 0 \\ 0 & \sigma_y - \frac{1}{3}(\sigma_x + \sigma_y + \sigma_z) & 0 \\ 0 & 0 & \sigma_z - \frac{1}{3}(\sigma_x + \sigma_y + \sigma_z) \end{bmatrix}. \tag{2.7}$$

The first term represents a hydrostatic stress with

$$\sigma_h = -p = \tfrac{1}{3}(\sigma_x + \sigma_y + \sigma_z), \tag{2.8}$$

while the second, called the *deviatoric stress* or *reduced stress* is

$$\sigma_d = \sigma_i - \sigma_h \quad (i = x, y, z). \tag{2.9}$$

The effect of the hydrostatic component is to cause a change in volume but no change of shape in the case of an elastically isotropic solid, i.e. negative dilatation (contraction) without distortion. Even at very high levels of stress, no plastic flow is caused by a hydrostatic stress. This is because there are no shear stresses on any planes – every plane is therefore identical and is, in fact, a principal plane.

The deviatoric stress term, however, produces shear strains and can therefore lead to plastic flow if the elastic limit is exceeded. A deviatoric stress causes no dilatation because the sum of the components is always zero.

2.1.1 *Special cases of stress*

There are many special cases in which most of the six independent stress components are zero:

(i) Uniaxial stress.

$$\begin{bmatrix} 0 & 0 & 0 \\ 0 & 0 & 0 \\ 0 & 0 & \sigma \end{bmatrix}. \tag{2.10}$$

This is the case for a vertical rod or wire in a state of tension produced by a hanging weight.

(ii) Hydrostatic pressure. If the same pressure, p acts perpendicular to each surface of the cube, the stress components are

$$\begin{bmatrix} -p & 0 & 0 \\ 0 & -p & 0 \\ 0 & 0 & -p \end{bmatrix} \tag{2.11}$$

as in eqn (2.6).

(iii) Pure shear. A vertical block subjected to pure shear in the horizontal plane has stress components

$$\begin{bmatrix} 0 & \sigma & 0 \\ \sigma & 0 & 0 \\ 0 & 0 & 0 \end{bmatrix}. \tag{2.12}$$

2.2 Generalized strain

We saw in § 1.8 that strain is a change in a dimension, measured as a fraction of another dimension. It can therefore be expressed as a measure of the displacement of a point in a body in relation to reference points. Moving or rotating an entire body does not change the distances between points within the body, and such movements are called *rigid body motions*. The overall displacement of points in a body can always be expressed as a sum of (i) rigid body motions and (ii) displacements associated with changes of shape or volume. Strain is only concerned with (ii).

In order to visualize the nature of a general distortion it is simplest to consider first the case of infinitesimal strain in a two-dimensional xy-plane. The results can then be generalized into three dimensions without difficulty. Under the action of the applied forces, a point P at initial position $\mathbf{r}$, coordinates x, y, is displaced to a point P' at position $\mathbf{r} + \mathbf{u}$ as shown in fig. 2.2. To exclude rigid body translations it follows that other points within the body are displaced by different amounts. The displacement $\mathbf{u}$ may be quite large even though the distortion of the material may be relatively small. For example, when a weight is hung from a long wire, its bottom end may be lowered through quite a large distance even though the percentage length increase of the wire is small. Suppose that a point Q, initially close to P at position $\mathbf{r} + \Delta\mathbf{r}$ is caused to move by the

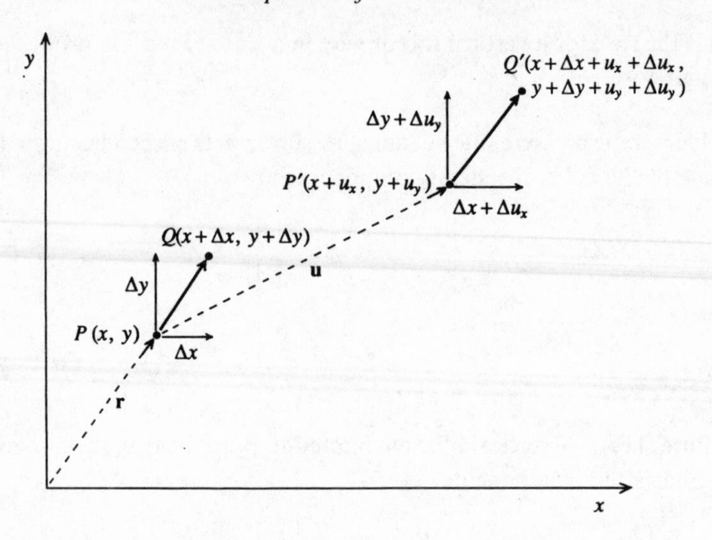

Fig. 2.2 Infinitesimal homogeneous elastic strain in two dimensions.

distortion to Q' at position $\mathbf{r} + \mathbf{u} + \Delta\mathbf{r} + \Delta\mathbf{u}$. For small relative distortions

$$|\Delta\mathbf{u}| \ll |\Delta\mathbf{r}|.$$

Since the components of **u** are, in general, slowly varying functions of x and y, the components of relative displacement are given by

$$\begin{aligned} \Delta u_x &= (\partial u_x/\partial x)\,\Delta x + (\partial u_x/\partial y)\,\Delta y, \\ \Delta u_y &= (\partial u_y/\partial x)\,\Delta x + (\partial u_y/\partial y)\,\Delta y. \end{aligned} \tag{2.13}$$

The linear relationships in eqns (2.13) correspond to the case of *homogeneous strain*, in which all straight lines remain straight after straining and parallel straight lines remain parallel, although their common directions may be altered.

The strain components are now defined in terms of the dimensionless displacement gradients:

$$\begin{aligned} e_{xx} &= \partial u_x/\partial x, \quad e_{xy} = \partial u_x/\partial y, \\ e_{yx} &= \partial u_y/\partial x, \quad e_{yy} = \partial u_y/\partial y. \end{aligned} \tag{2.14}$$

As in the example of the stretched wire, these quantities are independent of position for a homogeneous strain – see fig. 2.3. For a homogeneous extension

$$u_z'' > u_z',$$

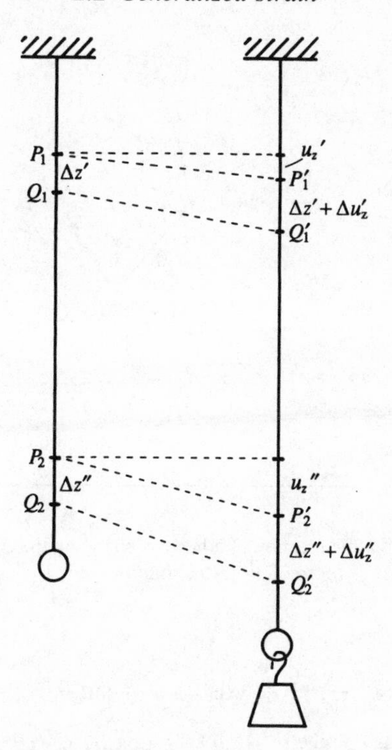

Fig. 2.3 Infinitesimal strain as it applies to the case of the elastic extension of a wire by a hanging weight.

but

$$\begin{aligned} e_{zz} &= \Delta u_z'/\Delta z' = \Delta u_z''/\Delta z'' \\ &= \partial u_z/\partial z, \end{aligned} \tag{2.15}$$

which is independent of z. We see from both figs. 2.2 and 2.3 that e_{xx}, e_{yy} and e_{zz} are compressive or tensile strains.

The quantities e_{ij} do not, however, quite determine the distortion of the cube of fig. 2.1 as intended, since a rigid rotation of a body without any deformation produces non-zero values of e_{xy} and e_{yx}. It is our intention to exclude rotations and rigid body displacements. Figure 2.4 shows the effect of a simple rotation of a body around some axis through O, perpendicular to the xy-plane. If P has coordinates x, y before rotation, the new position P' has coordinates

$$\begin{aligned} x' &= x + u_x = x\cos\theta - y\sin\theta, \\ y' &= y + u_y = x\sin\theta + y\cos\theta. \end{aligned} \tag{2.16}$$

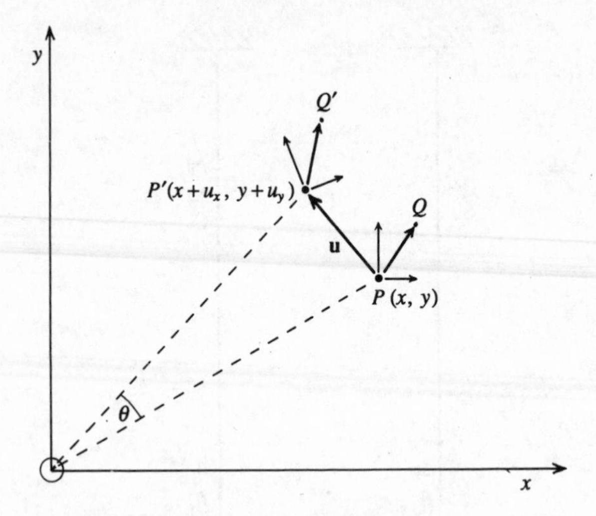

Fig. 2.4 The effect of a simple rotation of a body in changing the coordinates of an internal point P (before rotation) to those of P' (after rotation).

Thus

$$u_x = -x(1 - \cos\theta) - y\sin\theta \simeq -y\theta, \tag{2.17}$$

$$u_y = x\sin\theta - y(1 - \cos\theta) \simeq x\theta, \tag{2.18}$$

for small angles of rotation. Thus the displacement gradients, defined in eqn (2.14), have the values

$$e_{xx} = e_{yy} = 0, \quad e_{xy} = -\theta \quad \text{and} \quad e_{yx} = \theta. \tag{2.19}$$

Rigid rotations are characterized by the antisymmetry property

$$e_{xy} = -e_{yx} \tag{2.20}$$

of the shear displacement gradients. The components ε_{ij} of the *rational strain* ε are accordingly defined so that these rotational contributions are removed,

$$\varepsilon_{xx} = e_{xx} = \partial u_x/\partial x, \quad \varepsilon_{yy} = e_{yy} = \partial u_y/\partial y, \tag{2.21}$$

$$\varepsilon_{xy} = e_{xy} + e_{yx} = \frac{\partial u_x}{\partial y} + \frac{\partial u_y}{\partial x}. \tag{2.22}$$

It is obvious from this definition that

$$\varepsilon_{yx} = \varepsilon_{xy}. \tag{2.23}$$

These results for two dimensions generalize to three dimensions in a very obvious way. There are nine strain components, but analogous to eqn

(2.23),

$$\varepsilon_{yz} = \varepsilon_{zy}, \quad \varepsilon_{zx} = \varepsilon_{xz} \tag{2.24}$$

and only six components are independent. The components are conveniently written in a matrix array:

$$\begin{bmatrix} \varepsilon_{xx} & \varepsilon_{xy} & \varepsilon_{xz} \\ \varepsilon_{yx} & \varepsilon_{yy} & \varepsilon_{yz} \\ \varepsilon_{zx} & \varepsilon_{zy} & \varepsilon_{zz} \end{bmatrix}. \tag{2.25}$$

The components on the leading diagonal are normal strains which, since the ε_{ij} can be positive or negative, may be extensions or compressions. The off-diagonal elements ($i \neq j$) represent shear distortions. All ε_{ij} are $\ll 1$.

2.2.1 *Special cases of strain*

Many of the strain components vanish in some important practical cases when we use principal axes:

(i) Uniaxial extension. An extension parallel to the z-axis, with no change in the dimensions of the body in the xy-plane, has strain matrix

$$\begin{bmatrix} 0 & 0 & 0 \\ 0 & 0 & 0 \\ 0 & 0 & \varepsilon \end{bmatrix}. \tag{2.26}$$

(ii) Uniform expansion or compression. The strain matrix is

$$\begin{bmatrix} \varepsilon & 0 & 0 \\ 0 & \varepsilon & 0 \\ 0 & 0 & \varepsilon \end{bmatrix}. \tag{2.27}$$

The fractional change in volume of a strained material is the dilatation or the volume strain, previously defined in eqn (1.31). For a unit cube of an isotropic solid that has been strained, the dilatation is

$$\Theta = \Delta V/V = (1 + \varepsilon)^3 - 1 \simeq 3\varepsilon \tag{2.28}$$

since ε is small. This situation includes that of hydrostatic pressure, in which case all the strains are negative.

(iii) General dilatation. The strain matrix resembles case (ii), except that the strains are unequal, i.e.

$$\begin{bmatrix} \varepsilon_x & 0 & 0 \\ 0 & \varepsilon_y & 0 \\ 0 & 0 & \varepsilon_z \end{bmatrix}. \tag{2.29}$$

Since all the strains are small, the dilatation is

$$\Theta = (1 + \varepsilon_x)(1 + \varepsilon_y)(1 + \varepsilon_z) - 1 \simeq \varepsilon_x + \varepsilon_y + \varepsilon_x. \tag{2.30}$$

(iv) Volume-preserving extension. The volume of a stretched body can be preserved if the body contracts suitably in the plane perpendicular to the axis of tension. The strain matrix

$$\begin{bmatrix} -\frac{1}{2}\varepsilon & 0 & 0 \\ 0 & -\frac{1}{2}\varepsilon & 0 \\ 0 & 0 & \varepsilon \end{bmatrix} \tag{2.31}$$

preserves the initial volume of the material.

(v) Pure shear. The strain matrix

$$\begin{bmatrix} 0 & \varepsilon & 0 \\ \varepsilon & 0 & 0 \\ 0 & 0 & 0 \end{bmatrix} \tag{2.32}$$

corresponds to the shear deformation illustrated in fig. 2.5.

2.2.2 General state of strain

The relative magnitudes of the strain components vary with the axes that are used to measure them. There is always a set of principal axes for which all the components vanish except those on the leading diagonal. This does not mean that there is no shear strain in the material and fig. 2.6 shows pure shear in such a case. A lozenge *abcd* is inscribed in a square which is then elongated along x and compressed along y. If there is no change in dimension parallel to the z-axis, the strain matrix has the form

$$\begin{bmatrix} \varepsilon & 0 & 0 \\ 0 & -\varepsilon & 0 \\ 0 & 0 & 0 \end{bmatrix} \text{(unprimed axes)}. \tag{2.33}$$

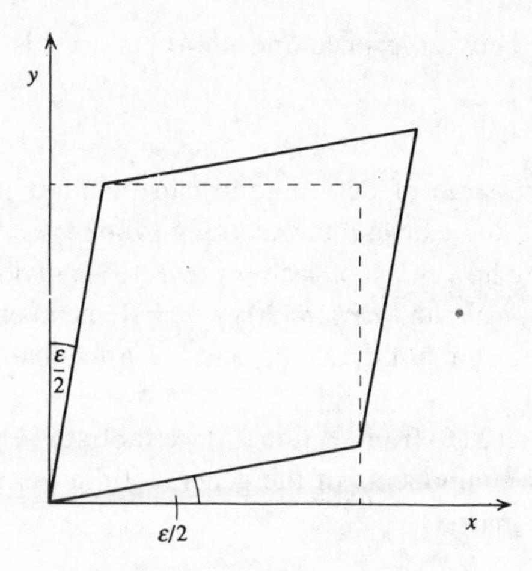

Fig. 2.5 Two dimensional representation of the effect of pure shear by the distortion of a square into a rhombus.

The matrix is diagonal and x, y and z are accordingly principal axes for the assumed strain. Consider however the set of axes x', y' and z'. The axes x' and y' bear the same relation to the strained material as do the axes x and y in fig. 2.5, and the strain matrix is

$$\begin{bmatrix} 0 & 2\varepsilon & 0 \\ 2\varepsilon & 0 & 0 \\ 0 & 0 & 0 \end{bmatrix} \text{(primed axes).} \tag{2.34}$$

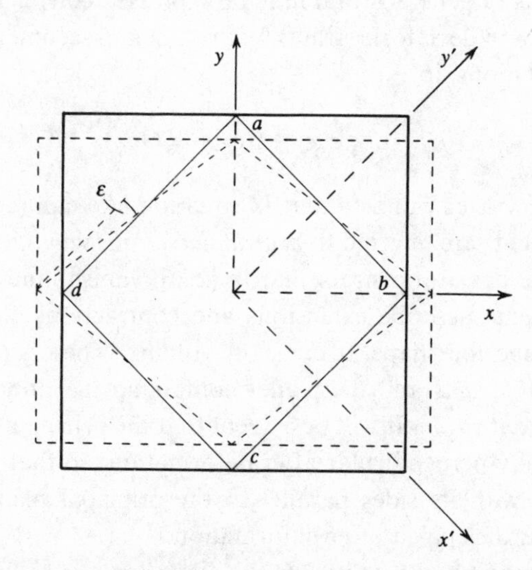

Fig. 2.6 Pure shear of an element *abcd* within a square deformed by compressive and tensile forces applied parallel to its edges in the plane of the diagram.

The relationship between primed and unprimed axes is

$$\varepsilon_{x'y'} = \varepsilon_{y'x'} = \varepsilon_{xx} - \varepsilon_{yy}. \tag{2.35}$$

Since the same state of deformation can manifest itself as different amounts of shear and elongation, according to the axes chosen, we need a way of identifying how much of each is present. We divide the total strain into *dilatational strain* and *deviatoric strain.* (Remember that the former changes the volume but not the shape, and the latter changes the shape but not the volume.)

Using the result (2.30) from section (iii) we subtract $\frac{1}{3}\Theta$ from each of the leading diagonal components of the general strain matrix to obtain the deviatoric strain matrix

$$\begin{bmatrix} \varepsilon_{xx} - \frac{1}{3}\Theta & \varepsilon_{xy} & \varepsilon_{xz} \\ \varepsilon_{yx} & \varepsilon_{yy} - \frac{1}{3}\Theta & \varepsilon_{yz} \\ \varepsilon_{zx} & \varepsilon_{zy} & \varepsilon_{zz} - \frac{1}{3}\Theta \end{bmatrix}. \tag{2.36}$$

By defining a symbol δ_{ij} with the properties that

$$\begin{aligned} \delta_{ij} &= 1 \quad \text{when} \quad i = j \\ \text{and} \quad \delta_{ij} &= 0 \quad \text{when} \quad i \neq j, \end{aligned} \tag{2.37}$$

the components of general strain may be expressed conveniently in terms of dilatational and deviatoric strains by the first and second terms respectively in the relationship

$$\varepsilon_{ij} = \tfrac{1}{3}\Theta\delta_{ij} + (\varepsilon_{ij} - \tfrac{1}{3}\Theta\delta_{ij}), \tag{2.38}$$

which may be written out as a matrix equation like eqn (2.7). When the strain components are referred to *principal axes of strain*, the off-diagonal elements in the deviatoric strain matrix (2.36) vanish, and the surviving diagonal elements describe extensions and contractions along principal axes that change the shape at constant volume. These strains are then called *principal strains* and the planes containing the principal axes are *principal planes of strain.* It will be evident that the principal axes of strain are still mutually perpendicular after deformation, so that a rectangular parallelepiped with its sides parallel to the principal planes remains a rectangular parallelepiped after deformation.

For the example shown in fig. 2.6, the dilatation is zero for both sets of axes, and both matrices (2.33) and (2.34) are entirely deviatoric, corresponding to the pure shear character of the strain that they represent.

2.3 Generalized elasticity

The stress and strain described in the previous two sections are proportional to one another in the manner of eqns (1.32) and (1.33). However, these equations seriously understate the complexity of the proportionalities for a general elastic solid. We have seen that there are six independent stresses σ_{ij} and six independent strains ε_{ij}. In general, any one component of the strain is proportional to all six components of the stress, and vice versa.

Because of the complexity of these relations between stresses and strains it is customary to reduce the number of subscripts by means of an abbreviated notation for pairs of coordinate directions, as follows:

$$\begin{array}{lcccccc} \text{full} & xx & yy & zz & yz & zx & xy \\ \text{abbreviated} & 1 & 2 & 3 & 4 & 5 & 6 \end{array} \tag{2.39}$$

The generalization of eqn (1.33) is written

$$\sigma_i = \sum_j c_{ij}\varepsilon_j, \tag{2.40}$$

where the subscripts i and j refer to the abbreviated notation in the second line of the conversions (2.39). There are, in general, 36 independent coefficients c_{ij} of elastic stiffness.

An example of the relation (2.40) written out in full for the normal stress in the x-direction is

$$\sigma_1 = c_{11}\varepsilon_1 + c_{12}\varepsilon_2 + c_{13}\varepsilon_3 + c_{14}\varepsilon_4 + c_{15}\varepsilon_5 + c_{16}\varepsilon_6 \,. \tag{2.41}$$

Further simplifications occur because it can be shown by consideration of the energy stored in a strained material (see for example Nye, 1957) that the stiffness coefficients have a symmetry property analogous to eqns (2.2) and (2.23) for the stress and strain, namely

$$c_{ji} = c_{ij}. \tag{2.42}$$

The six relations obtained from eqn (2.40) can thus be written in matrix form

$$\begin{bmatrix} \sigma_1 \\ \sigma_2 \\ \sigma_3 \\ \sigma_4 \\ \sigma_5 \\ \sigma_6 \end{bmatrix} = \begin{bmatrix} c_{11} & c_{12} & c_{13} & c_{14} & c_{15} & c_{16} \\ c_{12} & c_{22} & c_{23} & c_{24} & c_{25} & c_{26} \\ c_{13} & c_{23} & c_{33} & c_{34} & c_{35} & c_{36} \\ c_{14} & c_{24} & c_{34} & c_{44} & c_{45} & c_{46} \\ c_{15} & c_{25} & c_{35} & c_{45} & c_{55} & c_{56} \\ c_{16} & c_{26} & c_{36} & c_{46} & c_{56} & c_{66} \end{bmatrix} \begin{bmatrix} \varepsilon_1 \\ \varepsilon_2 \\ \varepsilon_3 \\ \varepsilon_4 \\ \varepsilon_5 \\ \varepsilon_6 \end{bmatrix}. \tag{2.43}$$

It is seen that because of the relation (2.42) only 21 of the 36 stiffness coefficients are different.

The physical significance of 21 different non-zero stiffness coefficients is that any one of the six independent strains that can occur in a material sample causes different non-zero forces corresponding to all six independent stresses. Conversely, any one applied stress produces non-zero contributions to all six of the strain components. Such extremely complicated elastic behaviour is, in practice, only found in the most unsymmetrical of crystal structures, those belonging to the triclinic class. In all materials of higher symmetry, it can be shown that some of the stiffness coefficients must vanish, and in many cases there are simple relations between different coefficients. The number of independent coefficients is considerably less than 21 for most materials of common interest. The various symmetries are systematically considered by Nye (1957), who gives tables of their compliance and stiffness matrices.

The properties of the elastic stiffness are emphasized above but it is a simple matter to obtain the corresponding results for the elastic compliance. The generalization of eqn (1.32) is

$$\varepsilon_i = \sum_j s_{ij}\sigma_j, \tag{2.44}$$

and the symmetry property

$$s_{ji} = s_{ij}, \tag{2.45}$$

analogous to eqn (2.42), reduces the number of independent coefficients s_{ij} of the elastic compliance from 36 to 21. The general relation (2.44) has a matrix form similar to eqn (2.43) but with σ and ε interchanged and c replaced by s.

Our further discussion of elasticity is confined to isotropic materials, where the compliance and stiffness matrices take their simplest possible forms. Some properties of the stiffness are obvious. Thus the x-, y- and z-directions are equivalent, so that

$$c_{11} = c_{22} = c_{33} \quad \text{and} \quad c_{23} = c_{31} = c_{12}. \tag{2.46}$$

Furthermore, it is clear that normal stresses only generate extensions or compressions, while shear stresses only generate shear strains, and many of the stiffness coefficients accordingly vanish. The complete symmetry analysis shows that the stiffness matrix has the form

$$\begin{bmatrix} c_{11} & c_{12} & c_{12} & 0 & 0 & 0 \\ c_{12} & c_{11} & c_{12} & 0 & 0 & 0 \\ c_{12} & c_{12} & c_{11} & 0 & 0 & 0 \\ 0 & 0 & 0 & \frac{1}{2}(c_{11} - c_{12}) & 0 & 0 \\ 0 & 0 & 0 & 0 & \frac{1}{2}(c_{11} - c_{12}) & 0 \\ 0 & 0 & 0 & 0 & 0 & \frac{1}{2}(c_{11} - c_{12}) \end{bmatrix}. \tag{2.47}$$

Table 2.1. *Values of elastic moduli, Poisson's ratio, stiffness coefficients and anisotropy factor for selected materials at room temperature and atmospheric pressure*

Substance	E	G	ν	c_{11}	c_{12}	c_{44}	A
Isotropic							
Granite	46	19	0.21	52	14	19	1
Concrete	15–35	6–15	0.2	16–40	4–10	6–15	1
Glass (Crown)	71.3	29.2	0.22	81.3	22.9	29.2	1
Fused silica	73.1	31.2	0.17	78.5	16.1	31.2	1
Polyethylene	1.2	0.4	0.45	0.1	0.03	0.04	1
(high density, unoriented fibres, at low strain rate)							
Rubber	0.01	0.01	0.45	–	–	–	–
(at low stress)							
Cubic							
Aluminium	72.5	27.6	0.31	107.5	60.8	28.5	0.82
Lead′	26	8.5	0.41	47.7	40.3	14.4	0.26
Iron′	215	77	0.29	228.1	133.5	110.9	0.43
Diamond′	900	360	0.24	950.0	390.0	430.0	0.65
Silicon	130	51	0.28	165.7	63.9	79.6	1.56
Sodium chloride	53.5	23.1	0.20	48.6	11.9	12.8	1.43
Argon	4.8	2.0	0.18	5.33	1.33		
Non-cubic							
Zinc				161	34.2	38.3	
Ice (−16° C)				13.8	7.1	3.2	
α-Quartz′	96	44	0.08	87.6	6.1	57.2	0.68
Calcite′	83	32	0.32	148.0	55.4	32.7	1.42
Olivine′	200	80	0.25	324.0	59.0	66.7	1.99
(91.7% Mg_2SiO_4)†							

The data come from various sources, but most are listed in Simmons, G. & Wang, H., *Single Crystal Elastic Constants and Calculated Aggregate Properties: a Handbook*, 2nd edn, M.I.T. Press, Cambridge, Mass., U.S.A., 1971. This book is also the source of the averages of bounds for E, G and ν, marked with a prime (′) and calculated as the properties of aggregates.
† This result is from Kumazawa, M. & Anderson, O.L. (1969). *J. Geophys. Res.* **74**, 5961–72. The units of E, G and c are GPa. (1 Pa is equal to 1 N m^{-2}).

Thus only 2 independent numbers are needed to characterize the elastic stiffness properties of an isotropic material. Table 2.1 shows some representative values.

It is often convenient to use a hybrid notation for the elasticity equations, in which the full Cartesian axis subscripts are used for the stress and strain components, but the abbreviated subscripts are used for the stiffness. With this convention, the stiffness matrix (2.47) generates the set of equations

$$\begin{aligned}
\sigma_{xx} &= c_{11}\varepsilon_{xx} + c_{12}\varepsilon_{yy} + c_{12}\varepsilon_{zz} \\
\sigma_{yy} &= c_{12}\varepsilon_{xx} + c_{11}\varepsilon_{yy} + c_{12}\varepsilon_{zz} \\
\sigma_{zz} &= c_{12}\varepsilon_{xx} + c_{12}\varepsilon_{yy} + c_{11}\varepsilon_{zz} \\
\sigma_{yz} &= \tfrac{1}{2}(c_{11} - c_{12})\,\varepsilon_{yz} \\
\sigma_{zx} &= \tfrac{1}{2}(c_{11} - c_{12})\,\varepsilon_{zx} \\
\sigma_{xy} &= \tfrac{1}{2}(c_{11} - c_{12})\,\varepsilon_{xy}.
\end{aligned} \tag{2.48}$$

The compliance matrix for an isotropic material has the form

$$\begin{bmatrix}
s_{11} & s_{12} & s_{12} & 0 & 0 & 0 \\
s_{12} & s_{11} & s_{12} & 0 & 0 & 0 \\
s_{12} & s_{12} & s_{11} & 0 & 0 & 0 \\
0 & 0 & 0 & 2(s_{11} - s_{12}) & 0 & 0 \\
0 & 0 & 0 & 0 & 2(s_{11} - s_{12}) & 0 \\
0 & 0 & 0 & 0 & 0 & 2(s_{11} - s_{12})
\end{bmatrix} \tag{2.49}$$

and the explicit relations between stress and strain are

$$\begin{aligned}
\varepsilon_{xx} &= s_{11}\sigma_{xx} + s_{12}\sigma_{yy} + s_{12}\sigma_{zz} \\
\varepsilon_{yy} &= s_{12}\sigma_{xx} + s_{11}\sigma_{yy} + s_{12}\sigma_{zz} \\
\varepsilon_{zz} &= s_{12}\sigma_{xx} + s_{12}\sigma_{yy} + s_{11}\sigma_{zz} \\
\varepsilon_{yz} &= 2(s_{11} - s_{12})\,\sigma_{yz} \\
\varepsilon_{zx} &= 2(s_{11} - s_{12})\,\sigma_{zx} \\
\varepsilon_{xy} &= 2(s_{11} - s_{12})\,\sigma_{xy}.
\end{aligned} \tag{2.50}$$

The sets of relations given in eqns (2.48) and (2.50) connect the same stress and strain variables and they must clearly be equivalent to one another. Solution of the eqns (2.48) to obtain the strains as functions of the stresses, and comparison with the eqns (2.50) shows that the compliance and stiffness coefficients must be related by

$$s_{11} = \frac{c_{11} + c_{12}}{(c_{11} - c_{12})(c_{11} + 2c_{12})} \quad \text{and} \quad s_{12} = -\frac{c_{12}}{(c_{11} - c_{12})(c_{11} + 2c_{12})}. \tag{2.51}$$

It follows that

$$2(s_{11} - s_{12}) = 2/(c_{11} - c_{12}) \tag{2.52}$$

in accordance with the equivalence of the shear relations in eqns (2.48) and (2.50). The fractional volume change of a stressed sample, given by the dilatation defined in eqn (2.30), can be expressed in the forms

$$\begin{aligned} \Theta &= \varepsilon_{xx} + \varepsilon_{yy} + \varepsilon_{zz} = (s_{11} + 2s_{12})(\sigma_{xx} + \sigma_{yy} + \sigma_{zz}) \\ &= (\sigma_{xx} + \sigma_{yy} + \sigma_{zz})/(c_{11} + 2c_{12}). \end{aligned} \tag{2.53}$$

The compliance s_{12} is negative or zero for most isotropic substances, while s_{11} is positive,

$$s_{11} > 0 \quad \text{and} \quad s_{12} \leqslant 0. \tag{2.54}$$

The stiffness coefficients similarly satisfy

$$c_{11} > 0 \quad \text{and} \quad c_{12} \geqslant 0. \tag{2.55}$$

The relations (2.51) between stiffness and compliance show that these inequalities are mutually consistent only if

$$c_{11} > c_{12}. \tag{2.56}$$

We shall henceforth confine the discussion to substances for which these inequalities are valid.

2.3.1 *Special cases of stress*

The relations between stress and strain simplify further for special kinds of applied stress:

(i) Uniaxial stress. If only σ_{zz} is non-zero (eqn (2.10)), as in the extension of a vertical wire by a hanging weight, the eqns (2.50) reduce to

$$\varepsilon_{xx} = \varepsilon_{yy} = s_{12}\sigma_{zz} \quad \text{and} \quad \varepsilon_{zz} = s_{11}\sigma_{zz}. \tag{2.57}$$

Extension in the z-direction is thus accompanied by contraction in the x- and y-directions. Young's modulus is defined to be

$$E = \frac{\text{normal stress}}{\text{normal strain}} = \frac{\sigma_{zz}}{\varepsilon_{zz}} = \frac{1}{s_{11}} = \frac{(c_{11} - c_{12})(c_{11} + 2c_{12})}{c_{11} + c_{12}}, \tag{2.58}$$

while Poisson's ratio is defined to be

$$\nu = \frac{|\text{transverse strain}|}{\text{normal strain}} = \frac{|\varepsilon_{xx}|}{\varepsilon_{zz}} = -\frac{s_{12}}{s_{11}} = \frac{c_{12}}{c_{11} + c_{12}}. \tag{2.59}$$

It is seen from the inequality (2.56) that Poisson's ratio lies in the range

$$0 \leqslant \nu < \tfrac{1}{2}; \tag{2.60}$$

for example ν has values close to $\frac{1}{3}$ for polycrystalline aluminium and nickel. The ratio of stiffness coefficients is given by

$$c_{12}/c_{11} = \nu/(1 - \nu). \tag{2.61}$$

The fractional change in volume caused by the uniaxial stress is obtained from eqn (2.53) as

$$\Theta = (s_{11} + 2s_{12})\sigma_{zz} = (1/E)(1 - 2\nu)\sigma_{zz}, \tag{2.62}$$

where eqns (2.58) and (2.59) have been used. It is seen from the allowed range of ν given in eqn (2.60) that, except for $\nu \approx \frac{1}{2}$, the volume of the stretched wire increases, despite the compensatory effects of its transverse contraction.

(ii) Hydrostatic pressure. If

$$\sigma_{xx} = \sigma_{yy} = \sigma_{zz} = -p \text{ and } \sigma_{yz} = \sigma_{zx} = \sigma_{xy} = 0, \tag{2.63}$$

as in eqns (2.6) and (2.11), then from eqn (2.50)

$$\varepsilon_{xx} = \varepsilon_{yy} = \varepsilon_{zz} = -(s_{11} + 2s_{12})p. \tag{2.64}$$

The sample is in a state of uniform compression as defined in eqn (2.27) and, using eqns (2.58) and (2.59), the dilatation given by eqn (2.53) is

$$\Theta = -3(s_{11} + 2s_{12})p = -\frac{3(1 - 2\nu)}{E}p. \tag{2.65}$$

The bulk modulus is defined (eqn (1.35)) as

$$K = \frac{p}{|\Theta|} = \frac{1}{3(s_{11} + 2s_{12})} = \frac{E}{3(1 - 2\nu)}. \tag{2.66}$$

(iii) Pure shear. If only σ_{xy} is non-zero, as in eqn (2.12), then the only non-zero strain component from eqn (2.50) is

$$\varepsilon_{xy} = 2(s_{11} - s_{12})\sigma_{xy}. \tag{2.67}$$

The sample is thus in a state of pure shear as defined in eqn (2.32). The rigidity modulus is defined to be

$$G = \frac{\text{shear stress}}{\text{shear strain}} = \frac{\sigma_{xy}}{\varepsilon_{xy}} = \frac{1}{2(s_{11} - s_{12})} = \tfrac{1}{2}(c_{11} - c_{12}) = \frac{E}{2(1 + \nu)}, \tag{2.68}$$

where eqns (2.58) and (2.59) have been used. Since the total angle of shear is equal to the shear strain, as shown in fig. 2.5, the definition of the rigidity modulus is equivalent to eqn (1.34).

The above considerations of special cases show how the various coefficients of elasticity are interrelated. It should be emphasized that only two of them are independent for a given elastically-isotropic substance. Different applications of elasticity theory conveniently make use of different pairs of coefficients, for example s_{11} and s_{12}, or c_{11} and c_{12}, or E and ν.

It should also be noted that although crystals belonging to the cubic class are optically isotropic, they are not in general elastically isotropic. Cubic crystals have three different non-zero elastic coefficients, whereas only two are needed for an elastically isotropic solid. This has important consequences in that, for example, the rigidity modulus depends on the plane and the direction of shear, so that shear on the $\{100\}$ cube planes in the $\langle 001 \rangle$ directions gives

$$G_{\{100\}} = c_{44},$$

while for the $\{110\}$ planes in the $\langle 1\bar{1}0 \rangle$ directions

$$G_{\{110\}} = \tfrac{1}{2}(c_{11} - c_{12}).$$

The ratio of these moduli is known as the *anisotropy factor* and is

$$A = (c_{11} - c_{12})/2c_{44}. \qquad (2.69)$$

The data in Table 2.1 enable the anisotropy factors for a few cubic crystals to be compared; the factor is of course equal to unity for elastically isotropic materials.

2.4 Elastic constants in relation to interatomic forces

The elastic constants of a material determine how it responds macroscopically to the application of forces. We can see how the macroscopic behaviour relates to the microscopic, or atomic, properties by examining a model solid whose atoms are arranged in a simple cubic array, as shown in fig. 2.7. We take the equilibrium separation of the atoms in the close-packed direction (i.e. parallel to the cell edge) to be r_0 and we proceed to look at the effect of applying balanced uniform tensile forces T to the solid, acting parallel to the x-axis, as shown in the figure. These forces affect all the rows of atoms parallel to the x-axis in a similar manner, so that the yz-planes of atoms become more widely spaced but remain planar. We may therefore ignore the forces between atoms in adjacent rows and restrict our discussion to what happens in a single row.

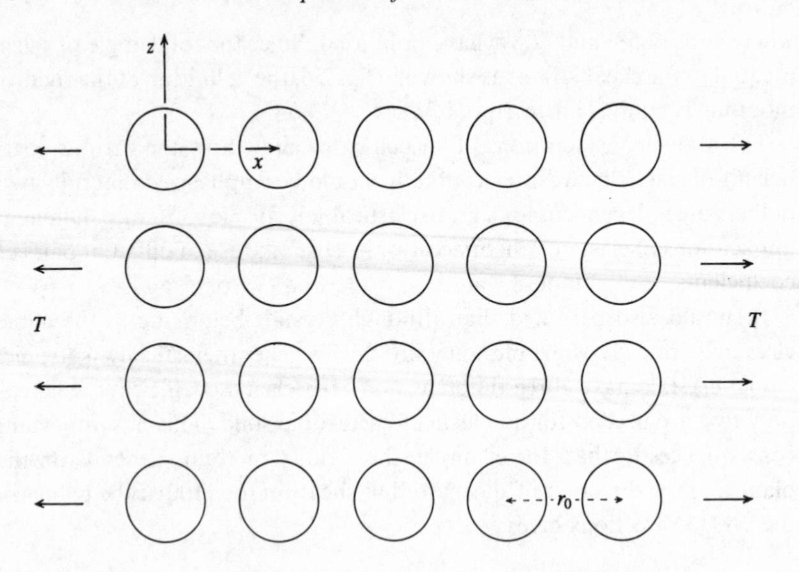

Fig. 2.7 The deformation of a simple cubic array of atoms under a uniform tensile stress field.

The effective cross-sectional area of a row of atoms, perpendicular to its length, is r_0^2, so that the stress on a row is

$$\sigma = T/r_0^2. \tag{2.70}$$

If the interatomic separation increases by δr, the normal strain is

$$\varepsilon = \delta r/r_0. \tag{2.71}$$

We assume a general form for the dependence of potential energy on interatomic distance for two atoms, called the *Mie potential*. (The Lennard-Jones potential discussed in § 1.1 is a special case of the Mie potential.) Thus the potential energy is

$$U = -\frac{A}{r^m} + \frac{B}{r^n}. \tag{2.72}$$

Then the interatomic force F is

$$F = -\frac{\mathrm{d}U}{\mathrm{d}r} = -\frac{mA}{r^{m+1}} + \frac{nB}{r^{n+1}}. \tag{2.73}$$

Using the condition that when $F = 0$, $r = r_0$, we find that

$$B = \frac{m}{n} A r_0^{n-m}. \tag{2.74}$$

The shape of the F versus r curve (see fig. 1.2(a)) in the neighbourhood of

r_0 determines, at the microscopic level, the amount of strain produced by a given macroscopic tension on the body. Increasing the interatomic separation by an amount δr produces a force δF between every adjacent pair of atoms in the row given by

$$\delta F = (\mathrm{d}F/\mathrm{d}r)_{r=r_0}\,\delta r, \tag{2.75}$$

where the gradient of the F–r curve is evaluated at $r = r_0$. T and δF must be opposed, so that using the sign convention for stress given in § 2.1 we see from eqns (2.70) and (2.71) that Young's modulus is given by eqn (2.58) as

$$E = \frac{\sigma}{\varepsilon} = \frac{T}{r_0\delta r} = -\frac{\delta F}{r_0\delta r} = -\frac{1}{r_0}\left(\frac{\mathrm{d}F}{\mathrm{d}r}\right)_{r=r_0}. \tag{2.76}$$

Differentiation of eqn (2.73) and elimination of B by the use of eqn (2.74) gives

$$\left(\frac{\mathrm{d}F}{\mathrm{d}r}\right)_{r=r_0} = \frac{m(m+1)A}{r_0^{m+2}} - \frac{n(n+1)B}{r_0^{n+2}} = \frac{m(m-n)A}{r_0^{m+2}}. \tag{2.77}$$

Thus substitution into eqn (2.76) gives

$$E = \frac{m(n-m)A}{r_0^{m+3}}. \tag{2.78}$$

This expression does not tell us very much, so we need to evaluate the constant A, which we can do as follows:

The energy of a bond between two atoms is the value of the potential energy U at the equilibrium atomic separation r_0, which from eqns (2.72) and (2.74) is

$$\begin{aligned} U_0 &= -\frac{A}{r_0^m} + \frac{B}{r_0^n} \\ &= \frac{A}{r_0^m}\left(\frac{m}{n} - 1\right). \end{aligned} \tag{2.79}$$

The power n in the repulsive force term can be determined by measuring the compressibility of a solid. For a rather ideal solid that behaves as though it is a close-packed structure of 'hard' spheres, with a weak van der Waals type of interaction (a good example is solid argon), appropriate values for m and n are 6 and 12 respectively (the Lennard-Jones potential, § 1.1), so that

$$U_0 = -A/2r_0^6. \tag{2.80}$$

The molar binding energy of solid argon is about 6.5 kJ mole^{-1} which, if divided by the product of one half the *coordination number*, 12 (the number of nearest neighbour atoms) and Avogadro's number, gives a binding energy (the energy of the bond between two atoms) of about 1.8×10^{-21} J. Taking the interatomic separation to be 0.37 nm, we are now able to estimate the constant A and hence, from eqn (2.78), Young's modulus E. The value of E derived in this fashion is about 2.6×10^9 Pa, which is of the same order as measured experimentally (see Table 2.1).

2.5 Stresses in thin shells subject to internal pressure from a fluid

Now that we have introduced some of the basic concepts about stress, and strain, and shown that the relationship between them (Hooke's law) has microscopic origins, it is appropriate to examine some practical situations. In general our aims are to show how to calculate the changes in the dimensions of a body resulting from the application of a given set of stresses. We start with two simple examples of thin shells of homogeneous material containing fluid under pressure.

(i) A thin spherical shell. We assume that the internal pressure exceeds the external pressure and denote the difference or excess pressure by p. We resolve the stress at some point on the shell into normal stress components σ_1 and σ_2 in the shell wall and σ_3 perpendicular to it, as shown in fig. 2.8(a). For a thin shell there is no gradient of stress in the shell thickness and the radial stress within the shell is zero, so that $\sigma_3 = 0$. By symmetry, the tangential components of stress in the shell must be equal so that

$$\sigma_1 = \sigma_2 = \sigma_T, \text{ say.} \tag{2.81}$$

Consider the equilibrium of one hemisphere of the shell, shown in fig. 2.8(b). The total force around the circumference must be produced by the internal pressure acting over the hemispherical interior surface. If the thickness of the shell wall is t and the radius of the shell is r, then

$$\sigma_T 2\pi rt = \int_{\pi/2}^{0} p \cos\theta \, 2\pi r \sin\theta \, r \, d\theta. \tag{2.82}$$

Simplifying this expression and integrating the right-hand side gives

$$\sigma_T = pr/2t. \tag{2.83}$$

Since the tangential stress, σ_T is constant throughout the shell wall and the radial stress in the wall is zero, the tangential strain ε in the wall is constant throughout the shell and has the value

$$\varepsilon = s_{11}\sigma_1 + s_{12}\sigma_2 = (1 - \nu)\sigma_T/E, \tag{2.84}$$

where eqns (2.58) and (2.59) have been used. Substituting from eqn (2.83)

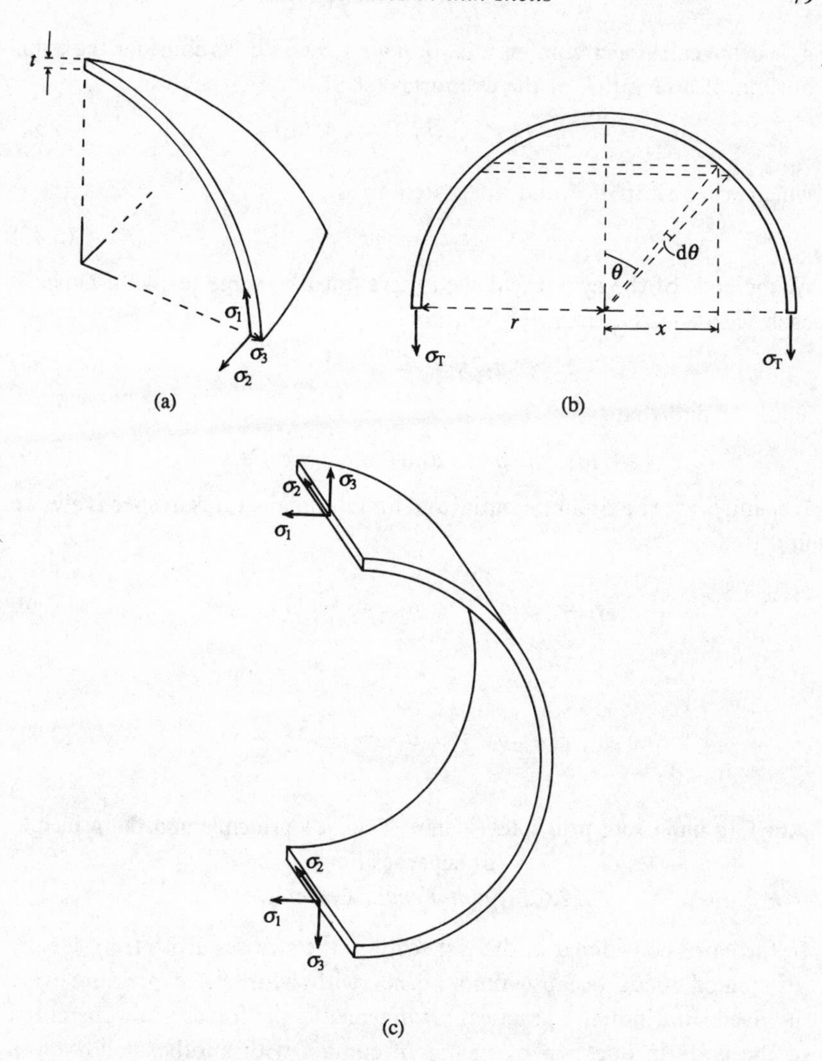

Fig. 2.8 The stresses in thin shells resulting from internal pressure: (a) fragment of spherical shell; (b) diametral section through half of a spherical shell; (c) half strip of cylindrical shell.

we obtain

$$\varepsilon = \frac{pr}{2Et}(1 - \nu). \tag{2.85}$$

(ii) A thin cylindrical shell. This case follows the principles of the preceding example very closely and fig. 2.8(c) illustrates the situation and the components of stress. For a thin shell again $\sigma_3 = 0$, but this time $\sigma_1 \neq \sigma_2$.

σ_1 is often called a circumferential or *hoop stress*. Let us consider the equilibrium of a length L of the cylindrical shell,

$$2\sigma_1 Lt = pLr \int_{-\pi/2}^{\pi/2} \cos\theta \, d\theta, \tag{2.86}$$

which when simplified and integrated gives

$$\sigma_1 t = pr. \tag{2.87}$$

At the ends of the cylindrical shell there must be some forms of closure, each with effective area πr^2, so that

$$\sigma_2 2\pi rt = p\pi r^2. \tag{2.88}$$

Thus we find that

$$\sigma_1 = pr/t \quad \text{and} \quad \sigma_2 = pr/2t. \tag{2.89}$$

If ε_1 and ε_2 are the circumferential and longitudinal strains, respectively, we find that

$$\varepsilon_1 = s_{11}\sigma_1 + s_{12}\sigma_2 = \frac{pr}{Et}(1 - \tfrac{1}{2}\nu) \tag{2.90}$$

and

$$\varepsilon_2 = s_{22}\sigma_2 + s_{21}\sigma_1 = \frac{pr}{Et}(\tfrac{1}{2} - \nu). \tag{2.91}$$

2.6 Two important principles – Saint-Venant's principle and the principle of superposition

2.6.1 *Saint-Venant's principle*

In the cases considered in the last section, the stresses arose from ideally distributed loads because fluid phases with hydrostatic pressure were involved. But in many practical arrangements, the forces are transmitted to the body in question by means of contact with another solid, e.g. a clamp, or a pivot pin. In many composite structures the forces are applied where bolts or rivets are attached and the stress distributions near these positions may be very complex. Stress concentrations usually occur and the problems may not at first appear to be amenable to analytical methods. A principle first propounded in 1855 by a French mathematician, Barré de Saint-Venant, fortunately often provides a way around these difficulties. The principle may be stated thus:

If the distribution of forces acting on a small† sector of the surface of

† 'Small' implies that the area of the sector is small in comparison with the total surface area of the body.

a body is changed in a manner such that any resultant force and couple remain the same then, at a sufficient distance† from the said sector, there is no change in the strain.

The significance of Saint-Venant's principle is that the net effect of some force on the main bulk of a body is independent of the means of applying the force. In solving problems, therefore, it may be possible to replace forces, whose effects are difficult to determine, by a set of forces which have the same resultant but produce a stress distribution that is calculable. The overall strain in the body will be the same for the two stress systems, although it will differ in detail at the positions where contact forces occur.

2.6.2 The principle of superposition

The solution of problems concerned with the elastic deformation of bodies subject to given forces requires, in general, that we find the stress components, or the resulting displacements, as determined by differential equations and the appropriate boundary conditions. Provided that the differential equation for a displacement is linear, its solutions for various arrangements may be superimposed. This is valid assuming that, when deriving equations and boundary conditions for an element of a body undergoing deformation, no distinction need be drawn between the position and shape of the element before loading, and its position and shape afterwards. In other words, the differential equations that we use, and their solutions, are valid as long as the displacements (i.e. the strains) are small and they do not substantially affect the action of the external applied forces. If this holds, then the principle of superposition may be invoked as a method of finding the displacements resulting from a complex system of loads.

2.7 Stresses and strains in elastically-bent beams

A *beam* is a structural member that has one dimension – its length – very much larger than the other two dimensions and, as a consequence, it deflects or *bends* when suitably supported and subjected to transverse forces. We shall consider some cases where the conditions of support are such that the *reactions* at the points of support can be found by the methods of *statics*. In effect we proceed by specifying what form the deformation can take and by showing that it is then possible to derive and

* The 'sufficient distance' is difficult to define precisely and the principle must be used with discretion. The proof of the pudding is in the eating!

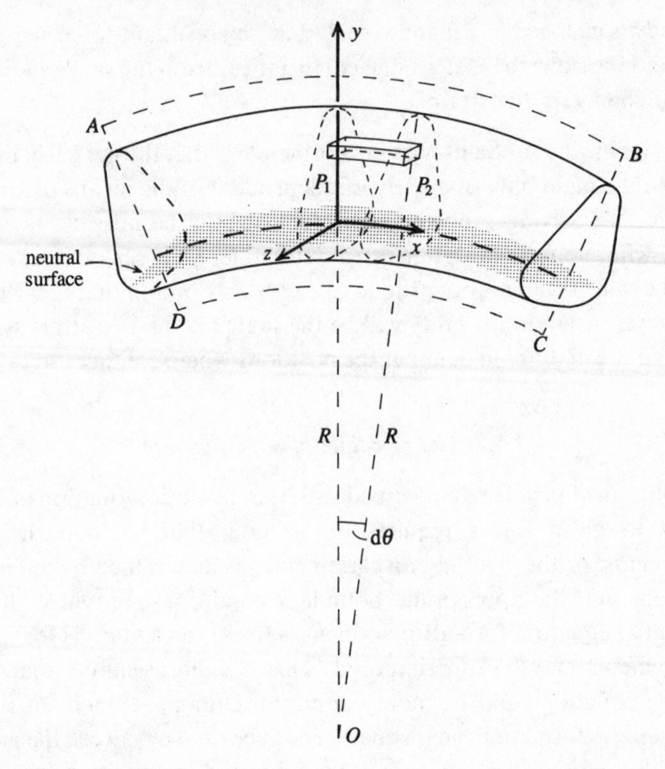

Fig. 2.9 The elastic deformation of a beam. (After Hall, I.H. (1968). *Deformation of Solids*. London: Nelson).

to solve a set of equations that determine the forces necessary to produce this deformation and also give the conditions within the beam. Consider a body in the form of a long, prismatic rod with a cross-section that need not be circular, but is symmetrical with respect to the plane *ABCD* (effectively a mirror plane) in fig. 2.9. We assume that the axis of the rod is straight before deformation, but that afterwards the axis is bent so that it still lies in the *plane of symmetry*, *ABCD*. We further assume that

(i) short elements of the rod axis approximate to the arcs of a circle;
(ii) planes that were normal to the rod axis before deformation remain normal to it afterwards;
(iii) cross-sections of the rod remain symmetrical about the plane of symmetry after deformation;
(iv) lines of the particles that constitute the rod, which were parallel to the axis before deformation, remain parallel afterwards. (We shall hereafter refer to these lines of particles as fibres.)
(v) the beam axis (which is the line through the centroids of the beam

cross-sections) coincides with the neutral axis (to be defined very shortly), which is the condition known as *pure bending*.

We now examine the state of strain in the bent beam by considering the behaviour of the small rectangular element depicted in fig. 2.9, using the coordinate axes shown. The end faces of the element are still rectangular after the bending, but they are no longer parallel because they are parts of planes, labelled P_1 and P_2, which are perpendicular to the rod axis. It follows that the upper face of the element is longer than the lower face in the direction of the rod axis. Since $ABCD$ is a plane of symmetry of the deformed rod, the strain in the rod must be independent of the z-coordinate and be solely a function of y. Because after deformation, the planes of the end faces of the element are still perpendicular to the rod axis and the long edges of the element remain parallel to it, we deduce that ε_{xy} and ε_{xz} are zero. Also since the end faces of the element can change their shapes and sizes without contravening the specification of the deformation given earlier, the values that can be taken by ε_{yy}, ε_{zz} and ε_{yz} are not restricted. This implies that the principal factor defining the deformation is the dependence of the strain component ε_{xx} on the y-coordinate, which we now explore.

We have seen that after bending the rod, the planes P_1 and P_2 are not parallel, so that their axes of symmetry will intersect at some angle, which we call $\mathrm{d}\theta$, at a point O, the centre of curvature of the element. It will be apparent that whether the deformed length of, say, the top surface of the element is greater or less than its undeformed length depends on its y-coordinate. There exists one value of the coordinate for which the two lengths are equal, which implies that there is a surface in which the length of any line drawn parallel to the rod axis is unchanged by the bending. This surface, shown in fig. 2.9, is known as the *neutral surface*. Its line of intersection with the plane of symmetry is called the *neutral axis* (which we have chosen to be the x-axis).

We now consider a length of the neutral axis $\mathrm{d}x$, whose ends are defined by the two planes P_1 and P_2 and call its radius of curvature R. If we also shrink the cross-section of the element so that it coincides with a single fibre of particles, then the undeformed length of this fibre is also $\mathrm{d}x$. The strain in the fibre is therefore

$$\varepsilon_{xx} = \frac{(R + y)\,\mathrm{d}\theta - \mathrm{d}x}{\mathrm{d}x}. \tag{2.92}$$

But

$$\mathrm{d}x = R\,\mathrm{d}\theta,$$

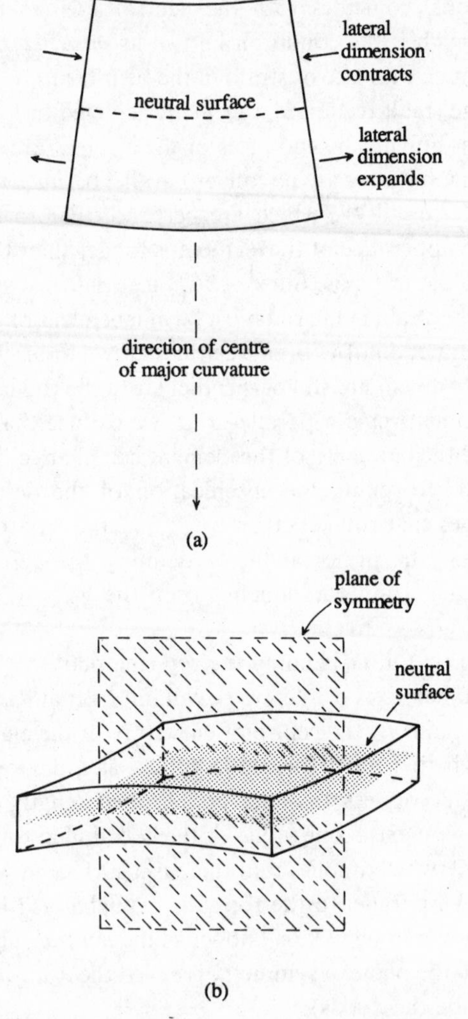

Fig. 2.10 (a) The effect of elastic bending on the cross-section of an initially-rectangular beam; (b) anticlastic curvature resulting from the change in the lateral dimensions of the beam.

so that

$$\varepsilon_{xx} = y/R. \tag{2.93}$$

This equation gives the tensional or compressional strain in fibres above or below the neutral axis, respectively, for the sense of bending depicted in fig. 2.9. For a fibre subject to simple elongation or contraction, the values of ε_{yy} and ε_{zz} could be found from ε_{xx} and Poisson's ratio. As indicated by fig. 2.10(a), the situation in bending is not so straightforward.

Fibres above the neutral axis are elongated and so have reduced cross-sections, while fibres below it are shortened and have enlarged cross-sections. These distortions cause the lateral dimensions of the rod to decrease above the neutral axis and to increase below it. Such dimensional changes can only be accommodated if the beam bends in a plane at right angles to the plane of symmetry, curving about an axis parallel to the x-axis, the sense of bending being opposite to that of the main bending. (The logic of this response will be evident if the principle of superposition is applied to the strains resulting from the two forms of bending.) The secondary bending is known as *anticlastic bending* and it is illustrated for the case of a beam of rectangular cross-section in fig. 2.10(b). It can be seen that the upper and lower surfaces of the beam have saddlelike forms, because of the additional *anticlastic curvature*. This is easily illustrated by the bending of a long rubber eraser. It can be shown that Poisson's ratio is given by

$$\nu = R_L/R_T, \tag{2.94}$$

where R_L and R_T are respectively the major, longitudinal radius of beam curvature and the radius of anticlastic, transverse curvature. Thus the pattern of interference fringes that may be obtained by using monochromatic light to illuminate a semi-silvered glass flat placed over the saddlelike, curved surface of an anticlastically-bent beam provides one means of determining Poisson's ratio for a material.

Anticlastic bending effects can be neglected in long, thin beams and we shall therefore not consider them in detail. In effect, this means that we shall treat the strain in a fibre within a long, thin beam as approximating to simple elongation and determine both the longitudinal stress and the transverse strains accordingly.

The only stress acting on the fibre is thus a normal stress, which from eqns (2.58) and (2.93) is

$$\sigma_{xx} = Ey/R. \tag{2.95}$$

Let the fibre have a cross-section $\mathrm{d}A$ and lie above the neutral axis. Then the outwardly-acting forces, $\mathrm{d}F$, on the ends of the fibre are

$$\mathrm{d}F = Ey\,\mathrm{d}A/R. \tag{2.96}$$

For such a fibre to be in equilibrium, these forces must be transmitted to its ends by virtue of its being attached to other fibres, all together constituting a filament that runs the length of the beam. For fibres below the neutral axis, the elongation and the strain are negative so that the internal forces on the ends of the corresponding filament are directed inwards.

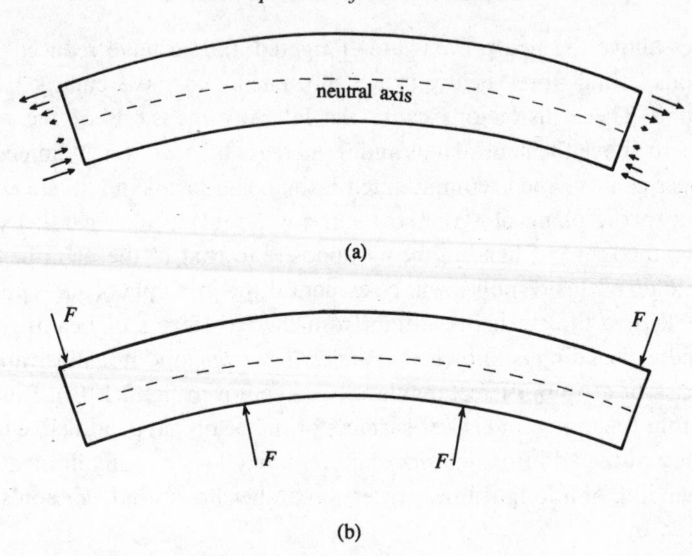

Fig. 2.11 (a) In principle, a beam may be bent by applying a distribution of forces to its ends; (b) achievement of bending similar to that shown in (a) by the application of a system of forces transverse to the length of the beam.

Together, these conclusions imply that a beam is bent by the application of the distribution of external forces as depicted in fig. 2.11(a).

In practice, however, beams are usually caused to bend by systems of forces applied in a direction transverse to the beam length. This apparent contradiction between our analysis of the problem and normal practice is resolved if we remember Saint-Venant's principle: any system of applied forces that has the same resultant as the distribution shown schematically in fig. 2.11(a) will produce the same deformation, provided that the beam is long and thin. The principle is illustrated by the set of forces shown in Fig. 2.11(b), and it indicates that the effect of the applied forces is to apply couples to the beam in the plane of symmetry. These couples are responsible for the bending and constitute what is called the *bending moment*. If the net force above the neutral axis was unequal to the net force below, there would also be a resultant axial force, but this would only affect the length of the beam, not its curvature. We shall only analyse the case where the resultant axial force is zero, which is the situation known as *pure bending*, mentioned in (v) in the list near the beginning of this section.

Since the normal forces acting on a filament depend only on its distance from the neutral surface, we may divide the beam into strips parallel to the neutral surface. Consider a strip of thickness $\mathrm{d}y$ at a distance y and let the width of the strip be b. The width is a function of y except for rectangular beams. Equation (2.96) applies to the strip if $\mathrm{d}A$ is replaced by $b\,\mathrm{d}y$, and the normal force applied to each end of the strip is

$$dF = Eby\,dy/R. \tag{2.97}$$

Its moment about the neutral axis is

$$dM = Eby^2\,dy/R. \tag{2.98}$$

The condition for pure bending, with a zero total axial force, is accordingly

$$F = (E/R)\int by\,dy = 0, \tag{2.99}$$

where the integral runs over the vertical dimensions of the beam. The total bending moment is

$$M = (E/R)\int by^2\,dy. \tag{2.100}$$

The integral in the last equation over the whole cross-section of the beam is analogous to a moment of inertia and is known as the *second moment of area* (or the *geometrical moment of inertia*) of the cross-section about the line $y = 0$, perpendicular to the plane of symmetry, i.e. the z-axis. We shall henceforth denote the second moment of area by the symbol I,

$$I = \int by^2\,dy. \tag{2.101}$$

Equation (2.100) then becomes

$$M = EI/R \tag{2.102}$$

in which the quantity EI is sometimes called the *flexural rigidity*.

The condition that the integral in eqn (2.99) vanishes implies that the plane $y = 0$ contains the centroid of the area of cross-section. Since the plane $y = 0$ is the neutral surface, the latter must contain the centroids of all the cross-sections, i.e. the beam axis. The neutral axis therefore coincides with the beam axis in pure bending, no matter what is the cross-section of the beam. The quantity I can usually be calculated simply from the dimensions of the beam, if necessary using theorems which pertain to moments of inertia. It is then a simple matter to calculate the radius of curvature of a beam in response to an applied couple M using the eqn (2.102).

2.8 Internal forces in beams subjected to distributed loads

The determination of the internal forces in a beam subjected to a general distribution of applied forces is a very complicated problem, which we do not consider here. However, many of the simpler cases are amenable to a method of calculation that we illustrate by means of a worked example.

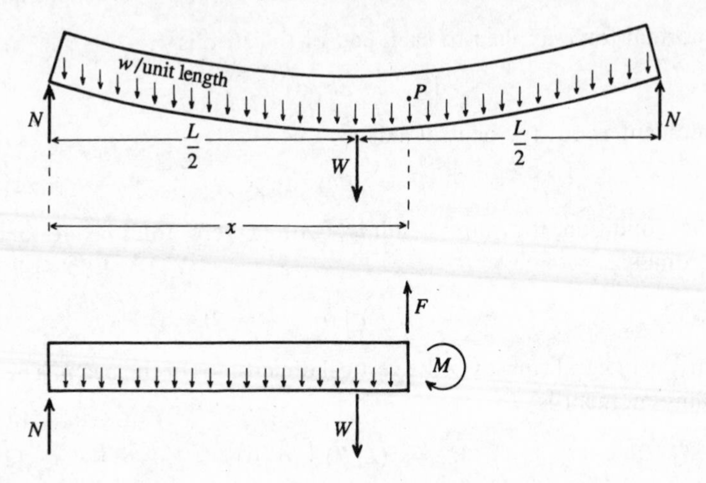

Fig. 2.12 The forces and moments acting on a loaded beam with a self-weight of w per unit length.

Consider the case of a beam, supported at its two ends, which is bent as a result of the combined effects of its own weight (uniformly-distributed) and a force W applied at the beam centre, as shown in fig. 2.12. Let the weight of the beam be equivalent to a downwards force of w per unit length of the beam and the reactions at the supports be N.

Let us imagine that the beam is now cut through vertically at the point P and consider the forces that must be applied to the cut face of the left-hand part of the beam in order to sustain equilibrium. It will be readily apparent that the forces must consist of a shear force acting vertically, which we shall call F, together with a couple (i.e. a bending moment) which we call M. We choose a sign convention for couples and forces such that a bending moment is taken as positive if it tends to produce bending with a centre of curvature below the beam, while a shear force will be considered positive if it tends to rotate the section in an anticlockwise direction. Thus for the system illustrated in fig. 2.12, F is positive, but M must have a negative value to produce the bending shown in the upper part of the figure.

For the left-hand part of the beam to be in equilibrium

$$F = W + wx - N$$

and

$$M = W(x - \tfrac{1}{2}L) + \tfrac{1}{2}wx^2 - Nx. \tag{2.103}$$

Since symmetry implies that the reactions at both supports are equal, static equilibrium requires that

$$N = \tfrac{1}{2}(W + wL), \tag{2.104}$$

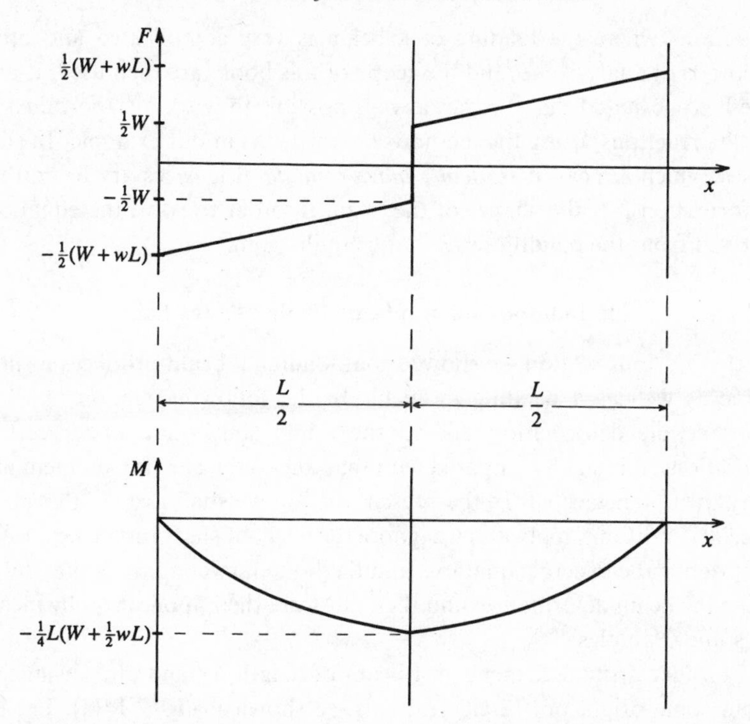

Fig. 2.13 The distributions of shear force and bending moment in the loaded beam shown in fig. 2.12.

so that

$$F = \tfrac{1}{2}W + w(x - \tfrac{1}{2}L)$$

and $\qquad\qquad (x \geqslant \tfrac{1}{2}L) \qquad (2.105)$

$$M = (x - L)\tfrac{1}{2}(W + wx).$$

Two comparable equations can be obtained by making the vertical cut at a point to the left of the beam centre, so that the variations of F and M along the whole length of the beam are easily found. The equations are

$$F = -\tfrac{1}{2}W - w(\tfrac{1}{2}L - x)$$

and $\qquad\qquad (x \leqslant \tfrac{1}{2}L). \qquad (2.106)$

$$M = -\tfrac{1}{2}x\{W + w(L - x)\}.$$

Not surprisingly, the four eqns (2.105) and (2.106) show that the distribution of bending moment within the beam is distributed symmetrically, as shown in fig. 2.13. However, there is a discontinuous change in F where the concentrated load W is applied, and this is generally the case. It is necessary to derive expressions for the shear force and the bending moment for each section of a beam between loading points (which includes supports). The approach that we have just illustrated can therefore be clumsy for situations where the loading of a beam is very complicated and other methods of analysis, beyond the scope of this book, are then used. It also needs to be stated that it is not always possible to work out the values of all the reactions, using the methods of statics, as in our example. In such cases, which are called *statically indeterminate*, it is necessary to find the deformation (i.e. the shape) of the beam in order to solve the equations derived from the conditions of static equilibrium.

2.9 Deformation of a beam by distributed loads

In the previous section we showed that loading a beam produces a shear force as well as a bending moment. In the following section we shall examine the deformation due to shear and show that, under certain conditions, it is small compared with that due to the bending moment and may then be neglected. In the present section we shall assume that these special conditions apply and we ignore the effect of shear on the beam. We first derive the general equations and the boundary conditions that determine the beam deformation and then illustrate their application by means of some examples.

Consider a small element of a beam of length $\mathrm{d}x$ and at a distance x from some origin on the neutral axis, as shown in fig. 2.14(a). Let the shearing forces and bending moments acting on the element at the face planes defined by the coordinates x and $x + \mathrm{d}x$ be F, M, and $F + \mathrm{d}F$, $M + \mathrm{d}M$, respectively, with directions and senses as shown in fig. 2.14(b). The conditions for equilibrium of the element are

$$(F + \mathrm{d}F) - F = w\,\mathrm{d}x$$

and

$$(M + \mathrm{d}M) - M = \tfrac{1}{2}F\,\mathrm{d}x + \tfrac{1}{2}(F + \mathrm{d}F)\,\mathrm{d}x.$$

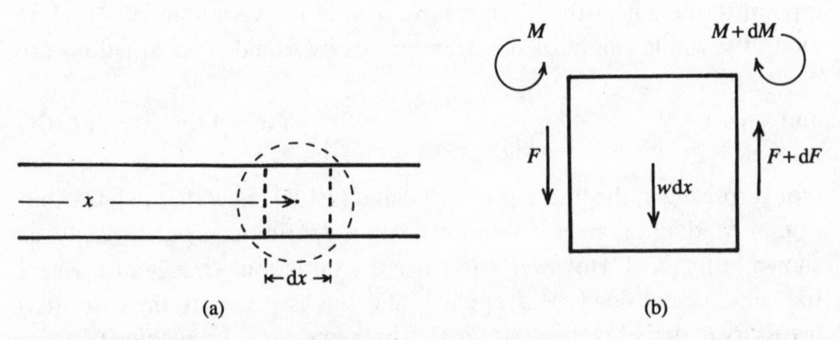

Fig. 2.14 (a) A small element of a thin beam, situated at a distance x from the origin; (b) enlarged view of the small element of the beam shown in (a) indicating the forces and moments acting on it when the beam is loaded.

Thus, neglecting second order terms, we obtain

$$dF/dx = w \tag{2.107}$$

and

$$dM/dx = F. \tag{2.108}$$

These general properties are consistent with the explicit eqns (2.105) and (2.106) derived for the beam problem illustrated in fig. 2.12. Combination of eqns (2.107) and (2.108) gives

$$d^2M/dx^2 = w. \tag{2.109}$$

The bending moment is determined by the radius of curvature R in accordance with eqn (2.102). A general expression for R is to be found in Appendix 1. The bent beams treated here have only small strains so that the gradient dy/dx is everywhere small, y being the displacement of the beam from its undistorted position parallel to the x-axis. We may therefore use the approximate result from eqn (A1.8) in Appendix 1,

$$1/R = d^2y/dx^2. \tag{2.110}$$

Our convention that a centre of curvature *below* the beam corresponds to a *positive* bending moment M makes it convenient to measure the y-coordinate vertically *downwards*. The second derivative in eqn (2.110) is then positive for positive M, and substitution in eqn (2.102) gives

$$M = EI d^2y/dx^2. \tag{2.111}$$

Then using eqns (2.108) and (2.109) we obtain

$$dM/dx = EI d^3y/dx^3 = F \tag{2.112}$$

and

$$d^2M/dx^2 = EI d^4y/dx^4 = w. \tag{2.113}$$

These equations determine the shapes of beams when deformed by various distributions of load.

The solution of eqn (2.113) to obtain y as a function of x requires values of four constants of integration, which are determined from knowledge of the conditions at the ends of the beam, i.e. whether the beam is clamped firmly, pinned, or rests upon a support. Four cases are easily recognized, with the following boundary conditions:

(i) Free end with no load applied to it.
In this situation $F = 0$ and $M = 0$, so that

$$d^3y/dx^3 = 0 \quad \text{and} \quad d^2y/dx^2 = 0. \tag{2.114}$$

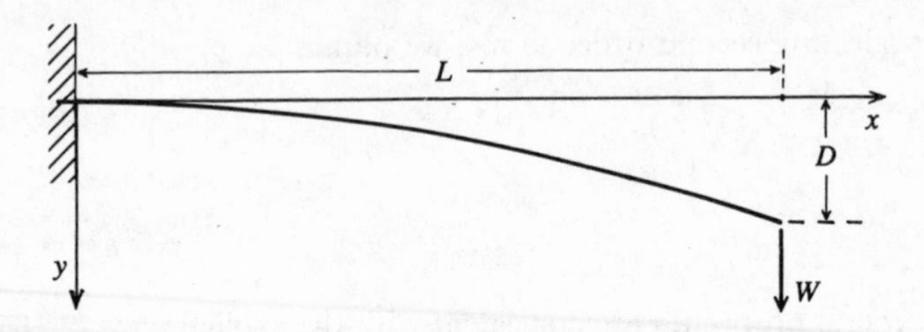

Fig. 2.15 The simple bending of a weightless beam, clamped at one end and loaded with a weight W at the other.

(ii) Free end with a load W applied to it.
Here $F = -W$ and $M = 0$ so that

$$EI\mathrm{d}^3y/\mathrm{d}x^3 = -W \quad \text{and} \quad \mathrm{d}^2y/\mathrm{d}x^2 = 0. \tag{2.115}$$

(iii) End supported but not clamped.
In this case y is known (we take the y-coordinate to be zero) and $M = 0$, so that

$$y = 0 \quad \text{and} \quad \mathrm{d}^2y/\mathrm{d}x^2 = 0. \tag{2.116}$$

(iv) End clamped.
Both y and $\mathrm{d}y/\mathrm{d}x$ are known, and with coordinates chosen to make them zero

$$y = 0 \quad \text{and} \quad \mathrm{d}y/\mathrm{d}x = 0. \tag{2.117}$$

We now apply these boundary conditions to several examples of simple bending.

(i) Beam clamped at one end, load W at the other. The coordinate system is shown in fig. 2.15. The general solution of eqn (2.113) is

$$EIy = C_0 + C_1x + C_2x^2 + C_3x^3 + (1/24)wx^4, \tag{2.118}$$

where the C_i are constants of integration to be determined. The boundary conditions (2.117) apply at the clamped end where $x = 0$ and accordingly

$$C_0 = C_1 = 0. \tag{2.119}$$

The general solution thus reduces to

$$EIy = C_2x^2 + C_3x^3 + (1/24)wx^4. \tag{2.120}$$

The boundary conditions (2.115) apply at the loaded end where $x = L$, and therefore

$$\begin{cases} 6C_3 + wL = -W \\ 2C_2 + 6C_3L + \frac{1}{2}wL^2 = 0 \end{cases}$$

with solutions

$$C_2 = \tfrac{1}{2}WL + \tfrac{1}{4}wL^2 \quad \text{and} \quad C_3 = -\tfrac{1}{6}W - \tfrac{1}{6}wL. \tag{2.121}$$

The equation of the deformed beam is therefore

$$y = \{12WL + 6wL^2 - 4(W + wL)x + wx^2\}x^2/24EI. \tag{2.122}$$

The maximum beam deflection, at the loaded end, is

$$D = (8W + 3wL)L^3/24EI. \tag{2.123}$$

(ii) Beam clamped horizontally at both ends. The boundary conditions (2.117), when applied to the end at $x = 0$, produce the same solution (2.120) as in example (i). Application of the same boundary conditions to the end at $x = L$ gives

$$\begin{cases} C_2L^2 + C_3L^3 + (1/24)wL^4 = 0 \\ 2C_2L + 3C_3L^2 + \tfrac{1}{6}wL^3 = 0 \end{cases}$$

with solutions

$$C_2 = wL^2/24 \quad \text{and} \quad C_3 = -wL/12. \tag{2.124}$$

The equation of the deformed beam is therefore

$$y = wx^2(L - x)^2/24EI \tag{2.125}$$

and its maximum deflection, at $x = \tfrac{1}{2}L$, is

$$D = wL^4/384EI. \tag{2.126}$$

Figure 2.16 shows the profile of the beam, together with the x-dependences of the bending moment

$$M = EI\,\mathrm{d}^2y/\mathrm{d}x^2 = w(L^2 - 6Lx + 6x^2)/12 \tag{2.127}$$

and of the shear force

$$F = EI\,\mathrm{d}^3y/\mathrm{d}x^3 = \tfrac{1}{2}w(2x - L). \tag{2.128}$$

Since I can be calculated from the cross-sectional dimensions of the beam ($I = ba^3/12$ for a rectangular cross-section of width b and thickness a), the quantities M, F and y are readily evaluated.

(iii) Beam supported at both ends, load W at the centre. This is the configuration illustrated in fig. 2.12, that we have already considered in § 2.8. The general methods used in the last example still apply, but we may combine the earlier results (eqns (2.105) and (2.106)) with eqn (2.111) and use the

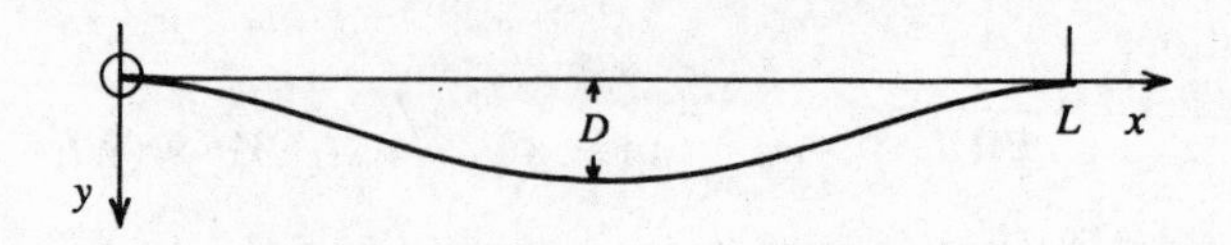

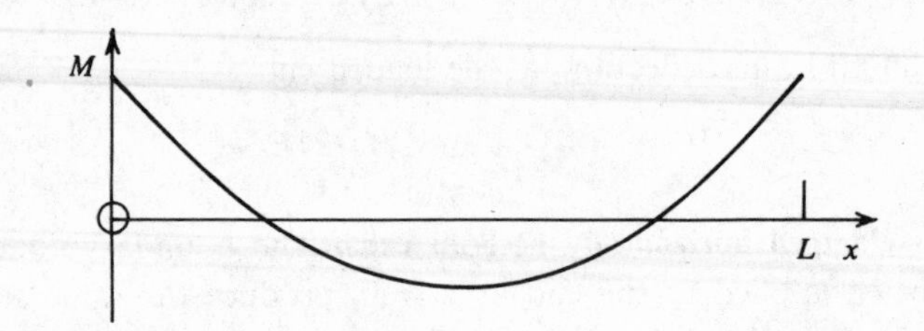

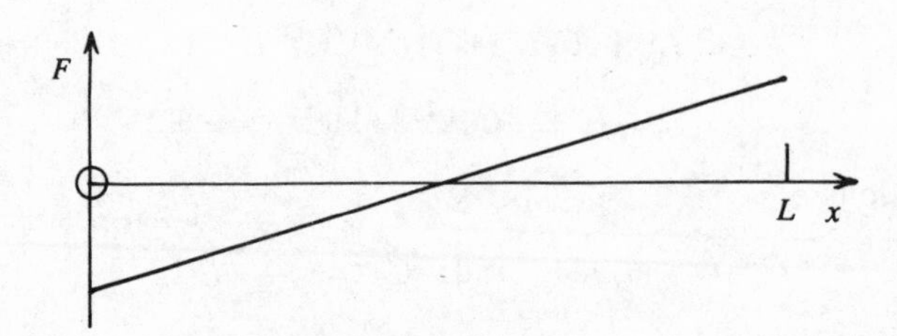

Fig. 2.16 The characteristics of a beam which is clamped at both ends: (a) shape of the bent beam; (b) distribution of bending moment; (c) distribution of shear force.

appropriate beam conditions to find the constants of integration. Using the second equation of (2.105) together with eqn (2.111) we obtain

$$2EI\,\mathrm{d}^2y/\mathrm{d}x^2 = (x - L)(W + wx) \quad \text{for } \tfrac{1}{2}L \leqslant x \leqslant L,$$

which, when integrated, gives

$$2EI\,\mathrm{d}y/\mathrm{d}x = \tfrac{1}{3}wx^3 + \tfrac{1}{2}x^2(W - wL) - WLx + C_1. \quad (2.129)$$

The beam will be symmetrical about the mid-point, since the load is applied at the centre, so that $\mathrm{d}y/\mathrm{d}x = 0$ at $x = L/2$. Using this fact in conjunction with eqn (2.129) gives

$$C_1 = (1/12)\,wL^3 + (3/8)\,WL^2.$$

Substituting C_1 in eqn (2.129) and integrating again, we obtain

$$\begin{aligned}4EIy &= \tfrac{1}{6}wx^4 + \tfrac{1}{3}x^3(W - wL) - WLx^2 \\ &\quad + \tfrac{1}{2}L^2x[(wL/3) + (3W/2)] + C_2.\end{aligned}$$

But $y = 0$ at the points of support, taken to be $x = 0$ and $x = L$. Since eqn (2.129) applies to values of x between $L/2$ and L, we must use the condition $y = 0$ when $x = L$, from which we obtain

$$C_2 = WL^3/12.$$

Thus the beam deflection is

$$y = \frac{1}{4EI}\left[\frac{wx^4}{6} + \frac{x^3}{3}(W - wL) - WLx^2 + \frac{L^2x}{2}\left(\frac{wL}{3} + \frac{3W}{2}\right) - \frac{WL^3}{12}\right]. \tag{2.130}$$

The deflection for $x < L/2$ can be obtained via the second of eqns (2.106) in a similar manner and both beam equations give the same value for the deflection at the mid-point, namely

$$D = \frac{L^3}{96EI}\left(2W + \frac{5wL}{4}\right). \tag{2.131}$$

The variations of the shear force F and the bending moment M were shown previously (fig. 2.13).

If the weight of the beam wL is small compared to the load we have a situation that is commonly encountered in deformation experiments. It is known as *symmetrical three-point loading* and is a simple way of testing the rheological properties of a solid. It is illustrated in fig. 2.17(a) which also gives the values of the bending moments on both sides of the centre. A superior system for the purposes of determining elastic or flow properties in bending, in which the loading is better distributed and stress concentrations are less severe, is *symmetrical four-point loading*. This is depicted in fig. 2.17(b) which again gives the values of the bending moments. The maximum deflections in the cases of three-point loading and four-point loading are, respectively

$$WL^3/48EI \quad \text{and} \quad W(2L^3 - 3La^2 + a^3)/48EI, \tag{2.132}$$

where a is the distance between the two inner, symmetrically-placed, loading points in the latter case.

2.10 Deformation of a loaded beam by shear

The shear force produced by a load was noted in § 2.8 but the contribution of shear to the deformation of a beam has so far been neglected. The effects of both bending and shearing will now be found in a particular case – the loaded cantilever. By comparing the two results we shall find the conditions under which the strain resulting from the transverse shear force is small, and the strain may therefore be neglected.

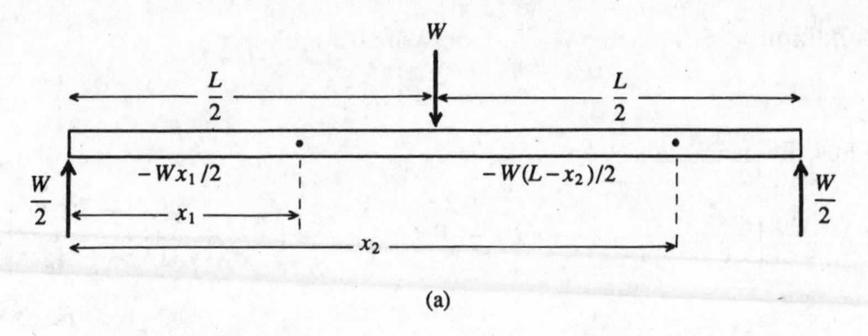

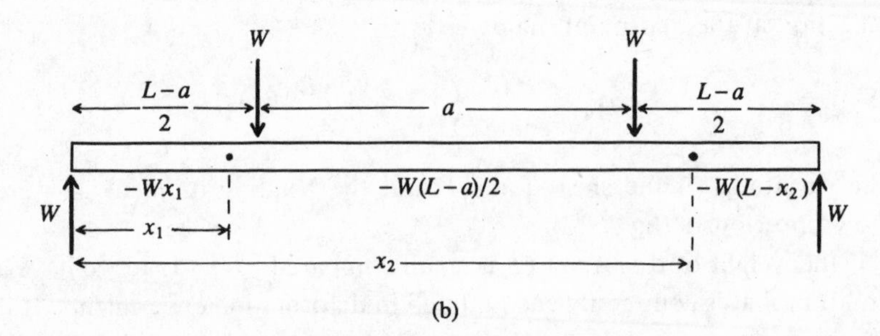

Fig. 2.17 (a) A beam under three-point loading; (b) a beam under four-point loading.

2.10.1 *Deflection produced by the bending moment*

The problem of the horizontal beam clamped at one end and loaded at the other end has been solved in example (i) of the previous section. The system is illustrated in fig. 2.15 and the equation of the beam is given in eqn (2.122). Consider the simplest case of a lightweight beam with $wL \ll W$, whereupon eqn (2.122) reduces to

$$y = W(3L - x)x^2/6EI \tag{2.133}$$

and the maximum beam deflection is

$$D_1 = WL^3/3EI. \tag{2.134}$$

For a rectangular bar of width b and depth a, the second moment of area is $ba^3/12$. Making this substitution in eqn (2.134) gives the deflection caused by the bending moment

$$D_1 = 4WL^3/Eba^3. \tag{2.135}$$

2.10.2 *Deflection produced by the shear force*

We now neglect the effect of the bending moment and just consider the shear due to the load W, acting over the cross-section of the beam at its end point in fig. 2.15. The stress is simply W/ab and this shear stress deflects the beam through an angle θ from its unloaded, horizontal position (the beam would have no curvature, of course). The deflection at the end point is then easily obtained from the shear strain θ, since for small stresses

$$\tan\theta \simeq \theta = y/x = D_2/L, \tag{2.136}$$

and the definition of the rigidity modulus in eqn (2.68) gives

$$G = (W/ab)/\theta \tag{2.137}$$

so that

$$D_2 = WL/Gab. \tag{2.138}$$

Therefore, taking the result for D_1 from eqn (2.135) gives

$$\frac{D_2}{D_1} = \frac{E}{4G}\left(\frac{a}{L}\right)^2. \tag{2.139}$$

Since $E \simeq G$ (see Table 2.1 for typical values), eqn (2.139) shows that the effect of shear forces on a beam is small provided that $a^2 \ll L^2$. This implies a long, thin beam. The dimension, b, in the direction perpendicular to the plane of symmetry (i.e. the width, parallel to the axis of curvature) is not critical.

2.11 Maximum stress in a loaded beam

If the stress within a beam reaches a sufficiently high value, the deformation ceases to be elastic and starts to be plastic, as explained in § 1.10. In practice it is important to allow a safety factor and to keep the maximum stress in a loaded structural member well below the yield stress. This factor should be adequate to overcome the effects of small inhomogeneities in the external surfaces or in the internal structure of the beam material (they act as concentrators of stress), or the possible effect of a severe environment on a member carrying a load for a long time (this may lead to a problem known as stress-corrosion). It is clearly important to be able to calculate the maximum stress in a loaded beam and to know where this value is attained. We obtain the normal stress in an elastically-bent beam from eqns (2.95) and (2.102), finding that

$$\sigma_{xx} = My/I. \tag{2.140}$$

Because I is uniform along the beam, the maximum stress on a filament occurs where both y and M are maximal, i.e. at the surface of the beam which is furthest from the neutral surface and at the filament in this surface where M is greatest. Let us now examine the situation in an earlier example, namely that shown in fig. 2.12. We see from fig. 2.13 that M has its maximum value at the point of application of the load W. Substituting the value $x = L/2$ in the first of the eqns (2.106) gives

$$M = -\tfrac{1}{4}L(W + \tfrac{1}{2}wL). \tag{2.141}$$

Using this result in eqn (2.140) gives

$$\sigma_{max} = Ly(W + \tfrac{1}{2}wL)/4I. \tag{2.142}$$

If the parts of the unbent beam above and below the neutral plane are not mirror images, the values of y corresponding to the outermost filaments of the upper and lower surfaces of the beam will differ in magnitude as well as in sign. This implies that the maximum values of the tensional and compressional stresses, above and below the neutral axis, will not be equal. However, if the beam has a rectangular cross-section, of width b and depth a, they are both numerically equal to

$$3L(W + \tfrac{1}{2}wL)/2ba^2.$$

2.12 Elastic energy stored in bending – Castigliano's theorem and its applications

2.12.1 Elastic energy stored in a bent beam

Stored elastic energy is used as the source of power for some mechanical devices (e.g. clockwork, a cross-bow) and we can calculate it in terms of the quantities that have already been introduced. Provided that no energy is lost in the production of heat, we expect that the energy stored in an elastically-strained solid, such as a bent beam, will be equal to the work done in deforming it. This is the simplest way of calculating the elastic energy. Consider a small element of a bent beam as shown in fig. 2.18, and assume that initially its axis was a straight line and the end faces were perpendicular to the axis. We assume also that after bending the end faces are mutually inclined at an angle $d\theta$, so that the element subtends the angle $d\theta$ at the centre of curvature O. Then

$$1/R = d\theta/ds, \tag{2.143}$$

where R is the radius of curvature of the element and ds is the length of the neutral axis.

Substituting for R from eqn (2.102) gives

$$d\theta = (M/EI)ds. \tag{2.144}$$

Let us now increase the bending moment by an amount dM, causing a corresponding increase $d^2\theta$ in the angle subtended at O.

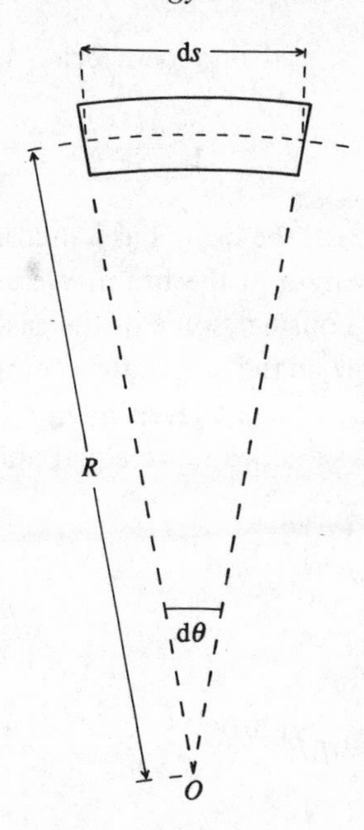

Fig. 2.18 A small element of an elastically bent beam.

Then

$$d\theta + d^2\theta = (1/EI)(M + dM)\,ds \qquad (2.145)$$

from which, subtracting eqn (2.144), we obtain

$$d^2\theta = (dM/EI)ds. \qquad (2.146)$$

This equation shows how the increase in the couple applied to the beam (which we saw in § 2.7 may be applied via the end faces) produces a further relative rotation of the end faces. The corresponding work done by the couple is just $M\,d^2\theta$ and this is equal to the additional elastic energy stored in the element on increasing M by dM. If the elastic energy stored in the bent element is dU when the applied bending moment has the value M, then

$$dU = \int M\,d^2\theta. \qquad (2.147)$$

Substituting for $d^2\theta$ from eqn (2.146) gives

$$dU = \frac{ds}{EI}\int_0^M M\,dM, \qquad (2.148)$$

so that

$$dU = (M^2/2EI)\,ds. \qquad (2.149)$$

The total elastic energy stored in a bent beam of length L is therefore

$$U = \int_0^L \frac{M^2\,\mathrm{d}s}{2EI}. \tag{2.150}$$

In the most common case, the beam has a uniform cross-section so that EI is constant over the length of the beam. We assume that the bending is small, so that we may consider some of the cases examined earlier and put $\mathrm{d}s = \mathrm{d}x$. For example, in the case of the centrally-loaded beam shown in fig. 2.12, with bending moment given by eqn (2.106), the total elastic energies in both halves of the beam are equal and have the value

$$U = \frac{1}{2EI}\int_0^{L/2} \frac{x^2}{4}\{W + w(L - x)\}^2\,\mathrm{d}x, \tag{2.151}$$

which when integrated gives

$$U = \frac{L^3}{3840EI}(20W^2 + 25WwL + 8w^2L^2). \tag{2.152}$$

2.12.2 Castigliano's theorem and its application in finding the deflection of a bent beam

The calculation of the strain energy in a stressed body is often a useful undertaking because it enables the displacement at particular points to be calculated through the application of Castigliano's theorem. The theorem, which we do not prove, applies to linearly elastic systems, i.e. those that obey the principle of superposition (§ 2.6.2). The strain energy of such an elastic system which is under the action of a set of concentrated forces F_1, F_2, F_3, . . . can be expressed as a quadratic function U of these forces, so that small changes in the forces produce an energy change

$$\delta U = (\partial U/\partial F_1)\,\delta F_1 + (\partial U/\partial F_2)\,\delta F_2 + \ldots . \tag{2.153}$$

By means of the principle of superposition it may be demonstrated that

$$\delta U = D_1\delta F_1 + D_2\delta F_2 + \ldots, \tag{2.154}$$

where D_1, D_2, . . . , are the displacement components associated with the forces F_1, F_2, . . . etc., so that it is clear that

$$D_1 = \partial U/\partial F_1, \quad D_2 = \partial U/\partial F_2 \ldots \quad \text{etc.} \tag{2.155}$$

Castigliano's theorem may thus be stated in the following general form:

> When the elastic strain energy is expressed as a quadratic function of the forces applied to a structure, the partial differential of the strain energy

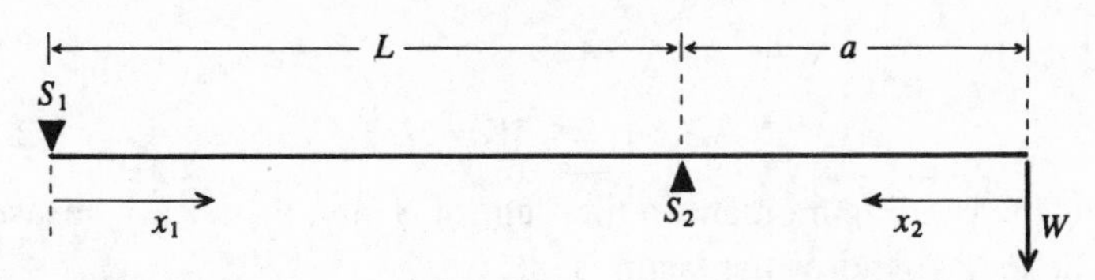

Fig. 2.19 A beam with two supports and a loaded overhanging section.

with respect to any one of the forces gives the corresponding component of displacement of the point of application of the force.

It is important to appreciate that we refer here to forces that are statically independent, in order that the principle of superposition should apply. Moreover, we should take the terms 'force' and 'displacement' in the generalized sense that, if the agent of an action is a couple rather than a force, the 'displacement' will be an angle of rotation in a sense dictated by the couple. The displacement corresponding to a true force will be a linear deflection in the direction of the force.

Armed with this useful theorem, we can now use the strain energy of a bent beam to calculate deflections at various points of loading. Clearly, the results should agree with those obtained by other means in earlier examples. This new approach is, however, often advantageous and, in particular, it may be used to treat statically-indeterminate problems. Let us first reconsider the problem of the centrally-loaded uniform beam with weight w per unit length. The elastic energy of one half of the beam is given by eqn (2.152) so that doubling the expression and differentiating it with respect to W gives the deflection at the centre as

$$D = \frac{L^3}{3840EI}(80W + 50wL) = \frac{L^3}{96EI}\left(2W + \frac{5wL}{4}\right),$$

which agrees with the result given in eqn (2.131) obtained from the equation for the beam.

Let us now consider the example of a lightweight beam that is loaded at one end with a weight W and held between two supports S_1 and S_2, distance L apart, the beam overhanging the underneath support S_2 as shown in fig. 2.19. We may use Castigliano's theorem to find the deflection of the overhanging part of the beam by first finding the strain energy. We consider in turn the parts of the beam on either side of S_2, using x_1 and x_2, respectively, to denote distances from the left- and right-hand ends of the beam (this avoids tedious algebra that follows if we work only from one end). Taking moments about S_2 we see that the reaction at S_1 must be equal to Wa/L, so that the bending moment at any point distance x_1 from

S_1 and to the left of S_2 is

$$M(x_1) = Wax_1/L \tag{2.156}$$

whereas the bending moment to the right of S_2 at a distance x_2 measured from the loaded end of the beam is

$$M(x_2) = Wx_2. \tag{2.157}$$

Using the general result for elastic strain energy of bending (eqn (2.150)), the total strain energy of the beam is

$$U = \int_0^L \frac{W^2a^2x_1^2}{2EIL^2}\,\mathrm{d}x_1 + \int_0^a \frac{W^2x_2^2}{2EI}\,\mathrm{d}x_2 = \frac{W^2a^2}{6EI}(L + a), \tag{2.158}$$

so that, applying eqn (2.155), the deflection at the loaded end is

$$D = \frac{\partial U}{\partial W} = \frac{Wa^2}{3EI}(L + a). \tag{2.159}$$

2.13 Buckling of columns – Euler's column formula

Let us now consider the case of a long thin rod, or slender column, AB, of length L, standing vertically with its lower end A firmly fixed into a solid floor, as shown in fig. 2.20(a). We assume that the rod is perfectly straight and uniform, and made of a material that is both homogeneous and perfectly elastic. The situation is that of a vertical cantilever, sometimes called a *cantilever column*. It can be shown that when a small load W is applied vertically to the top of the rod at B, the rod is compressed but at no point along its length is there any lateral deflection. Moreover, if a small lateral force is applied to the lightly loaded rod at the point B the top will be deflected to one side but the rod will return to the vertical as soon as the lateral force is removed. The compressed rod is said to be *laterally stable*.

Experiment shows that as the load W is gradually increased there comes a stage, corresponding to a particular value of W, where lateral stability is lost and the equilibrium state of the rod is no longer straight, as indicated in fig. 2.20(b). It is found that the rod, if displaced to one side by the application of a small lateral force F at some point along the length, remains out of the vertical line even when the force is removed. This instability in behaviour is called *lateral buckling* and the load at which it starts for a given rod is called the *critical load*, denoted hereafter by W_c.

It will be obvious that lateral buckling is not restricted to rods that are oriented in a vertical plane but will occur in all slender rods that are put in compression by loading the ends (unless, of course, they are laterally

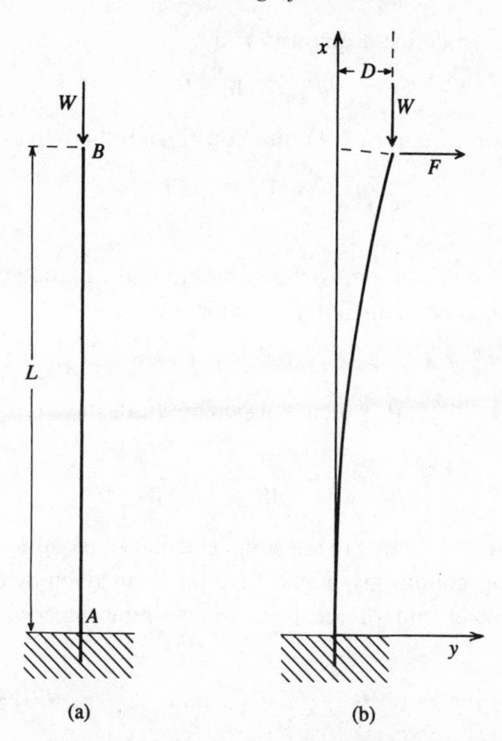

Fig. 2.20 (a) An axially-loaded strut, fixed in a vertical position at A; (b) showing the effect of applying a small horizontal force F to the free end B of the strut shown in (a).

constrained to remain straight). Such rods are often called *struts*, although the word strut is sometimes reserved for very short posts that do not readily suffer lateral instability.

In order to find the load W_c just sufficient to cause buckling, we consider the rod of fig. 2.20(a) in a condition of slight bending, produced by the lateral force F. The mass of the rod itself is assumed to be sufficiently small that the corresponding distributed load can be neglected. The coordinate system is shown in fig. 2.20(b), and the deflection D of the top of the rod from its unloaded position is assumed to be sufficiently small for only small bending strains to occur. Then eqn (2.111) applies in the form

$$d^2y/dx^2 = M/EI. \tag{2.160}$$

The bending moment produced by the applied forces at a general point x, y in the rod is

$$M = W(D - y) + F(L - x). \tag{2.161}$$

It is convenient to define a quantity k by

$$k^2 = W/EI \tag{2.162}$$

and the insertion of eqn (2.161) into eqn (2.160) then gives

$$\frac{d^2y}{dx^2} + k^2y = \frac{WD + FL}{EI} - \frac{F}{EI}x. \tag{2.163}$$

The solution of this inhomogeneous differential equation consists of the usual sum of the complementary function

$$y = C_1 \cos kx + C_2 \sin kx \tag{2.164}$$

and a particular integral, which can be chosen in the simple form

$$y = \frac{WD + FL}{W} - \frac{F}{W}x. \tag{2.165}$$

The two arbitrary constants in the complementary function are determined by the boundary conditions, which have the form given by eqn (2.117) at the clamped lower end of the rod. Some simple algebra provides the solutions

$$C_1 = -(WD + FL)/W \quad \text{and} \quad C_2 = F/Wk \tag{2.166}$$

and the complete general solution of eqn (2.163) is

$$y = \left(D + \frac{FL}{W}\right)(1 - \cos kx) - \frac{F}{Wk}(kx - \sin kx). \tag{2.167}$$

Note that in the absence of a vertical load, $W = 0$, this result reduces to

$$y = F(3L - x)x^2/6EI, \tag{2.168}$$

where care is needed in the evaluation of the trigonometric limits in eqn (2.167), bearing in mind the definition of k in eqn (2.162). This last result (2.168) has identical form to the result (2.133) obtained for the deflection of a weightless horizontal beam by a load W applied perpendicular to the undeformed beam, so that we now see that the spatial orientation of lightweight struts is not important, as mentioned at the beginning of this section.

The quantities D, F and W that occur in the general solution (2.167) are not mutually independent since the deflection of the top of the rod is itself determined by the applied forces. The relation between them is found from the condition that the point $x = L$, $y = D$ lies on the curve, that is

$$D = \left(D + \frac{FL}{W}\right)(1 - \cos kL) - \frac{F}{Wk}(kL - \sin kL),$$

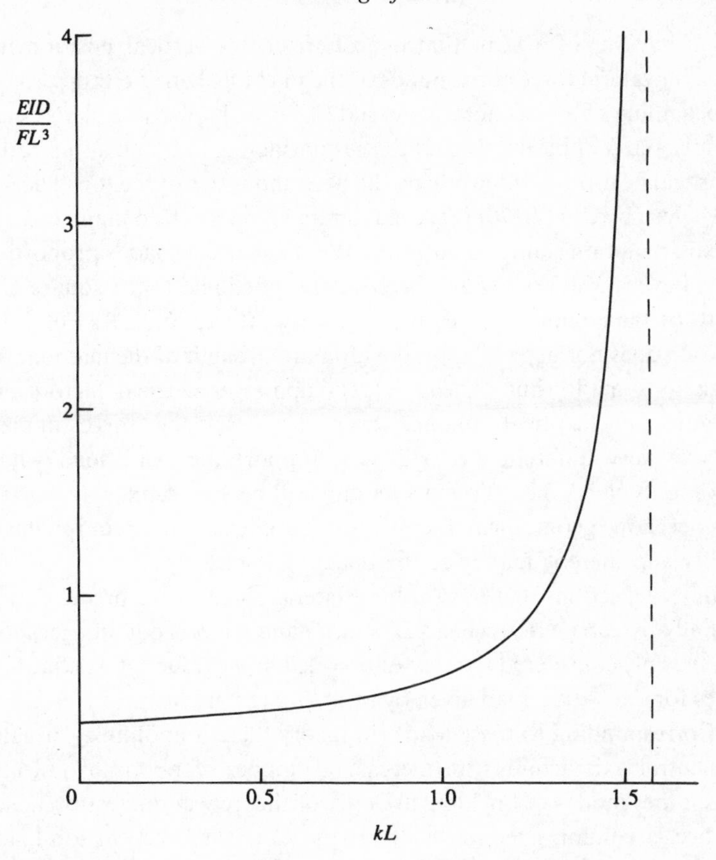

Fig. 2.21 The variation of the deflection of the top of the strut shown in fig. 2.20 as a function of kL for a constant horizontal force F.

which, by use of eqn (2.162) can be rearranged into the form

$$D = (\tan kL - kL)\,F/EIk^3. \tag{2.169}$$

Substitution of this expression for D into eqn (2.167) gives the equation for the curved rod in terms of the applied forces W and F. The boundary conditions with the forms given by eqn (2.115) for the free end of the rod are satisfied by the resulting expression.

Figure 2.21 shows how the deflection D of the top of the rod given by eqn (2.169) varies with kL for a fixed lateral force F. It is seen that D is independent of k, and hence of the load W, for $kL \ll 1$, but that as kL approaches $\pi/2$, the deflection increases rapidly. Thus as the load W is increased, the lateral force F required to produce a given deflection diminishes, and it falls to zero for critical values of k and W given by

$$k_c = \pi/2L \quad \text{and} \quad W_c = \pi^2 EI/4L^2. \tag{2.170}$$

This behaviour of a strut that deflects from the vertical position for a vanishing lateral force corresponds to the buckling behaviour described at the beginning of this section. The result (2.170) is known as Euler's column formula and it applies only to the case considered, in which a thin column (a vertical beam) is fixed firmly by the base and is free at the top. The value of W_c given by eqn (2.170) is the maximum load that the column can carry and still maintain stable equilibrium. We see that this load is proportional to the flexural rigidity, EI and inversely proportional to the square of the length of the column or rod. It follows that the failure of a column by buckling does not depend upon the ultimate strength of the material from which it is made, but instead largely upon geometrical factors. Two dimensionally identical columns, rods or tubes made, for example, of steels of very different strengths will support the same load without buckling, because their Young's moduli will be very similar. In §2.15 we consider how geometrical factors can be chosen to optimize flexural rigidity and thereby maximize the buckling load.

Finite deflections D for vanishing lateral force F are predicted more generally by eqn (2.169) when $kL = n\pi/2$ and n is *any* odd integer. So far we have only considered the case of $n = 1$. Larger values of n define larger values for the critical load given by $(n^2\pi^2 EI)/4L^2$, in the ratio $1:9:25:\ldots$ etc. Corresponding to these loads (in theory there is an infinity of values) the neutral axis acquires an increasing number of points of inflection. These other modes of buckling are only of theoretical interest because, in practice, a column fixed vertically at the base with vertical top-loading always buckles in the first mode.

2.14 Euler's result extended to struts with various end conditions

The equations obtained in the last section are often directly applicable to axially-loaded rods or struts with end conditions different from those of the column or vertical cantilever. We shall consider two examples and these are shown diagrammatically in figs. 2.22(a) and 2.22(b).

(i) Pinned strut. Both ends of the strut are pinned, not fixed, so that when it buckles, symmetry arguments dictate that the tangent at the midpoint will be parallel to the line between the pinning points. If the strut is light and the effect of its own weight may be neglected, the pins do not have to be in a vertical line. Then the strut may be considered in two halves, each of which corresponds to the system shown in fig. 2.20(b), so that eqn (2.163) applies. The critical load for this strut in the first buckling mode

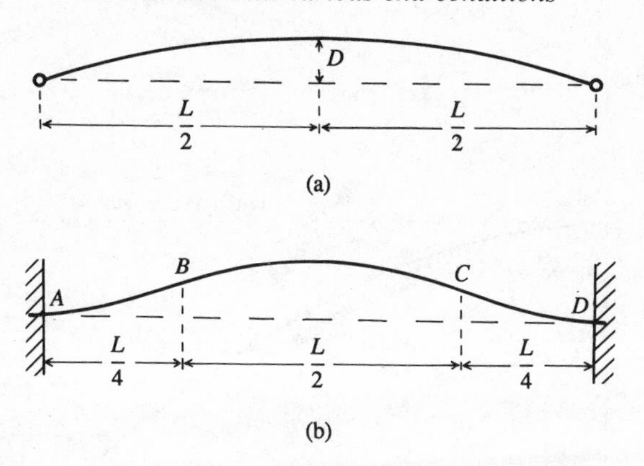

Fig. 2.22 (a) Strut pinned at both ends; (b) strut clamped at both ends.

is thus given by replacing L by $L/2$ in eqn (2.170), i.e.

$$W_c = \pi^2 EI/L^2. \tag{2.171}$$

(ii) Clamped strut. Both ends of a light strut are clamped; thus they correspond to point A in figs. 2.20(a) and (b). Symmetry considerations again dictate that the deflection of the strut will be a maximum at the midpoint and the tangent to the curve at that point will be parallel to the line through the ends of the strut. These conditions imply that the first buckling mode corresponds to a cosine wave with two points of inflection, B and C (fig. 2.22(b)). The two end portions AB and CD behave like the simple cantilever column described by eqn (2.170) so that eqn (2.170) again gives the critical load for buckling, provided that we replace L by $L/4$, i.e.

$$W_c = 4\pi^2 EI/L^2. \tag{2.172}$$

Cases of struts with yet different end conditions are also capable of solution using the approach outlined in § 2.13 and the treatments are generally known as Euler's theory of struts. The student may like to try the case of a light strut that is clamped at one end and pinned at the other (see Problem 2.12). It should be noted that in the cases which we have considered it is assumed that the approximation $1/R = \mathrm{d}^2y/\mathrm{d}x^2$ of eqn (A1.8) applies. From the above results it is clear that, to be effective, a strut should be short and thick, with a large moment of area. Then, in the limit, the critical buckling load exceeds the simple crushing strength of the strut, which is equal to the product of its cross-sectional area and either the yield stress (§ 1.9) or the fracture stress, according to the failure mechanism. Figure 2.23 shows how behaviour varies with the length of the strut. Very short

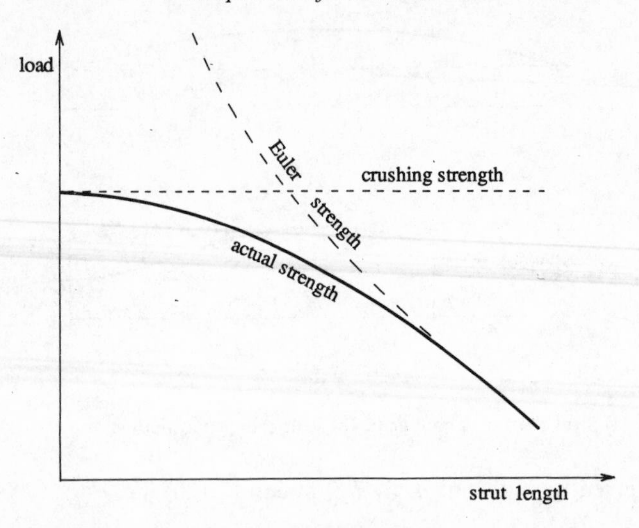

Fig. 2.23 The dependence of the compressive load at which a strut fails on the length of the strut, showing that short struts fail by crushing and longer struts fail by buckling, in accordance with the Euler theory.

struts (i.e. length $\lesssim 2 \times$ diameter) fail by crushing, while long ones (length $\gtrsim 2 \times$ diameter) fail by buckling. Intermediate struts fail by a combination of the two mechanisms.

2.15 Flexural rigidities of various cross-sections

The result for the critical buckling load obtained in § 2.14 (eqn (2.170)) shows that the strength of a long strut may be improved by increasing the second moment of area of its cross-section. In the case of structural members made of a single, homogeneous material, this is equivalent to increasing the moment of inertia. This improvement may be achieved without increasing the cross-sectional area by distributing the material as far as possible from the principal axes of the member. The following reasoning shows that a tubular strut is more resistant to buckling (i.e. stronger), weight for weight, than a solid strut of the same material and same length:

Consider a tube with outer and inner radii a and b, respectively, and wall thickness t. The second moment of area I' of the cross section about a diametral axis is

$$\begin{aligned} I' &= \tfrac{1}{4}\pi(a^4 - b^4) \\ &= \tfrac{1}{4}\pi a^4 - \tfrac{1}{4}\pi(a - t)^4. \end{aligned} \tag{2.173}$$

Let assume that $t \ll a$, so that we can neglect second order terms in t.

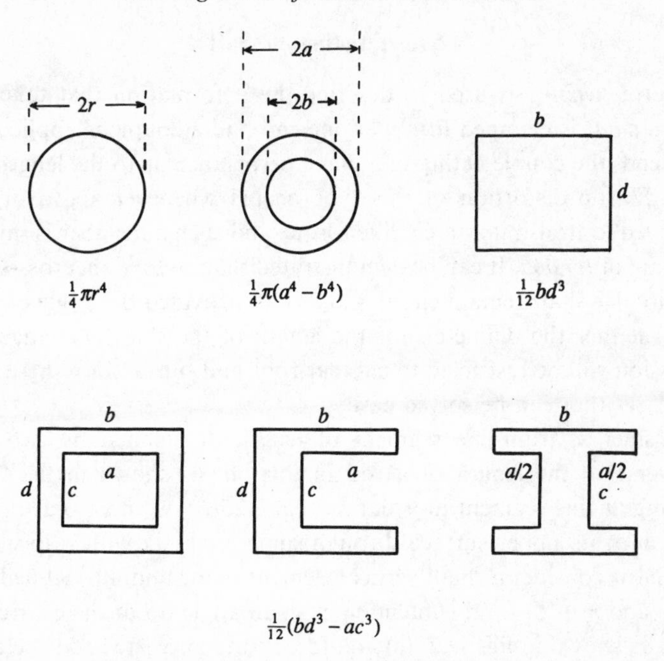

Fig. 2.24 The second moments of area about a horizontal axis through the centre of mass for various simple cross-sections.

Then

$$\begin{aligned} I' &= \tfrac{1}{4}\pi\{a^4 - (a^4 - 4a^3t + \text{smaller terms})\} \\ &= \pi a^3 t. \end{aligned} \tag{2.174}$$

The second moment of area I'' of a solid circular cross-section of radius r is

$$I'' = \tfrac{1}{4}\pi r^4. \tag{2.175}$$

If these cross-sections are respectively associated with a tube and rod with equal masses, then

$$\pi r^2 = 2\pi a t$$

so that comparing I' and I'' we find that

$$I' = aI''/t. \tag{2.176}$$

This means that the flexural rigidity EI of the tubular strut is greater than that of the solid strut. This result is fairly general, no matter what the actual shape of the hollow section, and everyday observation will show the principle in widespread use. Figure 2.24 gives the second moments of area for a number of simple cross-sections.

2.16 Twisting of shafts

The term *twisting* is used to describe the deformation that takes place when a shaft is clamped firmly at one end and a couple is applied to the other end, the couple acting in a plane perpendicular to the length of the shaft. Elastic distortion of this type occurs whenever a rod or bar is employed to transmit a screw-like torque and such a member is said to be in a state of *torsion*. It can be demonstrated that, in torsion, cross-sections of a circular shaft remain circular and that, provided the angle of twist is small, neither the diameter nor the length of the shaft is changed. Our discussion will be restricted to circular rods and tubes, since these are the only cases that can be solved easily.

Consider a drum-like segment of length $\mathrm{d}l$, defined by two planes transverse to the length of a rod in torsion, as shown in fig. 2.25(a). Looking at this segment in isolation, fig. 2.25(b), we note that there is a rotation of its upper surface through some angle $\mathrm{d}\alpha$ with respect to the base. Now consider a small vertical element in the annulus defined by the radii r and $r + \mathrm{d}r$, and subtending a small angle $\mathrm{d}\beta$ at the centre of the drum, as shown in figs. 2.25(c) and (d). In the untwisted rod the faces A and B of this element were perpendicular to the top and bottom surfaces of the drum. During twist of the parent rod they are inclined to the base at some angle ε, (the shear strain) indicated in fig. 2.25(d). The lengths of the faces A and B are essentially unchanged from their values when vertical, from which we may conclude that the element is in a state of strain approximating to pure shear. It is readily seen that the strain ε can thus be obtained from the fact that

$$\mathrm{d}s = r\,\mathrm{d}\alpha = \varepsilon\,\mathrm{d}l,$$

so that

$$\varepsilon = \mathrm{d}s/\mathrm{d}l = r\,\mathrm{d}\alpha/\mathrm{d}l = r\Psi, \text{ say.} \tag{2.177}$$

For a shaft or wire held at one end and twisted by a torque applied to the other, the angle of twist α at a particular cross-section is proportional to the distance from the fixed end. The angle of twist per unit length is also proportional to the applied torque Γ. The ratio of the applied torque to the resulting angle of twist in the shaft or wire is called the *torsional rigidity*.

Let the force acting on the top surface of the element, responsible for its shear, be $\mathrm{d}F$ (this force is transmitted via the adjacent material in the upper part of the rod). Then if the rigidity modulus of the rod material is

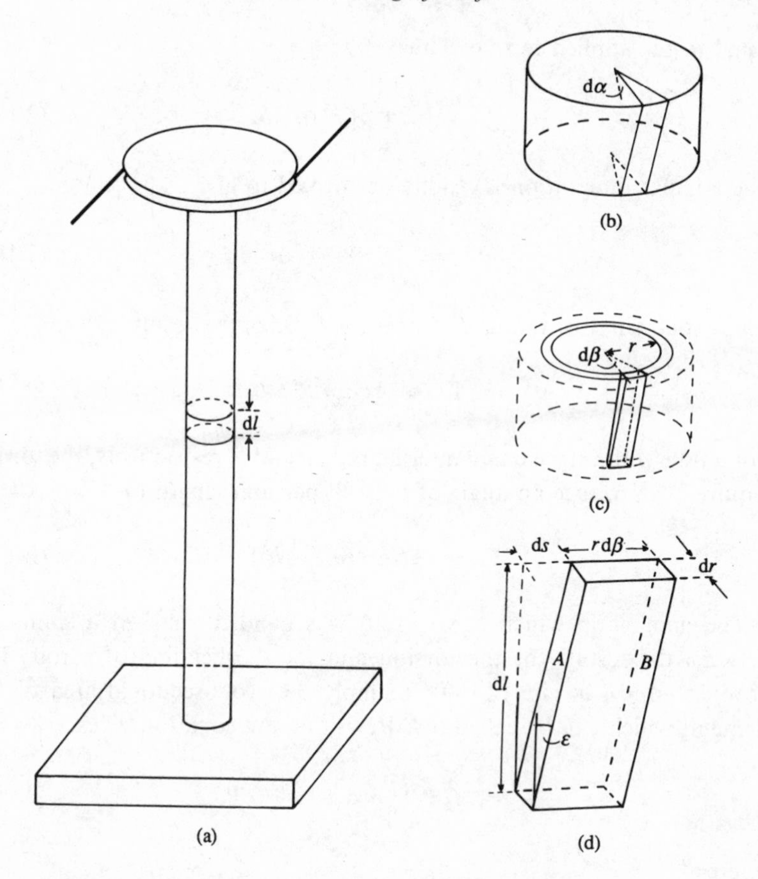

Fig. 2.25 The elastic torsion of a rod or wire: (a) a simple arrangement for applying a torque; (b) small element of the rod; (c) distortion of a vertical column in an annulus within the element shown in (b); (d) an enlarged view of the distorted column in (c).

G, the shear stress σ on the element is

$$\sigma = \mathrm{d}F/(r\,\mathrm{d}\beta\,\mathrm{d}r) = G\varepsilon = G\Psi r, \tag{2.178}$$

where eqn (2.177) has been used. The shear stress is thus directly proportional to the distance from the centre of the shaft and is a maximum at the outer surface. This is where plastic deformation starts if the yield stress is attained. Rearrangement of eqn (2.178) gives

$$\mathrm{d}F = G\Psi r^2\,\mathrm{d}r\,\mathrm{d}\beta. \tag{2.179}$$

There is therefore a moment, $r\,\mathrm{d}F$ about the rod axis, associated with the one small element and the sum of all the moments for the cross-section must be equal to the total torque about the axis of the shaft, resisting and equal to the applied torque. Thus

$$\Gamma = G\Psi \iint r^3 \, \mathrm{d}r \, \mathrm{d}\beta. \tag{2.180}$$

The angular integration is readily performed to give

$$\Gamma = 2G\pi\Psi \int r^3 \, \mathrm{d}r. \tag{2.181}$$

For a solid shaft of radius R, integration yields the result

$$\Gamma = \tfrac{1}{2}G\pi\Psi R^4. \tag{2.182}$$

For a tube with external and internal radii a and b, respectively, the torque required to produce an angle of twist Ψ per unit length is

$$\Gamma = \tfrac{1}{2}G\pi\Psi(a^4 - b^4). \tag{2.183}$$

The manner in which eqn (2.180) was handled masked a similarity between the results for the torsion and for the bending of a rod. The product $r \, \mathrm{d}r \, \mathrm{d}\beta$ in eqn (2.180) is simply the cross-sectional area of the element, which can be called $\mathrm{d}A$. Rewriting the equation gives

$$\Gamma = G\Psi \int_A r^2 \, \mathrm{d}A = G\Psi J, \tag{2.184}$$

where

$$J = \int_A r^2 \, \mathrm{d}A$$

and J is the *polar second moment of area* of the cross-section. A polar moment of area (or inertia) is the moment about an axis perpendicular to the plane of the section, where the product of each elemental area (or mass) with the square of its distance from the axis is integrated over the whole area of the section. (Polar moments of area for sections lacking rotational symmetry can often be calculated by finding the moments of area about two perpendicular axes passing through the same point in the section as the polar axis, and then applying the theorem of perpendicular axes.)

The appearance of the polar moment of area in eqn (2.184), which describes the twisting of a rod, may be compared with the result for the bending of a rod, eqn (2.102). In both cases, the magnitude of the torque or bending moment needed to effect the deformation is proportional to the appropriate second moment of area.

2.17 Suggestions for further reading

Anisotropic properties of crystals, tensors

Nye, J.F. (1957). *Physical Properties of Crystals*. Oxford: Oxford University Press.

Theory of elasticity

Landau, L.D. & Lifshitz, E.M. (1970). *Theory of Elasticity*. 2nd edn. Oxford: Pergamon Press.

Southwell, R.V. (1941). *An Introduction to the Theory of Elasticity*. 2nd edn. Oxford: Oxford University Press.

Timoshenko, S.P. & Goodier, J.N. (1970). *Theory of Elasticity*. 3rd edn. New York: McGraw–Hill.

Elastic constants

Huntington, H.B. (1958). The elastic constants of crystals. *Solid State Physics* **7**, 213.

Problems

1. Show that the gravitational energy of attraction between two molecules can be neglected when calculating the binding energy between them. (Take order of magnitude values for the properties of molecules in crystals, and see the end of the glossary for values of physical constants required.)

2. Assume that the Mie expression (equation (2.72)) can represent the potential energy of interaction between a sodium ion and a chlorine ion in a crystal of sodium chloride, NaCl. Assume also that the attractive term is the result of purely Coulombic forces between the ions while the repulsion is well represented by an exponent n equal to 9. Taking the smallest separation between Na^+ and Cl^- to be 0.281 nm, calculate the Young's modulus of NaCl. (See list of physical constants in glossary for the value of the electronic charge and other constants that you may need.)

3. The Lennard-Jones potential, discussed in §2.4 and illustrated in fig. 1.2(b), can be written in the convenient form

$$U(r) = U_0[(r_0/r)^{12} - 2(r_0/r)^6],$$

where $-U_0$ is its minimum value at $r = r_0$. Assume that at temperature T the mean potential energy of the two atoms is $k_B T$ higher than the minimum value, where $k_B T \ll U_0$. Hence find the mean atomic separation $\bar{r}$ at temperature T and show that the thermal expansivity of condensed material with such an interatomic potential is

$$(1/r_0)\,\mathrm{d}\bar{r}/\mathrm{d}T \simeq k_B/6U_0.$$

4. Working from the equations and data obtained in problem 2, estimate the theoretical fracture stress of a perfect single crystal of NaCl subject only to tensile forces. (Assume that the failure is ideally brittle, i.e. that there is no plastic strain prior to fracture.)

5. A steel wire of diameter 1 mm is wound around a cylindrical former of diameter 2 m. Find the maximum tensile force in the cross-section of the wire that results from bending, assuming that the Young's modulus of steel is 200 GPa. (Hint: examine the geometry of a small segment of arc and use similar triangles.)

6. The bulk modulus, K and the rigidity modulus, G of an isotropic solid are given by eqns (2.66) and (2.68). Assume that the bulk modulus, K remains finite, but the rigidity modulus, $G \to 0$ as a solid approaches its melting point. Hence show that Young's modulus, $E \to 3G$ and Poisson's ratio, $\nu \to 0.5$ as the melting point is approached.

7. A copper pipe, with internal and external diameters 10 mm and 20 mm respectively, is initially maintained horizontal by its ends which rest on two supports, 0.4 m apart. A load of 90 kg is then suspended from the centre of the pipe. Find the sag at the centre point, taking the Young's modulus of copper to be 100 GPa.

8. Find the equation for the shape of a beam subjected to four-point loading, using the parameters shown in fig. 2.17. Hence confirm the expression for the maximum beam deflection, given in eqn (2.132).

9. A uniform beam, of length L and weight w per unit length, is supported at its two ends and carries a load which is three times the total weight of the beam, and which is suspended from the centre of the beam. Show that the bending moment is equal to $w(x^2 + 2Lx - 3L^2)/2$, where the origin of coordinates is at one end of the beam. Derive an expression for the energy stored in the loaded, bent beam in terms of E, I, w and L, assuming that the beam undergoes only a small deflection, where E and I have the usual meanings.

10. A load W is applied to the free end of a uniformly-tapered cantilever beam of length L, circular cross-section, and negligible mass. The diameter D of the free end is half the diameter of the fixed end. Show that a cantilever of constant circular cross-section, with a diameter of $8^{\frac{1}{4}}D$ and length equal to that of the tapered beam, will undergo the same end deflection when it is subject to the same load, W. (Hint: take an origin of coordinates at a distance L beyond the free end of the tapered cantilever.)

11. Apply Euler's theory to find the equations giving (a) the deflection and (b) the buckling load for a lightweight strut that has one end fixed and the other end pinned, taking the origin of coordinates at the fixed end.

12. Consider the vertical strut of length L subjected to a load W and a horizontal force F illustrated in fig. 2.20(b). If the rod has a circular cross-section of diameter d, and W is smaller than the critical load for buckling, show that the material of the rod is nowhere in tension if

$$F < Wkd/8 \tan kL,$$

where $k^2 = W/EI$, and E and I have their usual meanings.

13. A propeller shaft must be manufactured to meet a specified torsional rigidity. Find the percentage of metal that can be saved if a hollow shaft is used with an internal diameter equal to one half its external diameter, rather than a solid shaft.

3

Dynamics and stability of elastic media

The dynamic manifestations of the elastic properties of media can be roughly divided into two categories that are distinguished by the length scale of the accompanying distortion of the material. Distortions whose length scale is short compared with at least one of the dimensions of the sample are produced when elastic waves are propagated through the bulk of the material or along its surface. Such waves may be generated by a transient event, for example an explosion or an earthquake, or by a continuously operating source, as in the case of waves generated in crystals by acoustic transducers. These waves are described variously as elastic waves, acoustic waves, or sound waves.

Distortions whose length scale is comparable to, or longer than the sample dimensions are associated with such dynamical phenomena as the vibrations of beams or the torsional oscillations of fibres. The static distortions of beams and shafts caused by applied stresses have been treated in §§2.7 to 2.12 and some consideration has been given to the conditions under which beams may buckle. The stability of a stressed beam is closely related to its frequency of vibration for small displacements from its static equilibrium position.

We consider in the present chapter both the small-scale elastic-wave properties of solids and their large-scale vibrations. Particular attention is paid to the links between the dynamical behaviour and the stability properties, where there are some parallels between the small- and large-scale elastic behaviour.

3.1 Vibrations of beams and shafts

It is convenient to begin the discussion of the stability of beams and their vibrational frequencies by considering the oscillations of some of the

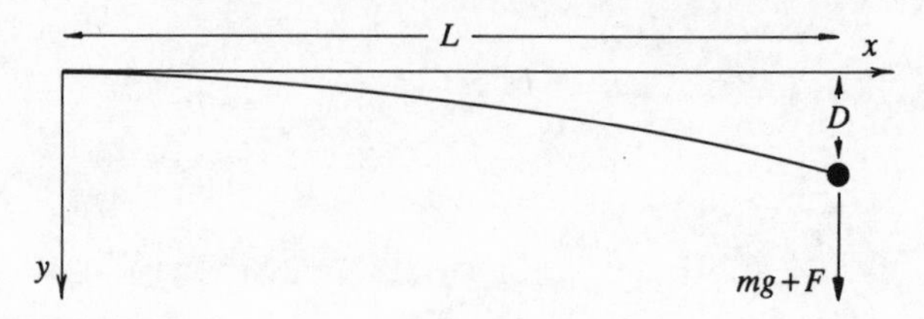

Fig. 3.1 Loaded horizontal beam subjected to deflecting force.

systems whose static equilibrium configurations have been determined in §2.9.

(i) Horizontal beam clamped at one end, load W at the other. This is the system illustrated in fig. 2.15 and treated as the first example in §2.9. Consider for simplicity the case of a lightweight beam with $wL \ll W$ so that w can be neglected and the beam eqn (2.133) is

$$y = W(3L - x)x^2/6EI. \tag{3.1}$$

It is convenient in treating the beam vibrations to assume that a mass m is attached to the free end of the beam, as shown in fig. 3.1. The load W is then made up of the weight of the mass and an applied force F,

$$W = mg + F. \tag{3.2}$$

The deflection D_0 under the influence of the weight alone is given by eqn (3.1)

$$D_0 = mgL^3/3EI. \tag{3.3}$$

The deflection D in the presence of the applied force F is given by eqn (3.1) as

$$D = (mg + F)L^3/3EI = D_0 + (D_0F/mg). \tag{3.4}$$

Thus the force needed to produce a static deflection D is

$$F = mg(D - D_0)/D_0 \tag{3.5}$$

and this must exactly balance the restoring force produced by the elasticity of the beam.

If the applied force F is now removed, the restoring force given by the expression on the right of eqn (3.5) is no longer balanced, and the end of the beam moves according to the equation of motion

$$m\,\mathrm{d}^2D/\mathrm{d}t^2 = -mg(D - D_0)/D_0. \tag{3.6}$$

This is an equation of simple harmonic motion, giving the angular frequency of vibration as

$$\omega = (g/D_0)^{1/2} = (3EI/mL^3)^{1/2}, \tag{3.7}$$

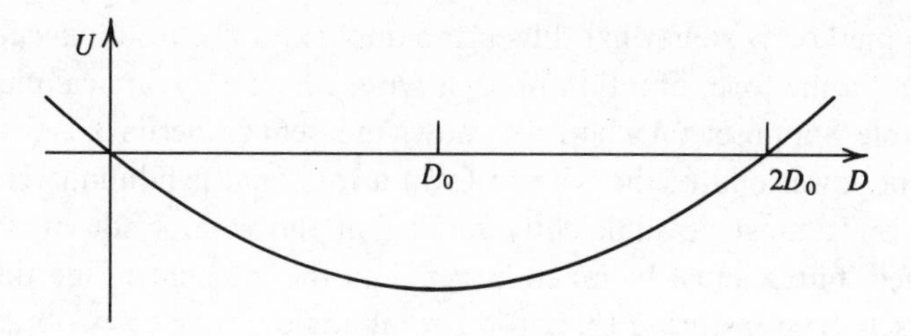

Fig. 3.2 Variation of energy of horizontal beam with end deflection.

while the period of vibration is

$$\tau = 2\pi(D_0/g)^{1/2}. \tag{3.8}$$

The period and frequency are thus the same as for a simple pendulum of length D_0.

The vibrational properties of the beam can also be approached via a consideration of its energy. The total energy U of the beam, at an arbitrary position relative to the horizontal, has an elastic strain contribution given by eqn (2.150) and a potential energy contribution determined by the descent of the mass, so that

$$U = (1/2EI)\int_0^L M^2\,\mathrm{d}x - mgD. \tag{3.9}$$

The bending moment obtained using eqn (3.4) is

$$M = (mg + F)(L - x) = 3EID(L - x)/L^3. \tag{3.10}$$

The integration in eqn (3.9) is easily performed to give a total energy

$$U = mgD(D - 2D_0)/2D_0, \tag{3.11}$$

where eqn (3.3) has been used. The dependence of the energy on beam deflection is illustrated in fig. 3.2. It is seen that the energy is minimized at the static deflection D_0. The parabolic dependence of the energy on the displacement from the minimum at D_0 leads in the usual way to harmonic vibrations with the frequency given in eqn (3.7).

(ii) Vertical wire clamped at upper end, free at lower end. The twisting of a shaft by an applied torque Γ has been treated in §2.16. The total angle of twist α in a rod of length L and radius R is

$$\alpha = \Psi L = 2\Gamma L/\pi G R^4, \tag{3.12}$$

where eqn (2.182) has been used. Measurement of the angle of twist produced by a known torque thus provides an experimental method for determining the rigidity modulus, G, of the material of the rod.

This method is somewhat difficult to apply and it is also inaccurate for material in the form of a thin fibre or wire, where a dynamical method is preferable. Suppose that a body of known moment of inertia $\mathscr{I}$ is suspended from the lower end of the wire to form a torsional pendulum. The mass of the body must be sufficiently small that the wire is not significantly stretched, but $\mathscr{I}$ must be much larger than the moment of inertia of the wire itself. The restoring force for a total angle of twist α is obtained by taking the negative of the expression on the right-hand side of eqn (2.182) with Ψ replaced by α/L. The suspended body thus oscillates in accordance with the equation of motion.

$$\mathscr{I}\,\mathrm{d}^2\alpha/\mathrm{d}t^2 = -\pi G R^4 \alpha/2L. \tag{3.13}$$

This again represents simple harmonic motion, and the angular frequency of the torsional oscillations is

$$\omega = (\pi G R^4/2\mathscr{I}L)^{1/2}. \tag{3.14}$$

Measurement of the oscillation period

$$\tau = (8\pi\mathscr{I}L/GR^4)^{1/2} \tag{3.15}$$

thus provides a value for the rigidity modulus G.

(iii) Vertical strut clamped at lower end. This is the Euler strut or column illustrated in fig. 2.20 and treated in § 2.13. The strut is now assumed to have a mass m attached to its upper end, and the mass of the strut itself is again assumed to be negligibly small. The results of § 2.13 apply with the identifications

$$W = mg \quad \text{and} \quad k^2 = mg/EI, \tag{3.16}$$

where eqn (2.162) has been used. The horizontal force F required to produce a static deflection D of the mass is obtained from eqn (2.169) as

$$F = EIk^3D/(\tan kL - kL). \tag{3.17}$$

The restoring force produced by a deflection D is given by the expression on the right-hand side, so that the equation of motion of the mass in the absence of the applied force F is

$$m\frac{\mathrm{d}^2 D}{\mathrm{d}t^2} = -\left(\frac{mgk}{\tan kL - kL}\right)D. \tag{3.18}$$

This is once more an equation for simple harmonic motion, and the angular frequency of vibration of the mass m about the undeflected vertical configuration of the strut is

$$\omega = \left(\frac{gk}{\tan kL - kL}\right)^{1/2}. \tag{3.19}$$

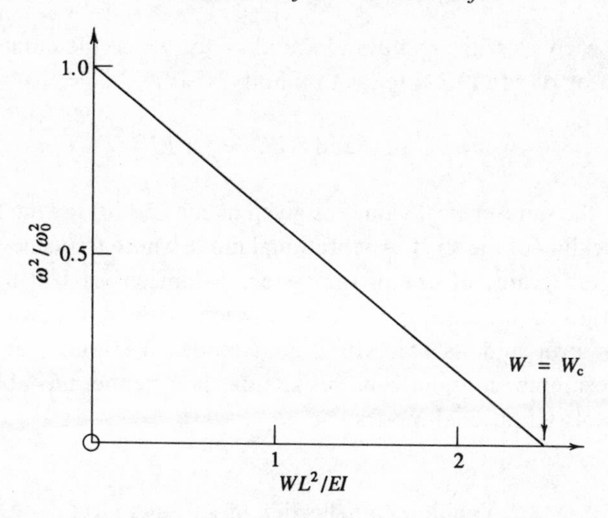

Fig. 3.3 Variation of squared vibrational frequency of an Euler strut with load, showing soft-mode behaviour.

For small values of kL we may use the expansion

$$\tan kL \simeq kL + \tfrac{1}{3}(kL)^3 \quad (kL \ll 1), \tag{3.20}$$

and a good approximation to the frequency is then

$$\omega \simeq \omega_0 = (3g/k^2L^3)^{1/2} = (3EI/mL^3)^{1/2}, \tag{3.21}$$

where eqn (3.16) has been used. This is the same frequency as was found in eqn (3.7) for the vibrations of the similar system of lightweight beam and mass, but with the undistorted position horizontal. We see that for a sufficiently small mass, the vibrational frequency is independent of the orientation of the beam.

Consider, however, the effect of an increased mass on the vibrational frequencies for the two systems. These now begin to differ, because the configurations of the deflected beam are different for the horizontal and vertical orientations. It is convenient to scale the 'vertical' frequency by the 'horizontal' frequency given by eqn (3.21) and write eqn (3.19) in the form

$$\frac{\omega}{\omega_0} = \left(\frac{k^3L^3}{3(\tan kL - kL)}\right)^{1/2}, \tag{3.22}$$

where from eqn (3.16)

$$k^2L^2 = WL^2/EI. \tag{3.23}$$

The square of the scaled frequency is plotted as a function of the load W in fig. 3.3. It is seen that the squared vibrational frequency tends to zero almost linearly at critical values of k and W for which the tangent in the denominator of eqn (3.22) tends to infinity, that is

$$k_c = \pi/2L \quad \text{and} \quad W_c = \pi^2 EI/4L^2. \tag{3.24}$$

These are the same critical values as given in eqn (2.170) for the load that causes buckling of the strut. A vibrational mode whose frequency tends to zero for some value of one of the system parameters is said to be *soft*. There is thus an intimate connection between the buckling instability of the Euler strut and its soft vibrational mode. We note that the two previous examples considered in this section show neither unstable behaviour nor soft vibrational modes.

3.2 Equilibrium deflection of an Euler strut

The behaviour of an Euler strut for loads that exceed the critical buckling value depends upon the size of the resulting deflection from the vertical and upon the rheological properties of the material from which the strut is made. The column may suffer some form of irreversible damage, for example by fracture or by plastic flow. Alternatively, the strut may retain its elasticity, and return to its original vertical position when the excessive load is removed. We consider this latter case in the present section, and show that there are analogies between the behaviour of the Euler strut and that of physical systems that show phase transitions.

The treatments given in §§2.13 and 3.1 provide values for the critical load but they do not give any information on the equilibrium deflection that sets in once the critical load is exceeded. In order to obtain this information, it is necessary to work to a higher order of approximation in the treatment of the distorted beam. For example, the curvature of the beam must be represented by the accurate formula given in Appendix 1 by eqn (A1.7) instead of the approximation given in eqn (A1.8), and account must be taken of the descent of the top of the strut below its undeflected height L. The more accurate theory is quite complicated, and the interested reader is referred to Thompson and Hunt (1973) for a clear presentation.

It is however possible to extract the essence of the critical behaviour of the Euler strut from the simpler model illustrated in fig. 3.4. The strut of elastic material is here replaced by a rigid rod, and the elasticity is transferred to a strip of spring metal that attaches the rod to a solid floor. The spring resists deflection of the rod away from the vertical position with a torque equal to k times the angle of deflection θ. This angle is assumed to be small, even for loads that exceed the critical value, so that the

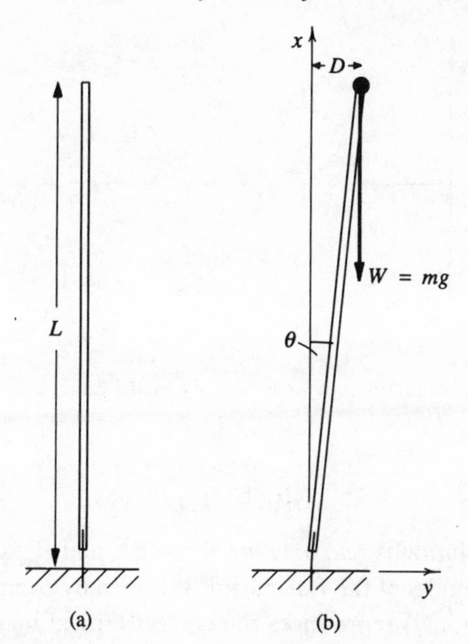

Fig. 3.4 Simple model of Euler strut showing (a) the vertical unloaded equilibrium position, and (b) the deflected position produced by loading.

approximations

$$\sin\theta \simeq \theta - (\theta^3/6) \quad \text{and} \quad \cos\theta \simeq 1 - \tfrac{1}{2}\theta^2 + (\theta^4/24) \tag{3.25}$$

are valid. The effect of any horizontal force is ignored for the present, and we consider the deflection of the rod caused by the vertical load

$$W = mg. \tag{3.26}$$

It is convenient in the present treatment to take θ as the primary variable that describes the deflection, rather than the displacement D of the top of the rod.

Consider first the condition for static equilibrium at a deflection with parameter values θ_0 and D_0, where the torque exerted by the spring must equal the bending moment at the bottom of the rod. We then obtain

$$\begin{aligned} k\theta_0 = WD_0 &= WL\sin\theta_0 \\ &\simeq WL\{\theta_0 - (\theta_0^3/6)\}. \end{aligned} \tag{3.27}$$

The solutions for θ_0 are

$$\begin{aligned} &\text{either} \quad \theta_0 = 0 \\ &\text{or} \quad \theta_0 = \pm\left\{6\left(1 - \frac{W_c}{W}\right)\right\}^{1/2}, \end{aligned} \tag{3.28}$$

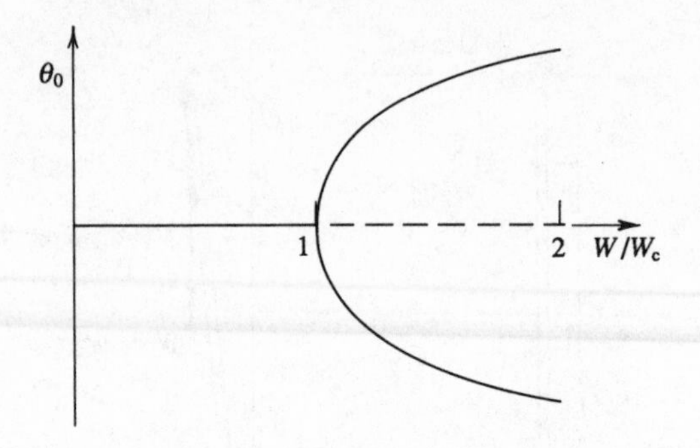

Fig. 3.5 Variation of equilibrium deflections of the model Euler strut with loading.

where

$$W_c = k/L. \tag{3.29}$$

The non-zero solution is real only for $W > W_c$, and W_c is the critical load for the simpler model of the Euler strut. It is seen by comparison with eqn (3.24) that eqn (3.29) reproduces the correct critical load of the vertical Euler strut for a spring constant

$$k = \pi^2 EI/4L. \tag{3.30}$$

The solutions (3.28) are illustrated in fig. 3.5. The curve for the non-zero solution is symmetrical about the horizontal axis since the rod can deflect to either side of the vertical when the critical load is exceeded. We shall show shortly that the solution for zero deflection corresponds to unstable equilibrium for $W > W_c$. This is indicated by the dashed portion of the horizontal axis in fig. 3.5. The transition at $W = W_c$ from one stable solution to two stable solutions is called a *bifurcation*.

The stability of the solutions is determined by the total potential energy of the system at a general deflection θ, given by a sum of the elastic energy stored in the spring and the change in gravitational energy of the descending mass,

$$\begin{aligned} U &= \tfrac{1}{2}k\theta^2 + WL(\cos\theta - 1) \\ &\simeq \tfrac{1}{2}(W_c - W)L\theta^2 + WL\theta^4/24, \end{aligned} \tag{3.31}$$

where eqns (3.25) and (3.29) have been used. The dependence of the potential energy on θ is plotted in fig. 3.6 for loads W that lie respectively below, at, and above the critical value W_c. It is seen that the minimum at $\theta = 0$ disappears as the load is increased through the critical value to become a position of maximum potential energy, and therefore unstable equilibrium. The new potential energy minima occur at the angles θ_0 given in eqn (3.28), as is easily verified by differentiation of eqn (3.31).

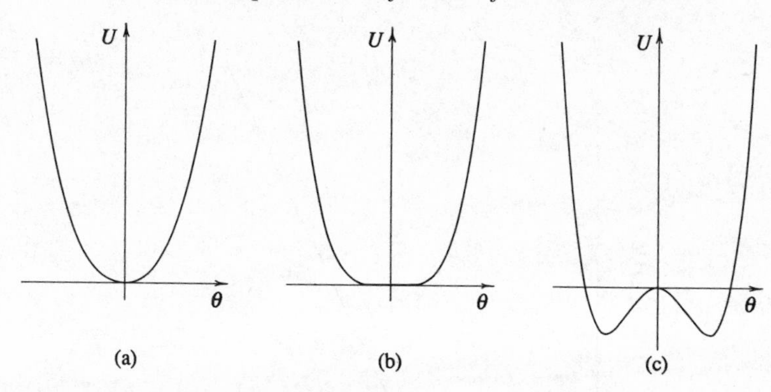

Fig. 3.6 Potential energies of the model Euler strut as functions of the deflection angle for loads W equal to (a) $\frac{1}{2}W_c$, (b) W_c and (c) $2W_c$.

Close to the potential minima, where $\theta - \theta_0$ is small, the potential energy can be approximately represented by means of a Taylor expansion. The low order terms give the parabolic form

$$U \simeq [U]_{\theta=\theta_0} + \tfrac{1}{2}(\theta - \theta_0)^2 \left[\frac{\mathrm{d}^2 U}{\mathrm{d}\theta^2}\right]_{\theta=\theta_0}. \tag{3.32}$$

This is a harmonic potential, and the rod undergoes simple harmonic motion when displaced from its static equilibrium position. The equation of motion of the mass at the top of the rod is approximately

$$mL\frac{\mathrm{d}^2\theta}{\mathrm{d}t^2} = -\frac{\mathrm{d}U}{L\,\mathrm{d}\theta} = -\frac{1}{L}(\theta - \theta_0)\left[\frac{\mathrm{d}^2 U}{\mathrm{d}\theta^2}\right]_{\theta=\theta_0} \tag{3.33}$$

and the vibrational frequency is given by

$$\omega^2 = \frac{1}{mL^2}\left[\frac{\mathrm{d}^2 U}{\mathrm{d}\theta^2}\right]_{\theta=\theta_0}. \tag{3.34}$$

The second derivatives of the potential given by eqn (3.31) at the equilibrium angles θ_0 given by eqn (3.28) are readily evaluated, and the resulting vibrational frequencies are

$$\begin{aligned} \omega &= \{(W_c - W)/mL\}^{1/2} \quad \text{for } W < W_c \\ \omega &= \{2(W - W_c)/mL\}^{1/2} \quad \text{for } W > W_c. \end{aligned} \tag{3.35}$$

It is convenient to scale these frequencies by the limiting frequency ω_0, for small loads, given by

$$\omega_0 = (W_c/mL)^{1/2}, \tag{3.36}$$

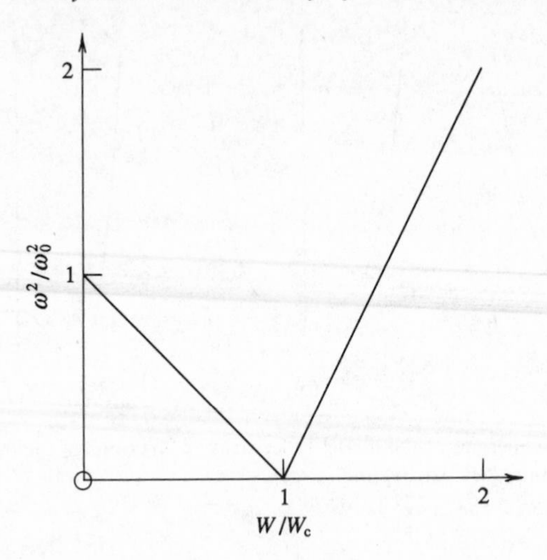

Fig. 3.7 Variation of squared vibrational frequency of the model Euler strut with load, showing soft-mode behaviour.

and write eqn (3.35) in the form

$$\frac{\omega}{\omega_0} = \left\{1 - \frac{W}{W_c}\right\}^{1/2} \quad \text{for } W < W_c$$

$$\frac{\omega}{\omega_0} = \left\{2\left(\frac{W}{W_c} - 1\right)\right\}^{1/2} \quad \text{for } W > W_c. \tag{3.37}$$

The square of the frequency is plotted as a function of W in fig. 3.7. It is seen that the part of the curve for loads smaller than the critical value reproduces the soft-mode behaviour shown in fig. 3.3. The more complete theory given here for the sprung rigid-rod model of the Euler strut shows that the vibrational frequency rises again from zero as the load is increased beyond the critical value. Of course the vibrations now take place around the deflected position of stable equilibrium. The more accurate theory of the original Euler strut, made from a rod of elastic material, predicts a very similar behaviour, with the vibrational frequency softening to zero as the critical load is approached from above or below.

The dependence of θ_0 on the load W is very similar to the dependence of the order parameter on the temperature in a second-order thermodynamic phase transition as defined in § 1.7. For example, the variation of the spontaneous magnetization of a ferromagnetic material with the temperature resembles the curve in fig. 3.5, except for a reversal of the direction of the horizontal axis. Thus the magnetization shows a bifurcation as the material is cooled through its critical temperature, known as the

Curie temperature. The equilibrium magnetization is zero in the paramagnetic phase above the Curie temperature, while below there are two stable states with equal and opposite magnetizations. The magnetic free energy develops in a manner similar to the sequence of curves shown in fig. 3.6 as the ferromagnet is cooled through the critical temperature. Second-order phase transitions are generally associated with some soft mode of excitation whose frequency in the neighbourhood of the critical temperature shows a form of variation similar to the vibrational frequency plotted in fig. 3.7.

3.3 The Euler strut and catastrophe theory

We now consider a slightly more general problem in which the sprung rigid rod shown in fig. 3.4 is subjected not only to the vertical load $W = mg$ but also to a horizontal force F applied to the mass at the upper end. The force nominally points in the x-direction but we allow F to take both positive and negative values. The total potential energy given in eqn (3.31) is thus generalized to

$$U = \tfrac{1}{2}k\theta^2 + WL(\cos\theta - 1) - FL\sin\theta. \quad (3.38)$$

We continue to assume that θ is small and we also take F to be much smaller than W, so that using eqns (3.25) and (3.29) an appropriate approximation to U is

$$U \simeq -FL\theta + \tfrac{1}{2}(W_c - W)L\theta^2 + WL\theta^4/24. \quad (3.39)$$

The equilibrium deflections are given by the turning points of the potential energy obtained from

$$dU/d\theta = -FL + (W_c - W)L\theta_0 + WL\theta_0^3/6 = 0 \quad (3.40)$$

or

$$\theta_0^3 + 6\left(\frac{W_c}{W} - 1\right)\theta_0 - 6\left(\frac{F}{W}\right) = 0. \quad (3.41)$$

This cubic equation is difficult to solve in general but there is a simple criterion for the number of real solutions, given by the conditions

$$\left.\begin{matrix}\text{1 real root}\\ \text{3 real roots}\end{matrix}\right\} \text{ for } \left(\frac{3F}{W}\right)^2 + 8\left(\frac{W_c}{W} - 1\right)^3 \left\{\begin{matrix} <0 \\ >0. \end{matrix}\right. \quad (3.42)$$

Thus for a given value of the load W, the regions where there exist 1 root and 3 roots are separated by a critical value of the applied force given by

$$F_c = (8/9W)^{1/2}(W - W_c)^{3/2}. \quad (3.43)$$

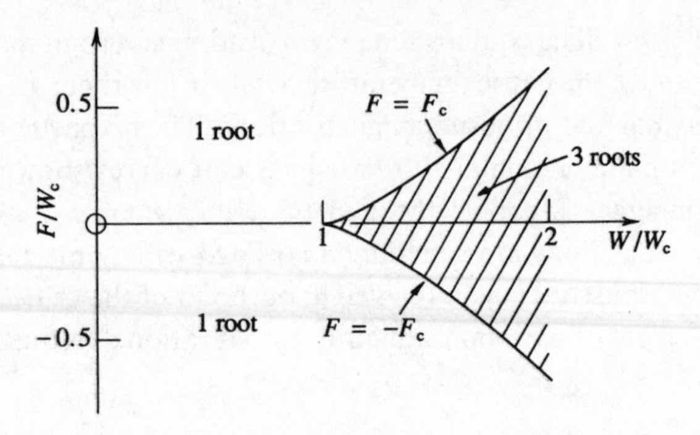

Fig. 3.8 Regions of WF-plane for which the model Euler strut subjected to vertical and horizontal loading has one or three equilibrium deflection angles.

The ranges of F and W for which there are three turning points or only one turning point are shown in fig. 3.8. The line $F = 0$ corresponds to the problem treated in § 3.2, and as has been shown there (see particularly figs. 3.5 and 3.6), two of the three turning points are potential energy minima and represent stable deflections of the rod.

We consider next what happens to the rod as F is varied while W is held constant at a value larger than W_c. It is convenient to write the relation (3.41) between F and θ_0 in the form

$$F = -(W - W_c)\theta_0 + W\theta_0^3/6. \tag{3.44}$$

Fig. 3.9 shows the form of this relation for $W = 1.02W_c$, where W is chosen only slightly larger than W_c in order for the approximations in eqn (3.25) to be reasonably well satisfied. There are either one or three values of θ_0 for each value of F. In the latter case the two stable configurations are indicated by the continuous curve while the unstable configuration is represented by the dashed curve. The part of the S-shaped curve corresponding to stable equilibrium deflections thus has two separate branches.

Suppose now that the rod is inclined at a positive angle to the vertical with F initially zero. This position corresponds to the intersection of the upper branch with the θ_0 axis in fig. 3.9. As the horizontal force F is increased from zero in the positive sense, the top of the rod is pulled further to the right as the equilibrium point moves up the upper branch in fig. 3.9. A reduction of F to zero returns the rod to its original position. If F is now increased again but in the negative sense, the deflection angle θ_0 is steadily decreased as the equilibrium point moves further down the upper branch until the upper bend in the S-curve is reached at $F = -F_c$. The position of positive deflection now becomes unstable and the rod suddenly flips across to take up a position of negative deflection with

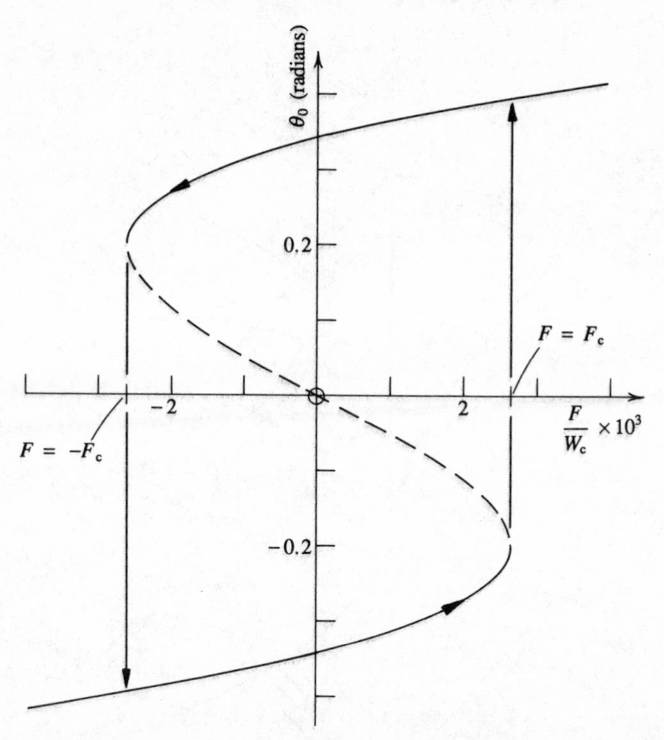

Fig. 3.9 Variation of the deflection angle of the model Euler strut with horizontal load F for a fixed value $W = 1.02W_c$ of the vertical load.

respect to the vertical. This discontinuous transition is represented by the downward vertical arrow in fig. 3.9. Any further increase in negative F pulls the rod further to the left as the equilibrium point moves down the lower branch in fig. 3.9. If the negative F is reduced to zero, the rod returns to a position of negative deflection that corresponds to the intersection of the lower branch with the θ_0 axis. An increasing F in the positive sense causes a steady decrease in the negative deflection angle as the equilibrium point moves up the lower branch until the lower bend in the S-curve is reached at $F = F_c$. The rod now flips across to the positive side of the vertical, as indicated by the upward vertical arrow in fig. 3.9.

These changes in θ_0 brought about by variations in F resemble the phenomena at a first-order thermodynamic phase transition. The distinguishing characteristics of a first-order transition described in § 1.7 are the discontinuous change in the configuration of the system, here represented by the deflection angle θ_0, and the hysteresis effect in that the transition occurs later for F increasing in one sense than it does for F increasing in the opposite sense. The sequence of events described in the previous

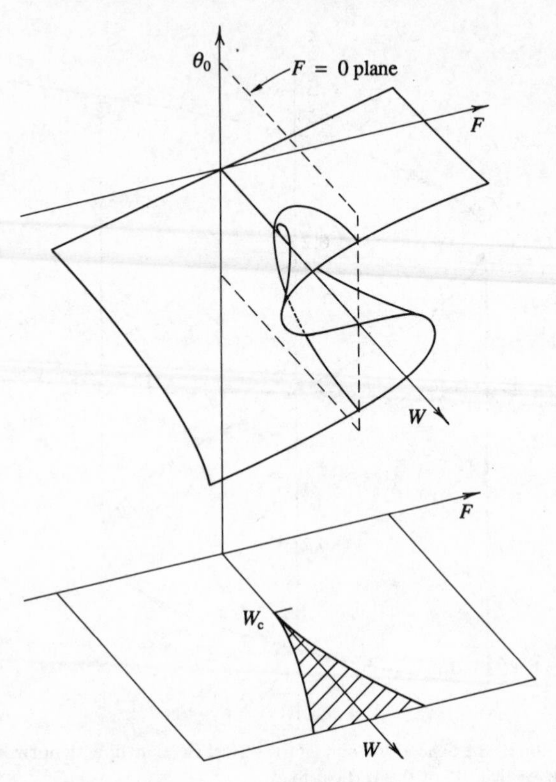

Fig. 3.10 Cusp catastrophe surface for the relation between deflection angle and applied horizontal and vertical loads for the model Euler strut.

paragraph takes the Euler strut around the *hysteresis cycle* or *hysteresis loop* indicated by the arrowed lines in fig. 3.9.

Most of the graphical information about the stable deflection of the Euler strut given in the present and previous sections can be collected together and displayed in a single figure. The relationship between θ_0, W and F given by eqn (3.40) represents a surface in a three-dimensional space. This surface, illustrated in fig. 3.10, is known as a *cusp catastrophe*, where the nomenclature reflects the catastrophic nature of the discontinuous collapse that occurs in mechanical systems at the points of instability. The section through the folded surface taken in the $F = 0$ plane reproduces the θ_0 versus W curve of fig. 3.5 together with the bifurcation that occurs for $W > W_c$. Sections through the surface at constant W for $W > W_c$ reproduce the S-shaped hysteresis curves similar to fig. 3.9, with discontinuous changes in θ_0 occurring at the bends in the curve. Finally, the projection of the cusp catastrophe surface into the WF-plane reproduces the ranges of F and W for which the system has two positions of stable equilibrium and one of unstable equilibrium.

Many different physical, chemical and biological systems show responses to applied stimuli that can be represented by cusp catastrophe surfaces, or more complicated surfaces in higher-dimensional spaces, and such phenomena are treated in general by a branch of mathematics known as *catastrophe theory*. The treatment of the Euler strut given here refers to the idealized model illustrated in fig. 3.4, and is valid only for small angles of deflection, corresponding to loads that cannot exceed the critical value W_c by more than a small amount. Nevertheless, the representation of the behaviour of the deflection of the top of the column by a cusp catastrophe surface is qualitatively correct for the column constructed from uniformly elastic material and rigidly clamped at its lower end, as illustrated in fig. 2.20. Very similar behaviour also occurs for the other kinds of strut treated in §2.14; for example the deflections of the mid-points of the struts illustrated in fig. 2.22 with a parallel load W and a perpendicular force F can be represented by cusp catastrophe surfaces similar to that of fig. 3.10.

3.4 Elastic energy and stability of bulk materials

The previous two sections describe a kind of instability that can occur for a beam under stress, in which the member as a whole distorts to a new configuration when the applied loads or forces exceed certain critical values. It has been pointed out that the behaviour of the beam may resemble that of a thermodynamic system undergoing a phase transition, either of the first order (fig. 3.9) or the second order (fig. 3.5), depending on the way in which the forces are applied.

True phase transitions analogous to these that we have described for mechanical systems do occur for bulk elastic media in which the material parameters are varied not by applied forces but by changes in temperature. The natures of the instabilities can be understood from a consideration of the elastic energy of a strained sample. The energy is calculated by a generalization of the method of §2.12 in which we find the work done in deforming the sample by means of an applied stress. Consider the unit cube of material shown in fig. 2.1 with its centre maintained in the same position while an applied stress σ_{xx} increases the strain ε_{xx} from its initially zero value to its equilibrium value. An increase $d\varepsilon_{xx}$ in this component, with no changes in the other strain components, corresponds to outward movements of the two faces perpendicular to the x-axis by equal amounts $\frac{1}{2}d\varepsilon_{xx}$. The work (force × distance) done on the two faces is

$$2\sigma_{xx} \times \tfrac{1}{2}d\varepsilon_{xx} \equiv \sigma_1 d\varepsilon_1 = \sum_j c_{1j}\varepsilon_j d\varepsilon_1, \qquad (3.45)$$

where the abbreviated notation of eqn (2.39) is used and the stress is expressed in terms of elastic stiffness coefficients by the use of eqn (2.40). Similar reasoning applies to the other stress components, and the change in energy associated with small increases in all the strain components is

$$dU = \sum_{i,j} c_{ij}\varepsilon_j \, d\varepsilon_i . \tag{3.46}$$

By using eqn (2.42) and integrating, we obtain

$$U = \frac{1}{2}\sum_{i,j} c_{ij}\varepsilon_i\varepsilon_j \tag{3.47}$$

as the energy per unit volume of an elastically strained sample.

We now specialize to the case of an elastically isotropic material where use of the explicit form of the stiffness matrix shown in eqn (2.47) gives, for the elastic energy, the expression

$$\begin{aligned} U = {} & \tfrac{1}{2}c_{11}(\varepsilon_{xx}^2 + \varepsilon_{yy}^2 + \varepsilon_{zz}^2) + c_{12}(\varepsilon_{yy}\varepsilon_{zz} + \varepsilon_{zz}\varepsilon_{xx} + \varepsilon_{xx}\varepsilon_{yy}) \\ & + \tfrac{1}{4}(c_{11} - c_{12})(\varepsilon_{yz}^2 + \varepsilon_{zx}^2 + \varepsilon_{xy}^2). \end{aligned} \tag{3.48}$$

The structure of the elastic material can be stable only if the energy is positive for any choice of strain components. When this condition is not satisfied, the material can lower its energy by spontaneous distortion to a new structure, analogous to the way in which the Euler strut considered in § 3.2 can lower its energy by deflection to a new equilibrium position. The stability condition places restrictions on the allowed values of the two independent stiffness coefficients. For example, eqn (3.48) must produce a positive energy when any one of the strain components is non-zero, so that the inequalities

$$c_{11} > 0 \quad \text{and} \quad c_{11} - c_{12} > 0 \tag{3.49}$$

must be satisfied. There is however no need for c_{12} to be positive, although this is the case for most substances. The above stability conditions have previously been assumed in eqns (2.55) and (2.56).

The elastic stiffness coefficients of a material usually vary with the temperature, and it sometimes happens that one of the quantities on the left-hand sides of the inequalities (3.49) tends to zero at a particular temperature. The spontaneous distortions to new atomic structures that result are examples of the structural phase transformations described in § 1.7. More generally, such structural changes associated with temperature-dependent elastic stiffness coefficients are known as *martensitic* phase transitions, and they are important in the technology of steels, and some non-metallic materials.

3.5 Elastic properties of liquids

Liquids have the property that they offer only a temporary or transient resistance to shear stresses (see § 1.6). Thus although the viscosity of a liquid has some resemblance to a rigidity, the liquid eventually flows in the direction of a stress, and there is no static equilibrium state of a free body of liquid subjected to shear. As with solids, we are not concerned with displacements produced by unbalanced forces. A contained liquid subjected to gravitational forces always takes up an equilibrium state in which it has a surface that is essentially flat, except that surface tension may produce curved boundary regions (discussed in § 4.8). These considerations imply that the shear stiffness coefficients of a liquid must vanish and, in accordance with eqn (2.47), which applies to an elastically isotropic material,

$$c_{44} = 0 \quad \text{or} \quad c_{11} = c_{12}. \tag{3.50}$$

The elastic properties of a liquid are therefore specified by a single quantity c_{11}. The energy per unit volume of a strained liquid from eqn (3.48) reduces to

$$\begin{aligned} U &= \tfrac{1}{2}c_{11}(\varepsilon_{xx} + \varepsilon_{yy} + \varepsilon_{zz})^2 \\ &= \tfrac{1}{2}c_{11}\Theta^2, \end{aligned} \tag{3.51}$$

where Θ is the dilatation defined by eqn (2.30).

As regards the elastic compliance, it is seen from eqn (2.51) that both the coefficients s_{11} and s_{12} are infinite for a liquid. Not only is the liquid infinitely compliant with respect to shearing stress, in accordance with the above discussion, but also with respect to normal stress. This property is also in agreement with common experience, since for example a uniaxial pressure applied parallel to the z-axis merely causes the liquid to flow outwards in all directions in the xy-plane, and there is no static equilibrium state. It follows from eqns (2.58), (2.59) and (2.68) that Young's modulus, Poisson's ratio, and the rigidity modulus have the values

$$E = 0, \quad \nu = \tfrac{1}{2} \quad \text{and} \quad G = 0 \tag{3.52}$$

for all liquids.

The single exception to the infinities in the compliance coefficients occurs for the combination

$$s_{11} + 2s_{12} = \frac{1}{c_{11} + 2c_{12}} = \frac{1}{3c_{11}}, \tag{3.53}$$

where eqns (2.51) and (3.50) have been used. This is the combination that relates strain to stress in the special case of hydrostatic pressure described by eqn (2.64). The liquid in this case is subjected to the same pressure in all directions, as for example when liquid in a cylinder is subjected to pressure via a well-fitting piston. No escape of liquid is possible, and its compression is determined by the bulk modulus, obtained from eqns (2.66) and (3.53) as

$$K = c_{11}. \tag{3.54}$$

3.6 Bulk elastic waves

The dynamical properties of a bulk elastic material determine the kinds of elastic or acoustic wave that can be propagated through it. In the present section we use the elasticity theory of §2.3 to obtain the elastic wave equation for an isotropic solid, and show how the stiffness coefficients can be determined from measured acoustic wave velocities.

Consider again the forces acting on an element of volume in an infinite elastic solid of density ϱ. The element, shown in fig. 3.11, has its faces oriented perpendicular to the Cartesian axes at coordinates x, $x + \Delta x$; y, $y + \Delta y$; and z, $z + \Delta z$. The stresses on opposite faces are equal and opposite when the material is in a state of static equilibrium. A displacement of the element from equilibrium produces non-balancing stresses that tend to return it to its equilibrium position. The net force in the x-direction resulting from the stresses on the first two faces is

$$\{\sigma_{xx}(x + \Delta x) - \sigma_{xx}(x)\}\,\Delta y\,\Delta z = \Delta x \frac{\partial \sigma_{xx}}{\partial x}\,\Delta y\,\Delta z. \tag{3.55}$$

Similar calculations for the other two pairs of faces give the total force in the x-direction as

$$F_x = \left\{\frac{\partial \sigma_{xx}}{\partial x} + \frac{\partial \sigma_{xy}}{\partial y} + \frac{\partial \sigma_{xz}}{\partial z}\right\}\Delta x\,\Delta y\,\Delta z. \tag{3.56}$$

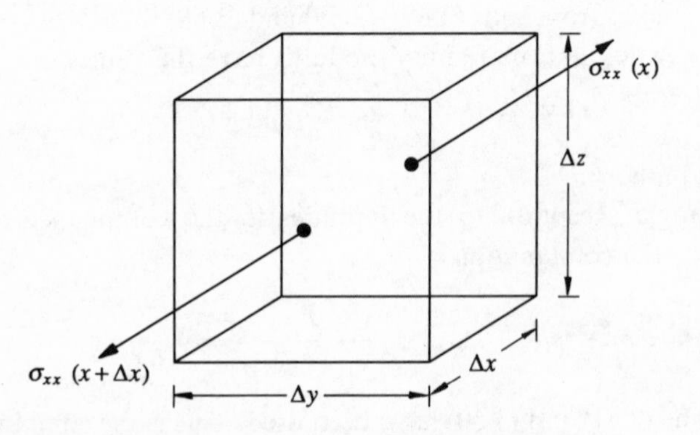

Fig. 3.11 Element of infinite solid.

The mass of the volume element is

$$\varrho \, \Delta x \, \Delta y \, \Delta z, \tag{3.57}$$

and Newton's equation of motion for the x-component of a displacement **u** is

$$\varrho \frac{\partial^2 u_x}{\partial t^2} = \frac{\partial \sigma_{xx}}{\partial x} + \frac{\partial \sigma_{xy}}{\partial y} + \frac{\partial \sigma_{xz}}{\partial z}. \tag{3.58}$$

Similar equations apply for u_y and u_z.

The stresses and strains in an isotropic material are related by eqn (2.48), so that eqn (3.58) can be written

$$\varrho \frac{\partial^2 u_x}{\partial t^2} = c_{11} \frac{\partial \varepsilon_{xx}}{\partial x} + c_{12} \left\{ \frac{\partial \varepsilon_{yy}}{\partial x} + \frac{\partial \varepsilon_{zz}}{\partial x} \right\} + \tfrac{1}{2}(c_{11} - c_{12}) \left\{ \frac{\partial \varepsilon_{xy}}{\partial y} + \frac{\partial \varepsilon_{zx}}{\partial z} \right\}. \tag{3.59}$$

The strains are in turn related to displacement gradients by eqns (2.21) and (2.22), so that the equation of motion becomes

$$\begin{aligned} \varrho \frac{\partial^2 u_x}{\partial t^2} &= c_{11} \frac{\partial^2 u_x}{\partial x^2} + c_{12} \left\{ \frac{\partial^2 u_y}{\partial x \, \partial y} + \frac{\partial^2 u_z}{\partial z \, \partial x} \right\} \\ &\quad + \tfrac{1}{2}(c_{11} - c_{12}) \left\{ \frac{\partial^2 u_x}{\partial y^2} + \frac{\partial^2 u_y}{\partial x \, \partial y} + \frac{\partial^2 u_z}{\partial z \, \partial x} + \frac{\partial^2 u_x}{\partial z^2} \right\}. \end{aligned} \tag{3.60}$$

This is recognized as a complicated form of 3-dimensional wave equation. The corresponding equations for u_y and u_z are obtained by cyclic permutation of x, y and z.

The solutions of the wave equation for an isotropic solid are expected to be independent of the direction of propagation. They can be found by use of some simple trial forms. Consider first a trial longitudinal solution of the form

$$u_x = u_{x0} \exp(-\mathrm{i}\omega t + \mathrm{i}qx), \quad u_y = u_z = 0 \tag{3.61}$$

where both displacement and propagation direction are parallel to the x-axis. Substitution in eqn (3.60) gives

$$-\varrho\omega^2 u_{x0} = -c_{11} q^2 u_{x0}, \tag{3.62}$$

where exponential factors have been removed from both sides. The longitudinal form (3.61) thus satisfies the wave equation if the phase velocity is

$$v_{\mathrm{L}} = \omega / q = (c_{11}/\varrho)^{1/2}. \tag{3.63}$$

Similarly, for a trial transverse solution

$$u_x = u_{x0} \exp(-\mathrm{i}\omega t + \mathrm{i}qy), \quad u_y = u_z = 0 \tag{3.64}$$

where the displacement is parallel to the x-axis and the propagation is parallel to the y-axis, substitution in eqn (3.60) gives

$$-\varrho\omega^2 u_{x0} = -\tfrac{1}{2}(c_{11} - c_{12})q^2 u_{x0}. \tag{3.65}$$

The transverse phase velocity must therefore be

$$v_T = \frac{\omega}{q} = \left(\frac{c_{11} - c_{12}}{2\varrho}\right)^{1/2} = \left(\frac{G}{\varrho}\right)^{1/2}, \tag{3.66}$$

where the rigidity modulus G is defined in eqn (2.68). The same velocity is also found for the independent transverse solution similar to eqn (3.64) but with the propagation direction changed from y to z.

We therefore find that three independent elastic waves, one longitudinal and two transverse, occur for any displacement direction or any propagation direction in an isotropic material. Typical values of the phase velocities are

$$v_L \simeq 5000\,\mathrm{ms^{-1}} \quad \text{and} \quad v_T \simeq 3000\,\mathrm{ms^{-1}}. \tag{3.67}$$

Measurement of the velocities enables the two independent stiffness coefficients to be determined from

$$\begin{aligned} c_{11} &= \varrho v_L^2 \\ c_{12} &= \varrho(v_L^2 - 2v_T^2). \end{aligned} \tag{3.68}$$

The ratio of the two velocities can be expressed in terms of Poisson's ratio ν with the help of eqn (2.61), giving

$$\frac{v_T^2}{v_L^2} = \frac{c_{11} - c_{12}}{2c_{11}} = \frac{1 - 2\nu}{2(1 - \nu)}. \tag{3.69}$$

The velocity ratio is plotted as a function of ν in fig. 3.12 over the allowed range of Poisson's ratio for solids (eqn (2.60)),

$$0 \leqslant \nu < \tfrac{1}{2}. \tag{3.70}$$

In circumstances where the stiffness coefficients c_{11} and c_{12} tend towards equality at a particular temperature, as in the martensitic phase transitions mentioned in § 3.4, the transverse velocity given by eqn (3.66) and the transverse wave frequencies tend to zero. These elastic vibrations represent soft modes associated with the material instability, analogous to the beam vibrational frequencies whose behaviour is shown in figs. 3.3 and 3.7. In the limit of a liquid, when $c_{11} = c_{12}$ and $\nu = \frac{1}{2}$,

$$v_L = (c_{11}/\varrho)^{1/2} \quad \text{and} \quad v_T = 0. \tag{3.71}$$

This result shows that transverse waves cannot be propagated through a liquid, as supported by experiment.

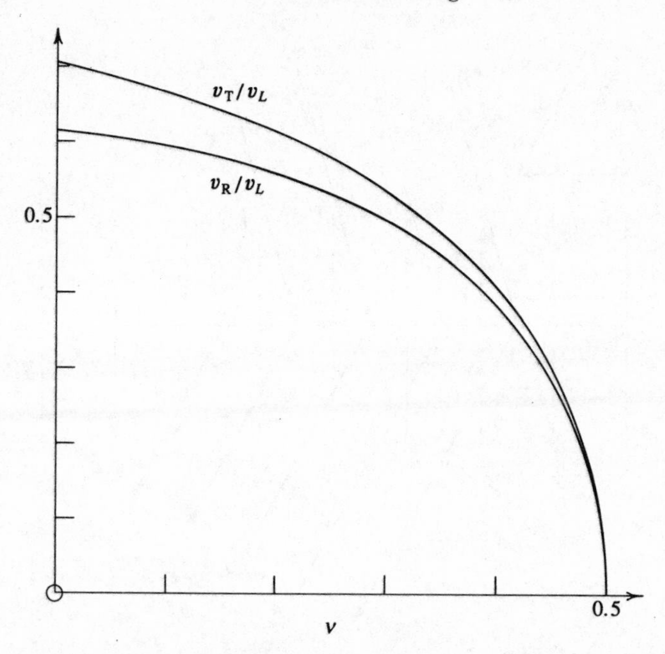

Fig. 3.12 Ratios of bulk transverse acoustic wave velocity and Rayleigh surface wave velocity to the bulk longitudinal velocity as functions of Poisson's ratio for isotropic solids.

The above results for solids are valid only for isotropic materials. The solution of the wave equation for crystalline materials is much more difficult. Thus even for crystals of cubic symmetry, the velocities vary with propagation direction, and the elastic waves have three distinct velocities and no simple polarization for a general direction of propagation.

3.7 Brillouin scattering

Elastic waves scatter light in a way that makes it possible to measure their velocities by optical techniques. Consider a wave of frequency ω propagating parallel to the x-axis, with a laser beam of frequency ω_l incident at some angle to the propagation direction, as shown in fig. 3.13. The periodically oscillating distribution of strain associated with the elastic wave produces an oscillation of frequency ω in the optical refractive index of the material appropriate to the laser frequency ω_l. The travelling elastic wave therefore acts like a moving diffraction grating, and for appropriate directions of incident light it can produce scattered light at frequencies

$$\omega_S = \omega_l \pm \omega. \tag{3.72}$$

The phenomenon is named after Brillouin who first predicted it. The

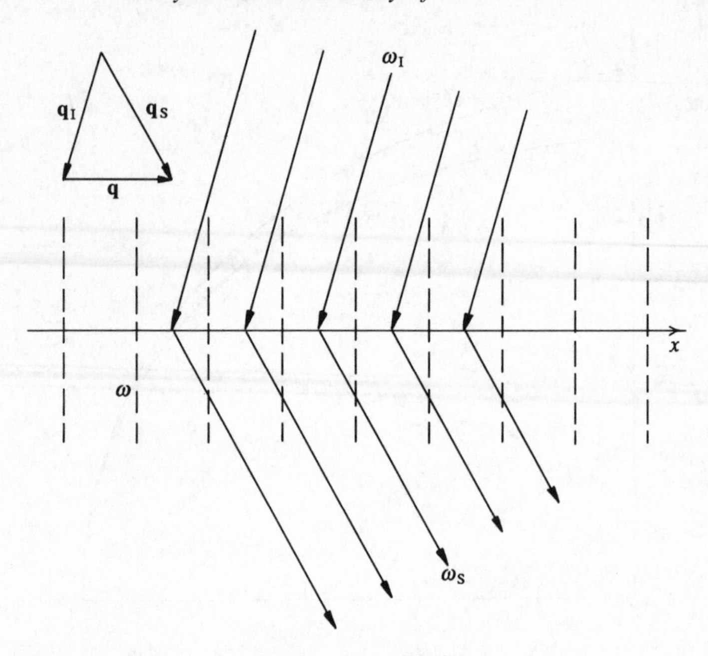

Fig. 3.13 Wavevector triangle for Brillouin scattering, and geometrical arrangement of incident and scattered light beams showing the orientation of the moving diffraction grating represented by dashed lines.

directions of the wavevectors $\mathbf{q}_I$ and $\mathbf{q}_S$ of the incident and scattered light beams for which the scattering is allowed are determined by a wavevector conservation rule

$$\mathbf{q}_S = \mathbf{q}_I \pm \mathbf{q}, \tag{3.73}$$

where $\mathbf{q}$ is the elastic wavevector.

The wavevector magnitudes must also satisfy

$$q_I = n\omega_I/c \quad \text{and} \quad q_S = n\omega_S/c, \tag{3.74}$$

where n is the refractive index of the elastic medium and c is the velocity of light. It is clear that measurements of ω_S, ω_I, $\mathbf{q}_S$ and $\mathbf{q}_I$ enable ω and q, and hence the velocity of the elastic wave and the stiffness coefficient to be found. In practice, for the Brillouin scattering of visible light with

$$\omega_I \simeq 3 \times 10^{15}\ \mathrm{s}^{-1}, \tag{3.75}$$

the conservation rules of (3.72) and (3.73) limit the measurable angular frequency shifts ω to be smaller than about $10^{11}\ \mathrm{s}^{-1}$ and thus the range of accessible wavevector shifts is approximately

$$10^4\ \mathrm{m}^{-1} < q < 2 \times 10^7\ \mathrm{m}^{-1}. \tag{3.76}$$

It is not necessary to generate the elastic waves externally in order to observe their Brillouin scattering, since all possible waves are thermally excited to some extent at temperature T. Consider first the energy contained in an elastic wave excited in a sample of volume V. According to eqn (3.48), the change in energy ΔE associated with the instantaneous strain ε of an elastic wave is symbolically

$$\Delta E = VU = \tfrac{1}{2} V c_{ij} \varepsilon^2, \tag{3.77}$$

where c_{ij} is the appropriate stiffness coefficient. Now according to the Boltzmann distribution, the relative probability of thermal excitation for a state of a system that requires an amount of energy ΔE is

$$\exp(-\Delta E/k_B T). \tag{3.78}$$

Thus the mean-square strain that accompanies the thermally-excited elastic wave of wavevector $\mathbf{q}$ is

$$\langle \varepsilon^2 \rangle_{\mathbf{q}} = \frac{\int_0^\infty \varepsilon^2 \exp(-\Delta E/k_B T)\, d\varepsilon}{\int_0^\infty \exp(-\Delta E/k_B T)\, d\varepsilon} = \frac{k_B T}{V c_{ij}}, \tag{3.79}$$

where eqns (A2.8) and (A2.12) from Appendix 2 have been used. This is independent of the magnitude and direction of $\mathbf{q}$ but it has different values for longitudinal and transverse waves because the appropriate stiffness coefficient c_{ij} differs in the two cases. In fact, the appropriate stiffness coefficient is that appearing in the relevant velocity (3.63) or (3.66), and the above result can be written

$$\langle \varepsilon^2 \rangle_q = k_B T / V \varrho v_L^2 \tag{3.80}$$

for longitudinal waves and

$$\langle \varepsilon^2 \rangle_q = k_B T / V \varrho v_T^2 \tag{3.81}$$

for transverse waves.

It follows from the above discussion that every possible elastic wave of a given polarization is thermally excited to produce the same strain fluctuations, irrespective of its frequency ω and wavevector $\mathbf{q}$. The strain fluctuations produce the refractive index fluctuations that are responsible for the Brillouin scattering. The strength of the scattering is thus proportional to the temperature, and is larger for less stiff materials, as might be expected. The form of scattered spectrum produced by the thermally-excited elastic waves in an isotropic solid is shown in fig. 3.14. There are contributions from the transverse and longitudinal waves, and these are symmetrically disposed on either side of the incident frequency in accordance with eqn (3.72). Measurement of the shifts of the scattered peaks from ω_I yields

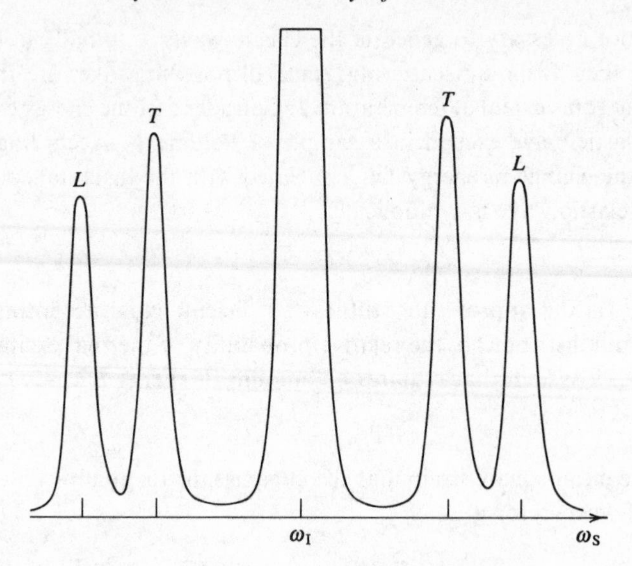

Fig. 3.14 Schematic Brillouin spectrum showing transverse and longitudinal contributions on either side of ω_I. The strong central peak at $\omega_S = \omega_I$ is caused by elastic scattering of the incident light.

the required frequencies ω. The wavevectors needed for substitution in eqn (3.73) are obtained from the geometrical arrangements of the incident and scattered beams. The widths of the spectral peaks can be used to obtain information on the rate of damping of the elastic waves.

Other high-resolution methods for measuring stiffness coefficients also rely on the determination of elastic wave velocities, followed by the use of eqn (3.68) or the analogous, more complicated relations for crystalline solids. Most of these methods use external sources of energy to generate elastic waves of known frequency in the sample. With a continuous source of variable frequency, it is possible to tune ω to excite standing-wave resonances where integral numbers of half wavelengths are matched to a given sample dimension. The elastic velocity can then be determined from the known frequency and measured wavelength. With a pulsed source of energy, usually in the ultrasonic frequency range, it is possible to observe echoes as the pulses reflect to and fro off the opposite ends of the sample. The velocity is determined directly by measurement of the time interval between successive echoes at one of the sample faces. The damping of the elastic waves can also be studied by observations of the rate of reduction in the energy of a train of echoes from the same initial pulse.

3.8 Rayleigh waves

All components of the stress must vanish at the free surfaces of a sample to which no forces are applied. Consider a semi-infinite sample that occupies the half-space $z < 0$ with its free surface in the $z = 0$ plane and a vacuum in the half-space $z > 0$. The boundary conditions at $z = 0$ are

$$\begin{aligned} \sigma_{xz} &= \tfrac{1}{2}(c_{11} - c_{12})\,\varepsilon_{zx} = 0 \\ \sigma_{yz} &= \tfrac{1}{2}(c_{11} - c_{12})\,\varepsilon_{yz} = 0 \\ \sigma_{zz} &= c_{12}\varepsilon_{xx} + c_{12}\varepsilon_{yy} + c_{11}\varepsilon_{zz} = 0. \end{aligned} \tag{3.82}$$

The conditions at a surface allow an additional solution of the elastic wave equations that corresponds to a ripple travelling along the surface, as shown in fig. 3.15. The amplitudes of the displacement u and strain ε fall off exponentially with distance into the material. The surface elastic wave on a solid is named after Rayleigh, who first predicted it in 1885.

We consider a surface wave whose displacements are confined to the zx-plane. The *Rayleigh wave* is equivalent to the superposition of a longitudinal wave

$$\mathbf{u}_{\mathrm{L}} \exp(\mathrm{i}\mathbf{q}_{\mathrm{L}} \cdot \mathbf{r} - \mathrm{i}\omega t) \tag{3.83}$$

whose displacement is parallel to the wavevector, so that

$$\mathbf{u}_{\mathrm{L}} \times \mathbf{q}_{\mathrm{L}} = 0, \tag{3.84}$$

and a transverse wave

$$\mathbf{u}_{\mathrm{T}} \exp(\mathrm{i}\mathbf{q}_{\mathrm{T}} \cdot \mathbf{r} - \mathrm{i}\omega t), \tag{3.85}$$

whose displacement is perpendicular to the wavevector, so that

$$\mathbf{u}_{\mathrm{T}} \cdot \mathbf{q}_{\mathrm{T}} = 0. \tag{3.86}$$

The two waves have the same frequency ω and they also have the same wavevector component parallel to the surface, which we denote simply q, so that

$$q_{\mathrm{L}x} = q_{\mathrm{T}x} \equiv q. \tag{3.87}$$

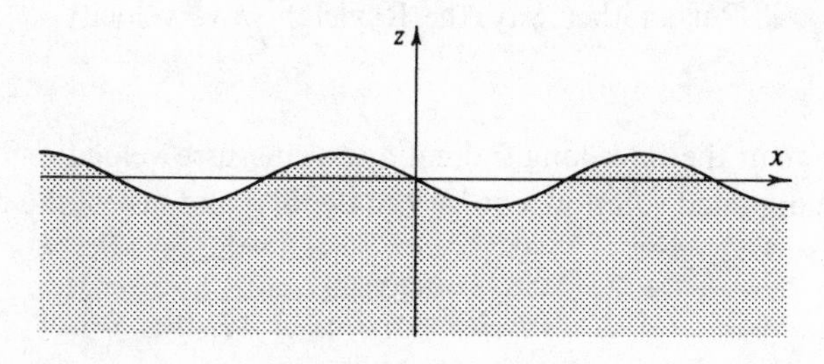

Fig. 3.15 Coordinate system for Rayleigh surface wave.

The longitudinal and transverse conditions (3.84) and (3.86) can therefore be written

$$\begin{aligned} q_{Lz}u_{Lx} - qu_{Lz} &= 0 \\ qu_{Tx} + q_{Tz}u_{Tz} &= 0. \end{aligned} \qquad (3.88)$$

The z-components of the wavevectors have different values determined by

$$\begin{aligned} q_L^2 &= q^2 + q_{Lz}^2 = \omega^2/v_L^2 \\ q_T^2 &= q^2 + q_{Tz}^2 = \omega^2/v_T^2 \end{aligned} \qquad (3.89)$$

where v_L and v_T are the elastic wave velocities given by eqns (3.63) and (3.66).

The superposition that represents the Rayleigh wave thus has the form

$$\mathbf{u} = \{\mathbf{u}_L \exp(iq_{Lz}z) + \mathbf{u}_T \exp(iq_{Tz}z)\} \exp(iqx - i\omega t). \qquad (3.90)$$

In order that the wave should disturb only the surface region, it is necessary that **u** must diminish with increasing negative z. This happens only if the wavevector z-components are negative imaginary quantities

$$q_{Lz} = -i|q_{Lz}| \quad \text{and} \quad q_{Tz} = -i|q_{Tz}|, \qquad (3.91)$$

so that eqn (3.90) becomes

$$\mathbf{u} = \{\mathbf{u}_L \exp(|q_{Lz}|z) + \mathbf{u}_T \exp(|q_{Tz}|z)\} \exp(iqx - i\omega t). \qquad (3.92)$$

The conditions (3.89) also become

$$\begin{aligned} q^2 - |q_{Lz}|^2 &= \omega^2/v_L^2 \\ q^2 - |q_{Tz}|^2 &= \omega^2/v_T^2 \end{aligned} \qquad (3.93)$$

and it is clear that

$$\omega < v_L q \quad \text{and} \quad \omega < v_T q. \qquad (3.94)$$

The frequency of the Rayleigh wave is therefore smaller than the frequencies of the bulk longitudinal and transverse waves that have the same surface wavevector q. Put another way, the Rayleigh wave velocity

$$v_R = \omega/q \qquad (3.95)$$

is smaller than the bulk longitudinal and transverse velocities.

For the assumed solution, eqn (3.92), the first and last of the boundary conditions (3.82) give

$$\begin{aligned} iq(u_{Lz} + u_{Tz}) + |q_{Lz}|u_{Lx} + |q_{Tz}|u_{Tx} &= 0 \\ ic_{12}q(u_{Lx} + u_{Tx}) + c_{11}(|q_{Lz}|u_{Lz} + |q_{Tz}|u_{Tz}) &= 0, \end{aligned} \qquad (3.96)$$

and the middle boundary condition is automatically satisfied. It is seen that eqns (3.88) and (3.96) provide four simultaneous equations for the four unknowns u_{Lx}, u_{Lz}, u_{Tx} and u_{Tz}. The eliminant of these four equations provides a condition on the wavevector components that appear in the trial solution. The result can be put into a convenient form if the wavevector z components are removed by the use of eqns (3.91) and (3.93) and the Rayleigh wave velocity is introduced via eqn (3.95), when after some algebra we obtain

$$4v_T^3(v_L^2 - v_R^2)^{1/2}(v_T^2 - v_R^2)^{1/2} = v_L(2v_T^2 - v_R^2)^2. \qquad (3.97)$$

This equation can in principle be solved to find the Rayleigh wave velocity for a material with known stiffness coefficients, and hence known longitudinal and transverse velocities.

The solution of eqn (3.97) must in general be found numerically, and fig. 3.12 shows the variation of v_R/v_L with Poisson's ratio ν. The Rayleigh velocity is typically about 0.9 of the transverse velocity, and both these velocities tend to zero as ν tends to the limiting value of $\frac{1}{2}$, characteristic of a liquid. An analytical solution of eqn (3.97) does exist in the middle of the allowed range of values of ν, where it can be verified that for $\nu = \frac{1}{4}$

$$v_T/v_L = 1/3^{1/2} \quad \text{and} \quad v_R/v_L = \tfrac{1}{3}(6 - 12^{1/2})^{1/2}. \qquad (3.98)$$

For the same value of Poisson's ratio,

$$|q_{Lz}| = 0.85q \quad \text{and} \quad |q_{Tz}| = 0.39q. \qquad (3.99)$$

The Rayleigh wave causes the solid surface to ripple in the manner shown in fig. 3.15. The surface displacement obtained from the $z = 0$ value of eqn (3.92) is

$$\mathbf{u} = (\mathbf{u}_L + \mathbf{u}_T)\exp(iqx - i\omega t). \qquad (3.100)$$

We again quote results for the middle of the range of Poisson's ratio where

$$\left.\begin{aligned} \mathrm{Re}\,[u_x] &= 0.71u_0 \sin(qx - \omega t) \\ \mathrm{Re}\,[u_z] &= u_0 \cos(qx - \omega t) \end{aligned}\right\} \nu = 1/4, \qquad (3.101)$$

and u_0 is the arbitrary wave amplitude perpendicular to the surface. The Rayleigh wave thus displaces a particle at the surface along an elliptical path as shown in fig. 3.16. The long axis of the ellipse is perpendicular to the surface, reflecting a greater freedom of particle motion in this direction. Note that the particle motion is anticlockwise for the Rayleigh wave travelling to the right in fig. 3.16. The amplitude of the motion falls off in an exponential fashion with increasing distance into the solid in accordance with eqn (3.92), the penetration depth being comparable to or less than the surface wavelength, $2\pi/q$.

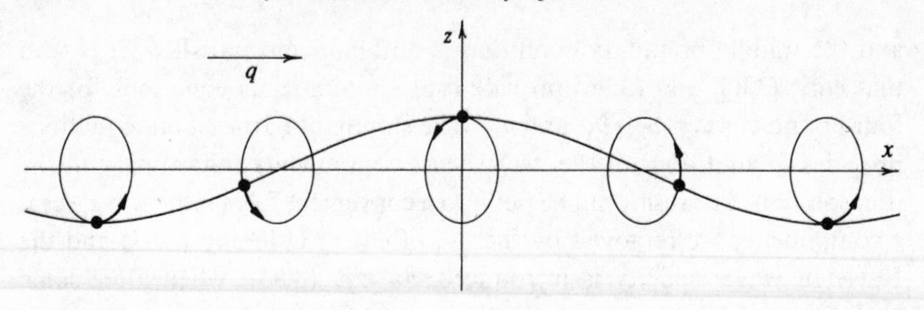

Fig. 3.16 Elliptical paths followed by surface elements in the presence of a Rayleigh wave.

Rayleigh wave theory is applied in seismology, where earthquakes and explosions generate surface waves in addition to the bulk elastic waves. In accordance with the ordering of velocities shown in fig. 3.12, the times of reception at a distant observation station are such that the bulk longitudinal waves arrive first, followed by bulk transverse waves, followed by surface Rayleigh waves. The three kinds of wave can be identified by the different characteristic earth motions to which they give rise. This was noted by Sir Harold Jeffreys when lying in his bed in Cambridge during an earthquake on 7 June, 1931 . . . 'I was lying approximately northeast to southwest, when the waves from the Dogger Bank earthquake arrived, and the motion began with rapid oscillation along the bed. After some time it changed to a larger but slower motion across the bed. The change was just what would be expected if the first part of the motion was Pg' (primary, longitudinal wave) 'and the second SHg' (secondary, horizontally-transverse shear wave). These observations show that one does not always need sensitive equipment to detect the different wave characteristics!

Rayleigh waves have also found important applications in electronics, particularly in electronic signal processing. They can readily be generated and detected on solid surfaces by electrical devices known as interdigital transducers. As Rayleigh waves travel along the surface, they can be sampled, manipulated or modulated in various useful ways. With a Rayleigh velocity of about 3000 m s^{-1}, processing on a microsecond timescale corresponds to path lengths of order 3 mm, which is a convenient dimension in terms of easily available crystal sizes. The electronic components that use Rayleigh waves are known as surface-acoustic-wave devices. The theory of isotropic materials given above needs generalization to cover the crystalline materials used in practice.

3.9 Scattering of light by surface waves

Surface ripples on both solids and liquids cause light to be scattered in a similar way to the Brillouin scattering by bulk elastic waves described in

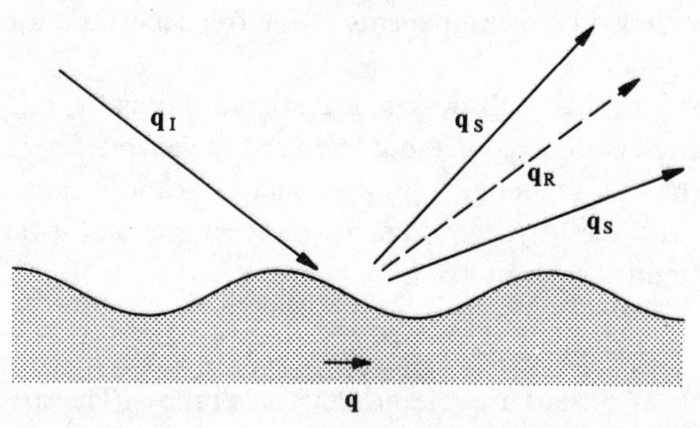

Fig. 3.17 Wavevector orientations for reflection and scattering of incident light by a surface wave. The magnitudes of the scattered wavevectors are almost the same as those of the incident and reflected wavevectors on account of the small size of the surface wave velocity compared to the velocity of light.

§3.7. Observations of such scattering provide values of the frequency ω and wavevector $\mathbf{q}$ of the surface waves. Brillouin spectroscopy forms an important technique for investigating surface wave properties and for measuring the material parameters that control the wave motion.

Consider again a surface wave of frequency ω and wavevector $\mathbf{q}$ propagated parallel to the x-axis, and suppose that a monochromatic light beam of frequency ω_I and wavevector $\mathbf{q}_I$ is incident on the surface, as shown in fig. 3.17. There is the usual reflected light beam, whose frequency ω_R and wavevector $\mathbf{q}_R$ must satisfy

$$\omega_R = \omega_I \quad \text{and} \quad q_{Rx} = q_{Ix}; \tag{3.102}$$

the reflected beam is indicated by the dashed arrow in fig. 3.17. The light can also exchange energy with the surface wave in such a way that two scattered beams are generated, disposed as shown by the continuous arrows in the figure. The frequency ω_S and wavevector $\mathbf{q}_S$ of the scattered light must satisfy

$$\omega_S = \omega_I \pm \omega \tag{3.103}$$

and

$$q_{Sx} = q_{Ix} \pm q. \tag{3.104}$$

These requirements are similar to eqns (3.72) and (3.73) for bulk Brillouin scattering, but only the travelling wave components parallel to the surface are now involved in the wavevector conservation. With free space above the surface, the optical wavevector magnitudes are given by

$$q_I = \omega_I/c \quad \text{and} \quad q_s = \omega_S/c. \tag{3.105}$$

Clearly, as in the bulk case, the surface wave frequency and wavevector can be determined by measurements of the frequency and wavevector of the scattered light.

Surface waves, like bulk waves, are excited thermally, and their light scattering can be observed without any need to generate them artificially. The strength of the scattering is proportional to the mean-square amplitude associated with the Rayleigh wave of wavevector **q**, and it can be shown that this quantity is given by

$$\langle u_0^2 \rangle_{\mathbf{q}} = k_B T(1 - \nu)/A \varrho v_T^2 q \tag{3.106}$$

for a sample of surface area A and Poisson's ratio ν. The strength of the scattering increases with the temperature T, like the bulk scattering proportionality in eqns (3.80) and (3.81), but there is an additional inverse dependence on the wavevector q for the surface scattering. The strength of scattering is thus proportional to the Rayleigh wavelength, and the strongest scattering occurs for directions close to that of the reflected beam. The scattered light is in all practical cases very weak compared to the incident and reflected beams, and considerable experimental skill is needed to measure the properties of the scattered light.

3.10 Suggestions for further reading

Catastrophe theory

Poston, T. & Stewart, I.N. (1978). *Catastrophe Theory and Its Applications*. London: Pitman.

Zeeman, C. (1976). Catastrophe theory. *Scientific American*, **234** (April), 65.

Elastic waves

Jeffreys, H. (1962). *The Earth*, 4th edn. Cambridge: Cambridge University Press.

Landau, L.D. & Lifshitz, E.M. (1970). *Theory of Elasticity*. 2nd edn. Oxford: Pergamon Press.

Euler's strut

Thompson, J.M. & Hunt, G.W. (1973). *A General Theory of Elastic Stability*. London: Wiley.

Pippard, A.B. (1980). Demonstration experiments in critical behaviour and broken symmetry. *European Journal of Physics*, **1**, 13–18.

Pippard, A.B. (1985). *Response and Stability*. Cambridge: Cambridge University Press.

Light scattering by elastic waves

Hayes, W. & Loudon, R. (1978). *Scattering of Light by Crystals*. New York: Wiley.

Problems

1. Obtain an expression for the vibrational frequency of a mass m attached to the centre point of a horizontal beam of length L and negligible mass which is supported, but not clamped, at both ends. Calculate the value of the vibrational period for the loaded copper pipe whose parameters are given in problem 8 of Chapter 2.

2. The *susceptibility* χ of the Euler strut considered in § 3.3 can be defined as

$$\chi = (\mathrm{d}\theta_0/\mathrm{d}F)_{F=0},$$

where θ_0 is the equilibrium deflection produced by a horizontal force, F. It provides a measure of the ease with which the strut can be deflected. Obtain expressions for χ for the cases $W < W_c$ and $W > W_c$, and hence sketch the variation of χ with W.

3. Consider the Brillouin scattering of visible light by the elastic waves in aluminium. If the light is scattered through 180°, calculate the angular frequency shifts between scattered and incident light caused by the longitudinal and transverse acoustic waves. (Use $\varrho = 2700\,\mathrm{kg\,m^{-3}}$ and the values of c_{11} and c_{12} given in Table 2.1, ignoring the departure of the anisotropy factor A from unity for aluminium.)

4. Use eqn (3.106) to obtain an expression for the total mean-square surface displacement amplitude by summing the contributions of the different wavevectors $\mathbf{q}$. It can be assumed that the distinct wavevectors on a surface of area A have a density $A/(2\pi)^2$ and that they are isotropically distributed with a maximum magnitude π/a where a is the length of the edge of the unit cell.
Show that the root-mean-square surface displacement of aluminium at room temperature is about 1.2×10^{-11} m. (Use $a = 4 \times 10^{-10}$ m, and values of c_{11} and c_{12} from Table 2.1, ignoring the anisotropy of aluminium.)

5. Consider the propagation of longitudinal elastic waves parallel to the axis of a rod of density ϱ whose transverse dimensions are much smaller than the elastic wavelength. Show that the wave velocity is

$$v'_{\mathrm{L}} = (E/\varrho)^{1/2} = [(c_{11} - c_{12})(c_{11} + 2c_{12})/\varrho(c_{11} + c_{12})]^{1/2}.$$

(Hint: use the method of § 3.6 but with all the stress components except the normal stress parallel to the axis set equal to zero in eqn (2.48) or (2.50)).
Show that $v_{\mathrm{L}} \geqslant v'_{\mathrm{L}}$, where v_{L} is the bulk longitudinal velocity from eqn (3.63), and give the physical explanation of this result.

6. Consider the propagation of elastic waves in the plane of a plate of density ϱ, whose thickness is much smaller than the elastic wavelength. The displacement vector $\mathbf{u}$, assumed also to lie in the plane of the plate, can be oriented either parallel or perpendicular to the wavevector $\mathbf{q}$. Show that the corresponding longitudinal and transverse velocities are

$$v''_{\mathrm{L}} = [(c_{11}^2 - c_{12}^2)/\varrho c_{11}]^{1/2}$$

and

$$v''_{\mathrm{T}} = [(c_{11} - c_{12})/2\varrho]^{1/2}.$$

(Hint: use the method of § 3.6 but with only the stress components entirely within the plane of the plate not equal to zero in eqn (2.48).)
Show that

$$v_L \geqslant v_L'' \geqslant v_L',$$

where v_L is the bulk longitudinal velocity given in eqn (3.63) and v_L' is the longitudinal velocity in a rod given in problem 5.

7. A hypothetical spherical planet is made of homogeneous material for which Poisson's ratio, $\nu = \frac{1}{4}$. An explosion at a point on the planet's surface generates Rayleigh surface waves, together with bulk longitudinal and transverse acoustic waves. What is the angular coordinate, relative to the position of the explosion, of the points on the planet's surface that Rayleigh waves take twice as long to reach as do the bulk longitudinal waves? What is the relative time interval between the arrivals of the longitudinal and transverse waves? (Ignore the effect of the curvature of the planet's surface on the Rayleigh wave velocity.)

8. The propagation of the seismic waves through the Earth is sometimes described by the equation

$$\varrho \frac{\partial^2 u}{\partial t^2} = c \frac{\partial^2 u}{\partial x^2} + c\tau \frac{\partial^3 u}{\partial x^2 \partial t},$$

where u is the displacement, ϱ is the density and c is the appropriate stiffness coefficient. The second term on the right models damping of the wave and τ is the relaxation time. Assume a plane wave solution for u with displacement proportional to $\exp(-i\omega t + iqx)$, where ω is a real frequency and q is a complex wavevector. Determine the dependences of the real and imaginary parts of q on the other parameters in the limits of
(i) weak damping with $\omega\tau \ll 1$
(ii) strong damping with $\omega\tau \gg 1$.

4

Static properties of liquids

The shapes of the surfaces of static liquids – a raindrop on a window pane or water in a glass – are amongst the best-known and most familiar characteristics of materials. However, the understanding of the nature and fundamental origin of the property – surface tension – that determines the shapes of liquid surfaces requires quite subtle considerations of the effects of the interatomic forces close to the surface. Thus although the main concerns of this chapter are the macroscopic properties of liquids at rest, it is helpful to preface these treatments with brief descriptions of the underlying molecular behaviour that produces the macroscopic effects. We accordingly begin with a discussion of the atomic or molecular structure of liquids.

4.1 Nature of the liquid state

The development of understanding of the behaviour of liquids has been slow compared with progress on gases and solids. The kinetic theory of gases works well because it assumes that molecules are far apart, on average. It also assumes that they interact with one another only during collisions, which affect their momenta and trajectories, but otherwise allow the forces between the molecules to be ignored. In contrast, and as indicated in Chapter 1, theories of solids are generally based on the simplifying assumption that the atoms in solids are arranged in orderly patterns, so that the ranges over which the forces act between them are known and the effects can be calculated. Most liquids fall uncomfortably between the extremes of these two approaches. Their molecules are close together, but long-range order is usually absent and the molecules are in a constant state of motion. Such an assembly is most easily specified by an average intermolecular separation and an average energy of motion. Unlike gases, therefore, the structure of liquids can be studied by diffraction methods, usually with neutrons or X-rays. Liquids formed by the condensation of monatomic gases exhibit diffuse diffraction peaks, but the gases do not. It is scarcely surprising that the first advances in understanding of the properties of liquids largely arose from treatments that considered them to be intermediate between gases and solids, and applied statistical methods to them. Some such treatments view the liquid as a condensed form of gas, while others model liquids as disordered forms of solids.

The properties of all states of matter are, to different extents, manifestations of the characteristics of the individual atoms and the forces which act between them. Thus the electronic structures of the constituent atoms are apparent in chemical properties, spectral lines, etc., while the strength of solids (§ 1.10) and the viscosity of liquids (§ 5.12) depend on interatomic and intermolecular forces. The most striking difference betwen fluids (i.e. liquids and gases) and solids is that the forces between the ordered, relatively close-packed atoms in solids cause them to keep their shape, unless subject to external forces. In contrast, liquids 'deform' under their own weight (i.e. they flow if not contained), while gases expand in volume and diffuse away through small openings in a container. Solids and liquids exhibit the property of *cohesion* (conserving volume by the constituent particles 'sticking together'). This characteristic is fundamental to the properties that are demonstrated by static liquids.

The intermediate nature of liquids is most clearly illustrated by their transformations into the other phases of matter at high and low temperatures, and/or high and low pressures. The phase transformations of many substances occur for specific conditions and, except for these conditions, the physical interface between two phases is generally quite sharp, so that its position can be determined with accuracy. (Glasses, many polymers and some other materials show more complex and less well-defined behaviour. Most of the simple descriptions and models given here are therefore inappropriate.) The transformation of a liquid to a gas takes place at the *critical temperature*, T_c, where the cohesion of the liquid is totally destroyed and the densities of the liquid and gaseous phases become equal, so that the liquid–gas interface no longer exists. The effects of this phase transformation on liquid surface properties are described in § 4.11, but most of the derivations assume a well-defined liquid–vapour interface at temperatures removed from T_c.

Since our main concern is with macroscopic properties, consideration of the molecular structure of the liquid state is limited to a simple model, which is described only briefly. We do not attempt to cover the behaviour and uses of liquid crystals (§ 1.5) and other unusual liquid-like compounds whose structures have stimulated much research interest.

4.2 Surface of a liquid – the phenomenon of surface tension

The suspected presence of a thin film or a transitional layer on a liquid can be investigated optically using polarized light. This approach, which has been used to examine films of superfluid helium creeping up metal mirrors at low temperatures, was used by Lord Rayleigh to examine the nature of the surface of a normal liquid, namely water.

When unpolarized light is reflected from a dielectric (e.g. glass) it becomes partially plane-polarized. At a particular angle, characteristic of the reflecting medium, reflected rays are completely plane-polarized. The reflected and refracted rays are then exactly 90° apart, so that the angle for the occurrence of total plane-polarization is related to the refractive index of the medium. This angle is given by Brewster's law

$$n = \tan \phi, \tag{4.1}$$

where n is the refractive index and ϕ the angle between the incident (and/or reflected) ray and the normal to the surface.

The above behaviour is expected when the refractive index changes abruptly at the dielectric surface. If however n changes more gradually at the surface, the reflected light should be elliptically-polarized, a fact that Rayleigh knew well. He found that light reflected from a clean water surface showed no evidence of ellipticity; it was plane-polarized. Rayleigh deduced from his result that the liquid–vapour interface was well-defined and that any transitional layer that might exist could, at the most, be only a few molecular diameters in thickness. This conclusion is valid for many simple liquids, except when close to their critical temperatures.

For a liquid surface (i.e. a liquid–vapour interface) to be in equilibrium, the numbers of atoms arriving and leaving the surface from the vapour must be equal. A similar condition must apply to the liquid side of the boundary. The property of liquids that is known as *surface tension* is a logical consequence of these conditions, as may be demonstrated from simple models of the liquid state. One such model is outlined in the next section. Surface tension causes a liquid to behave as if it has a contractual skin, thus giving a smooth profile to the liquid between the lines or curves defining its contact with another substance.

As shown first by the work of Rayleigh, any 'skin' or surface layer on a liquid with different physical properties from the bulk must be very thin and so, when considering the dynamical properties of liquids, as in Chapter

5, we shall mostly assume that it does not exist. In this chapter, however, it is fundamental to our assumption of the existence of tensile forces in the surface that the nature of the liquid surface differs from the bulk, although the dynamic structure of the surface is not explored in detail. For many years this was a contentious point to many scientists, so that respected textbooks of the time tend to explain effects and properties dependent on surface tension purely in terms of arguments about the minimization of surface energy. In the following sections we shall show that surface tension and surface energy are closely related, but we shall also argue the reality of a tension in the tangent plane of a liquid surface.

4.3 Origin of surface contraction

When a solid melts there is typically a volume expansion of about 10 per cent which, assuming that the atoms do not increase in size, implies the existence of holes or 'vacancies' in the resulting liquid. We now assume that this picture applies to a liquid of simple, near-spherical molecules and base our considerations on ideas about the structure of simple liquids put forward by Bernal and Eyring. The holes in the liquid are sufficiently large to accommodate a molecule and the energy to produce them comes from the latent heat of fusion. We assume that holes exchange places with the adjacent molecules randomly and continuously, as indicated in fig. 4.1. There is one hole for approximately every nine molecules and at any

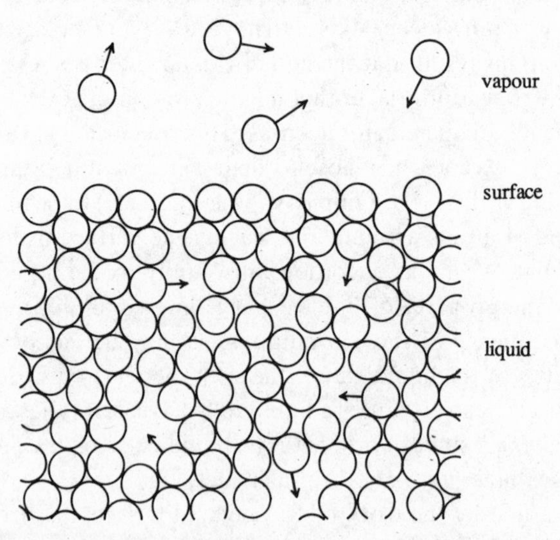

Fig. 4.1 The interface region between a simple liquid and its vapour, based upon the Bernal and Eyring models of a liquid.

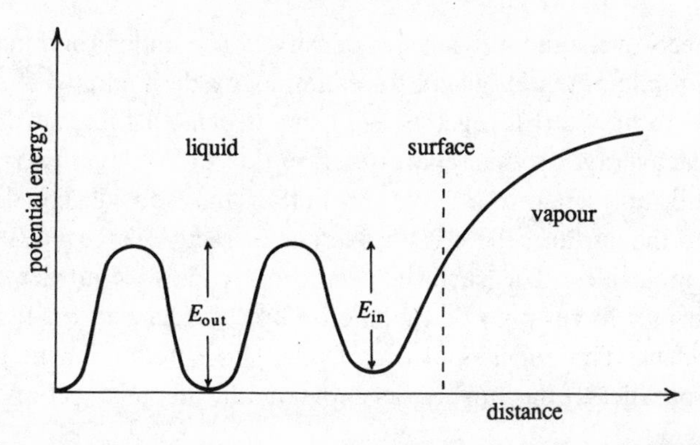

Fig. 4.2 The variation of the potential energy as a function of distance in the vicinity of a liquid surface.

moment the number of molecules involved in exchanges is roughly equal to the number of holes. Such interchanging molecules therefore exhibit a gas-like behaviour, as do also their partner holes, while the remainder of the molecules are fairly static and may be considered as solid-like. Except near to the liquid surface, the spatial distribution of holes is statistically uniform and each hole is in a spherically-symmetric environment most of the time. However, a hole at the liquid surface has a hemispherical distribution of molecules. A hole at the liquid–vapour interface is more likely to exchange with a molecule in the interface than with a molecule just below it, for the latter kind of exchange is impeded by the net inward force on the molecules. The consequence of the asymmetry of forces experienced by near surface molecules is that the concentration of holes, when time-averaged, is greater at the surface of the liquid than in the bulk. Thus the surface of a liquid differs in structure from the bulk, but, in the light of Rayleigh's results, this surface 'skin' can only be a few, up to perhaps a dozen or so, molecular diameters in thickness.

Because of the connection between force and potential energy, the same conclusions can be reached by considering the shape of the potential energy wells for atoms in the liquid interior and at the liquid surface, see fig. 4.2. The asymmetrical environment of a near-surface molecule produces an activation energy E_{in} for migration towards the interior that is less than the activation energy E_{out} for migration towards the surface. However, since the liquid is in a state of dynamic equilibrium, there must be some compensating adjustment to equalize the flows of the molecules of the liquid towards and away from the surface. This is achieved by a reduction in the density of the liquid very close to the liquid–vapour interface, again indicating a density at the liquid surface which is smaller than the normal bulk value.

The effects of the reduced surface density can be understood in terms of the interatomic potential given, for example, by the Lennard-Jones form illustrated in fig. 1.2(b), together with the associated force shown in fig. 1.2(a). The average atomic separation r_s in the surface layer is larger than the equilibrium separation r_0 in the bulk. Thus considering directions parallel to the surface, there is an average attractive force $F(r_s)$ between pairs of molecules. The surface layer thus tends to contract, and the intermolecular forces provide the microscopic mechanism for the surface tension effect. The effect is also equivalent to a reduction in the local pressure parallel to the surface, as explained in the next section.

4.4 Definition of surface tension

It is a simple matter to demonstrate the existence of tension parallel to a liquid surface, by using a flat soap film stretched across a frame of thin wires, as depicted in fig. 4.3. This shows a U-shaped hoop and another, movable, straight wire. If there is no friction where the wires touch and a film is formed by dipping the frame into soap solution, it is found that an outward perpendicular force, F must be applied in the plane of the frame to the slidable wire in order to prevent the film contracting. This implies that the film to the left of the slide wire is exerting a force on it. If this force is investigated using several different frames, we find that it acts in the plane of the soap film, and it is proportional to the length of the slide wire but it does not depend on the thickness of the film.

This experiment confirms that there is a tension in the film surface

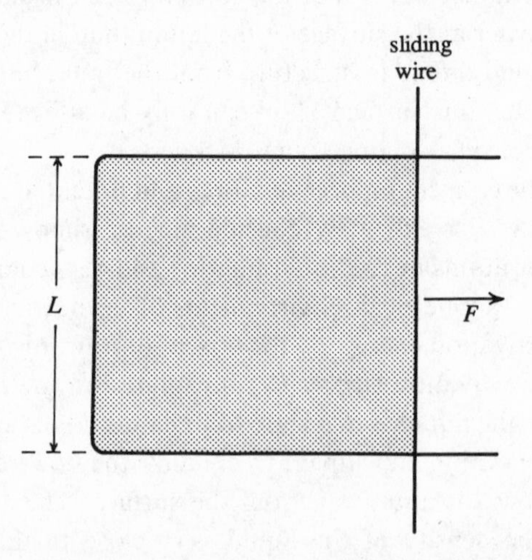

Fig. 4.3 An expandable wire frame consisting of a U-shaped hoop and frictionless slide wire. A thin liquid film fills the aperture and is maintained in equilibrium by a force F.

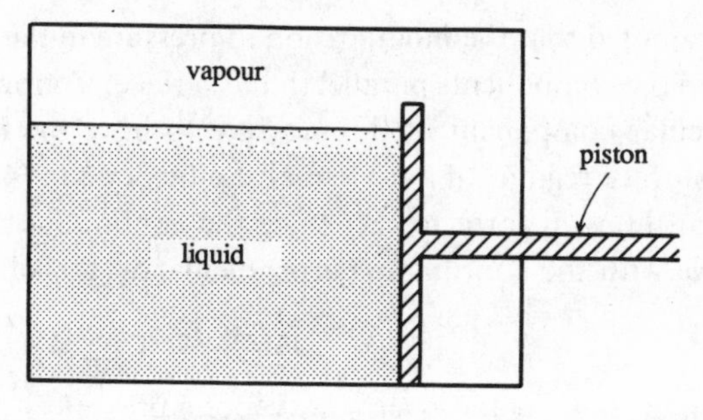

Fig. 4.4 Hypothetical apparatus for demonstrating the relationship between surface tension and the mean internal pressure, p_S of a thin layer below the surface of a liquid, whose density is less than that of the bulk liquid (after Brown, R.C. (1947). *Proceedings of the Physical Society, London*, **59**, 429).

(surfaces in this case, since there are clearly two), which acts on any member where the film terminates. Surface tension is defined as the force acting across a unit length of the contacting surface and therefore has units of $N\,m^{-1}$. The force cannot originate in the volume of the film since this would imply that (i) the pressure deep within the volume of the film is less than in the surrounding atmosphere, (ii) the liquid could maintain internal shear forces, which is contrary to both experiment and theory (see § 3.5). The first point, which focuses our attention on the pressure, suggests a way of visualizing the surface tension effect. The reasoning in § 4.3 showed that the density of a thin layer below the liquid surface is less than that of the liquid interior, and a reduced density implies a reduced local pressure. There is no sharp transition in density and pressure as one moves from the surface into the liquid interior, but for the sake of simplicity we will assume that the surface layer is distinct, has thickness t, and has a mean horizontal internal pressure p_s, lower than the pressure p_0 of the bulk liquid. We now apply these assumptions to the system illustrated in fig. 4.4, which depicts a piston contacting a liquid in a part of a vessel otherwise containing the vapour of the liquid. The liquid surface is here assumed to meet the face of the piston at right angles, hence avoiding the need to account for meniscus effects. We also neglect any hydrostatic effects from gravity and assume that the pressure in the bulk of the liquid is uniform and equal to the overlying vapour pressure, p_0. If the piston is frictionless, the force per unit length of surface that must be applied to the piston rod in order to prevent the piston moving to the left is

$$\gamma = (p_0 - p_s)\,t, \qquad (4.2)$$

where γ is the *surface tension*.

It should be noted that the modification in pressure in the surface layer applies only to the components parallel to the surface. Continuity requires the perpendicular component of the pressure to have the bulk value p_0 through the surface region. In a more realistic theory eqn (4.2) should be replaced by an integral form to recognize the continuous nature of the variation of p_s with the coordinate perpendicular to the surface, so that

$$\gamma = \int \{p_0 - p_s(z)\}\,dz. \tag{4.3}$$

Note that any change in the overlying vapour pressure p_0 produces an equal change in p_s, so that γ in eqn (4.2) or (4.3) has a constant value.

The numerical value of the surface tension can, in principle, be calculated if the detailed form of the interatomic potential is known for a given liquid. It is first necessary to determine the average atomic or molecular spacing r_s in the surface layer by satisfying the condition for equilibrium between the molecular flows towards and away from the surface, as discussed in §4.3. The surface tension, given by the force acting per unit length of surface, is then simply

$$\gamma = F(r_s)/r_s, \tag{4.4}$$

where $F(r)$ is the interatomic force at separation r, as illustrated in fig. 1.2(a) for the example of a Lennard-Jones potential. This way of estimating γ gives results in reasonable agreement with experiment, but it is too lengthy for our purposes. We shall therefore estimate values for γ using more approximate methods, after examining the relationship between surface tension and other important parameters.

4.5 Surface tension and surface energy

Since every molecule or atom in the surface of a liquid has both potential and kinetic energy, a unit area of surface has a characteristic energy which will reflect the packing density of its constituent particles. The potential energy term is a manifestation of the integrated effect of the bonding between the particles at their equilibrium average spacing. The existence of this potential energy is clearly seen when a bubble blown from soap solution bursts – liquid drops fly in many directions as the potential energy is converted to kinetic energy.

Consider again the U-shaped wire hoop and slider shown in fig. 4.3. The force F which is necessary to prevent the soap film contracting in area is given by

$$F = 2\gamma L. \tag{4.5}$$

The factor of two appears because the film has two surfaces. The area of the film of liquid is increased if the straight wire is pulled to the right, but if we assume that there is a fixed volume of liquid, more molecules must become exposed at the surface as the film is stretched and the thickness of the film is reduced. Therefore the total surface energy of the film is increased when the slide wire moves to the right and most of this energy is supplied by the work done against the forces of surface tension. If the stretching is done rapidly, the liquid is cooled because surface energy is created at the expense of internal (kinetic) energy. This is adiabatic stretching. Suppose instead, that the film is stretched slowly and isothermally, providing sufficient heat, δQ to maintain the liquid at a constant temperature as the wire is moved a distance δx to the right. If the potential energy of a unit area of the film is E_s, then the energy balance in moving the slide wire is described by

$$E_s \cdot 2L\delta x = F\delta x + \delta Q \tag{4.6}$$

so that using eqn (4.5) gives

$$E_s \cdot 2L\delta x = 2\gamma L\delta x + \delta Q,$$

that is

$$E_s = \gamma + \delta Q/2L\delta x = \gamma + \delta Q/\delta A, \tag{4.7}$$

where $\delta A = 2L\delta x$ is the area of film created. This equation shows that surface tension is equivalent to a *surface energy per unit area*, and the two quantities have the same dimensions of $\mathrm{N\,m^{-1}}$ or, equivalently, $\mathrm{J\,m^{-2}}$.

The final term in the last equation represents the heat that must be supplied as the film surface is increased by unit area, in order to keep the liquid at a constant temperature. Its form can be determined by a consideration of the thermodynamics of the stretched liquid film. In thermodynamic nomenclature, E_s is the total internal energy per unit area of surface while the surface tension contribution in eqn (4.7) is the Gibbs free energy per unit area of surface, usually called simply the *surface free energy*. It can be shown that the final term in eqn (4.7) is given by

$$\delta Q/\delta A = -T(\partial\gamma/\partial T)_A. \tag{4.8}$$

The rate of change of surface tension with temperature at constant area is negative for the great majority of liquids (see eqn (4.48)), so that an isothermal increase in surface area requires a transfer of heat to the liquid from its surroundings. Thus combining eqns (4.7) and (4.8), we obtain

$$E_s = \gamma - T(\partial\gamma/\partial T)_A, \tag{4.9}$$

and we see that the total surface energy E_s is normally larger than its free energy component γ.

4.6 Surface tension, latent heat and binding energy

A rough estimate of the surface tension of a liquid can be made if one assumes that the energy of a unit volume of a liquid in the form of a thin film is the same as that of a bulk liquid. In other words, we choose to take no account of our knowledge that the structure and bonding of a liquid close to its surface differ from those of the interior. The surface energy of a liquid can then be calculated from the *latent heat of evaporation*, a quantity that can be measured experimentally.

From ideas discussed earlier, we know that every atom or molecule has potential energy resulting from the forces exerted by near neighbours. We designate ε_B as the energy that must be supplied to an atom to overcome the potential energy of a neighbouring atom when the separation between the two is increased from its value in the liquid to its value in the vapour. This quantity ε_B is approximately equal to the binding energy of a pair of atoms, but only approximately, because the binding energy is, strictly speaking, the work done in separating the atoms by an infinite distance.

A value for the latent heat can be derived by considering a system of N atoms, each of which has m nearest neighbours, and neglecting any interactions between (i.e. contributions to the energy from) next-nearest and more distant atoms. (The quantity m, the *coordination number*, has already been encountered in §2.4.) Thus there is one bond for every neighbouring pair of atoms, and the total energy L required to vapourize N atoms of liquid is

$$L = \tfrac{1}{2} N m \varepsilon_B . \tag{4.10}$$

If we let N be equal to Avogadro's number, then L is the latent heat of evaporation. For a dense liquid m is about 10, while for the most closely-packed solids it is 12. Of course, for a gas it is zero. These differences in the value of m for the three states of matter explain why the *latent heat of melting* is small compared to the latent heat of evaporation. For example, to convert a kilogram of ice to water requires $\sim 3.3 \times 10^5$ J, whereas $\sim 2.5 \times 10^6$ J is needed to change one kilogram of water into steam. Strictly speaking, eqn (4.10) only applies to atoms, not molecules, but it may be applied to molecules provided that their shapes are more or less spherical. We can 'bend the rules' slightly to include diatomic molecules like N_2 but clearly not long chain molecules.

The surface energy of the liquid can be found by breaking a column of the liquid which has a unit cross-section into two parts, creating two new surfaces, each of unit area. After separation of the column, each molecule residing in the new surfaces has only $\tfrac{1}{2}m$ nearest neighbours. If there are

N' molecules per unit area of cross-section of the column, the number of bonds broken in separating the column into two parts is $\frac{1}{2}N'm$, which implies an energy input of $\frac{1}{2}N'm\varepsilon_B$. Since two new surfaces have been formed, the surface energy per unit area is

$$E_s = \tfrac{1}{4}N'm\varepsilon_B. \tag{4.11}$$

The quantity N' can be expressed in terms of the molecular weight, M and the density, ϱ of the liquid, using the fact that the volume, v of a molecule is

$$v = M/N\varrho. \tag{4.12}$$

This allows the effective area of surface occupied by a molecule to be estimated, giving

$$N' = (N\varrho/M)^{2/3}. \tag{4.13}$$

Combining eqns (4.10), (4.11) and (4.13), we obtain

$$E_s = \frac{L}{2N}\left(\frac{N\varrho}{M}\right)^{2/3}. \tag{4.14}$$

Evaluating eqn (4.14) for liquid nitrogen, using the values $\varrho = 810\,\mathrm{kg\,m^{-3}}$, $L = 5.44 \times 10^3\,\mathrm{J\,mole^{-1}}$, $M = 28$, gives

$$E_s \simeq 3 \times 10^{-2}\,\mathrm{J\,m^{-2}}.$$

The measured value of surface tension, γ for liquid nitrogen well below its boiling point is about $1.0 \times 10^{-2}\,\mathrm{N\,m^{-1}}$, i.e. $1.0 \times 10^{-2}\,\mathrm{J\,m^{-2}}$. The discrepancy between the estimated and measured values is not surprising considering the rather crude assumptions that have been made. Nitrogen is not an ideal liquid, since its molecules are not spherical. For liquid argon, the agreement is somewhat better: $E_s = 2.2 \times 10^{-2}$, $\gamma = 1.3 \times 10^{-2}\,\mathrm{J\,m^{-2}}$. For liquid neon, the agreement is better still: $E_s = 6.0 \times 10^{-3}$, $\gamma = 5.5 \times 10^{-3}\,\mathrm{J\,m^{-2}}$.

From eqn (4.9) it is seen that the surface tension γ is the sum of two terms E_s and $T(\partial\gamma/\partial T)_A$. No attempt has been made to calculate the contribution from the second term, which represents the additional heat necessary to keep the temperature constant when the liquids are separated. Discrepancies between the calculated values of E_s and values for γ are therefore to be expected for any temperature above 0 K.

4.7 Excess pressure at a curved surface of a liquid

In many of the practical cases where surface tension is important, the surface of the liquid is not flat but curved. In order for a curved surface

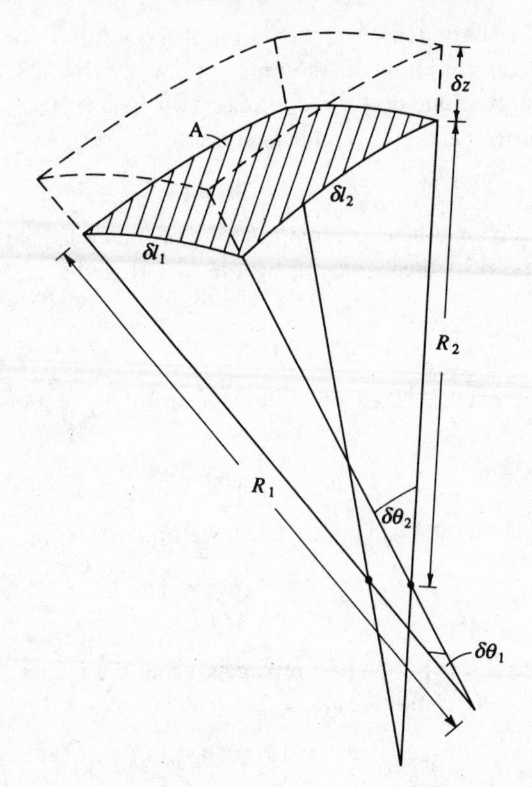

Fig. 4.5 The curvatures and displacement of a small elemental area of liquid surface.

to be stable, the pressures on the two sides of the surface must be unequal. The difference in pressure across the surface, or the *excess pressure* acting on the concave side can be calculated easily by considering the work done in causing the area of surface to be slightly increased.

Consider a small elemental area A of a liquid film with edge lengths equal to δl_1 and δl_2 as shown in fig. 4.5. The radii of curvature of the two edges are R_1 and R_2 respectively, and they subtend angles $\delta\theta_1$ and $\delta\theta_2$ at their centres of curvature. Thus

$$\delta l_1 = R_1\delta\theta_1 \quad \text{and} \quad \delta l_2 = R_2\delta\theta_2. \tag{4.15}$$

Any slight differences in the lengths and radii of curvature of opposite sides of the curvilinear rectangle are ignored. The area of the element of film, which has *two surfaces*, is therefore

$$A = 2\delta l_1\delta l_2 = 2R_1\delta\theta_1 R_2\delta\theta_2. \tag{4.16}$$

Now let the surface, which we assume to be in equilibrium, be displaced by a small distance δz in a direction normal to the elemental area. The new area of the element of the film is

$$A + \Delta A = 2(R_1 + \delta z)\,\delta\theta_1 \times (R_2 + \delta z)\,\delta\theta_2. \qquad (4.17)$$

Thus if terms of order δz^2 are neglected, the increase in area is

$$\Delta A = 2(R_1 + R_2)\,\delta\theta_1\,\delta\theta_2\,\delta z, \qquad (4.18)$$

where eqn (4.16) has been used. Further use of this equation puts the increase in area in the form

$$\Delta A = (1/R_1 + 1/R_2)\,A\delta z. \qquad (4.19)$$

Furthermore the work done must equal the increase in surface energy

$$\Delta W = \gamma \Delta A. \qquad (4.20)$$

We can take the change in pressure needed to move the film through δz to be small compared with the excess pressure p across the film, so that

$$\begin{aligned} \Delta W &= \text{force} \times \text{distance} \\ &= p\,\delta l_1\,\delta l_2\,\delta z. \end{aligned} \qquad (4.21)$$

Therefore combination of the last three equations with eqn (4.16) gives

$$p = 2\gamma(1/R_1 + 1/R_2). \qquad (4.22)$$

It has been assumed that the centres of curvature are on the same side of the film. If they are not then

$$p = 2\gamma(1/R_1 - 1/R_2), \qquad (4.23)$$

where p is the excess pressure on the same side of the film as the centre of curvature of the edge δl_1.

Equation (4.22), here derived by considering virtual work, is a special form of a slightly more general relationship for the pressure at a point within a liquid, which was obtained by Laplace by calculating the attractive force exerted on a small column of liquid normal to a surface by the liquid outside the column. The equation is accordingly known as *Laplace's formula*.

4.8 Special cases of simple curved liquid surfaces

4.8.1 A bubble in a liquid

Consider a bubble of gas within a liquid, depicted in fig. 4.6(a). The bubble is assumed to be small and spherical, with a radius R. These assumptions allow the effect of gravity upon the shape of the bubble to be neglected and any variations in pressure within it to be ignored. The excess pressure

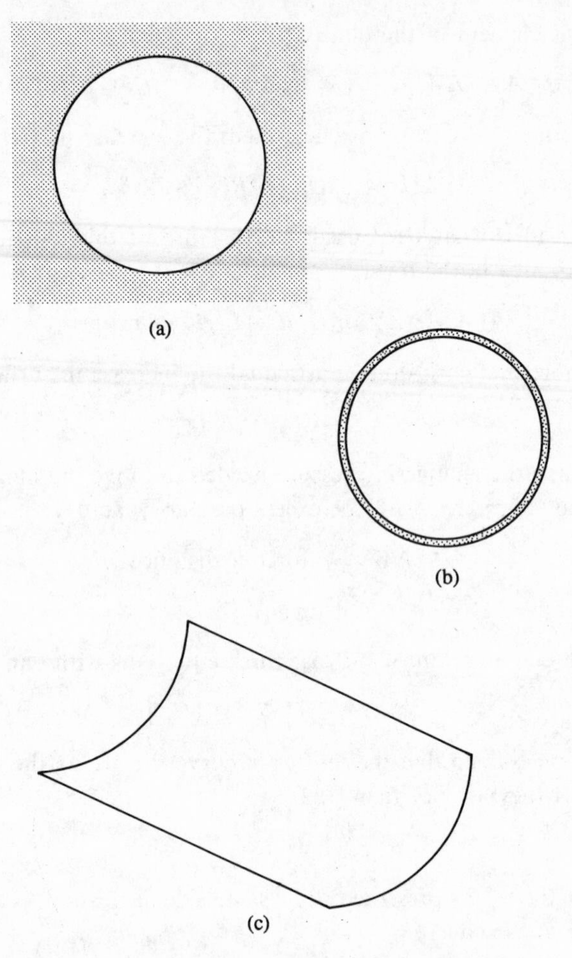

Fig. 4.6 Special cases of curved liquid surfaces: (a) a vapour bubble within a liquid; (b) a free-floating bubble; (c) a portion of a cylindrically-curved surface.

across the surface of the bubble is then, from eqn (4.22)

$$p = 2\gamma/R. \tag{4.24}$$

The factor of two in eqn (4.24) arises from the bubble having two equal radii of curvature; a bubble in a liquid has only one surface so that eqn (4.22) has to be modified to apply to this case. A bubble may be considered as two hemispherical halves in equilibrium by virtue of the equal surface tension forces acting perpendicular to the boundary planes. The excess pressure across a single-sided hemispherical surface is therefore also given by eqn (4.24).

4.8.2 A blown, free bubble

Now consider a free-floating soap bubble, as in fig. 4.6(b), which is sufficiently small to remain essentially spherical, despite the influence of gravity. It has two film surfaces, inner and outer, and two radii of curvature, both approximately equal to R. From eqn (4.22), the pressure inside the bubble exceeds the pressure outside by

$$p = 4\gamma/R. \tag{4.25}$$

4.8.3 A cylindrical liquid surface

If the liquid surface has a cylindrical form with a radius of curvature R and is single-sided, as in fig. 4.6(c), the excess pressure on the concave side is simply

$$p = \gamma/R. \tag{4.26}$$

The foregoing cases are the simplest examples of curved liquid surfaces. Slightly more complex shapes will be considered when discussing methods of measuring surface tension and the angles of contact between liquid surfaces and solids.

4.9 The interface between two liquids – cohesive energy

Two liquids which do not mix with one another and have different densities exhibit a distinct boundary or interface when in static equilibrium, as illustrated in fig. 4.7. Such liquids are said to be *immiscible*. Molecules of the liquid labelled A which are very close to the interface must be subject to forces exerted by molecules of the other liquid, labelled B, and vice versa. The potential energy of liquid A when it is contact with another liquid B, therefore differs from the potential energy of A when it is in contact with its own vapour. One may deduce, then, that a two-component liquid system has a characteristic (or intrinsic) total *interfacial energy*, which we designate by E_{AB}. The corresponding free energy contribution, or interfacial tension, is denoted γ_{AB}.

Consider an interface of unit area and assume that it is possible to separate the two liquids cleanly and isothermally at their surface of contact by pulling the liquids in opposite directions normal to the interface, and performing an amount of work W_{AB}. Since this process creates new exposed surfaces on the liquids A and B which are in contact with their respective vapours, it is necessary to supply heat in order to maintain isothermal conditions. Let us call the quantities of heat for A and B, ΔQ_A

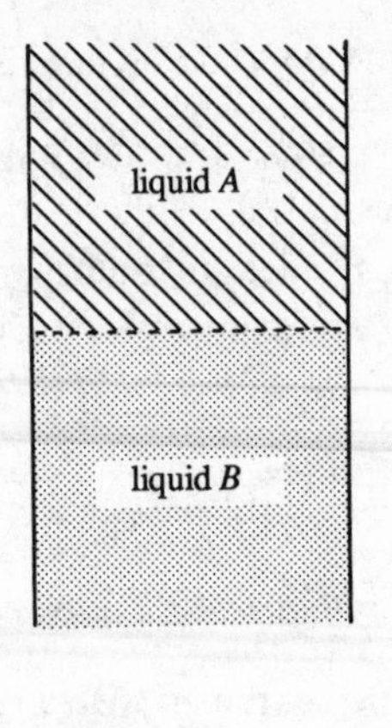

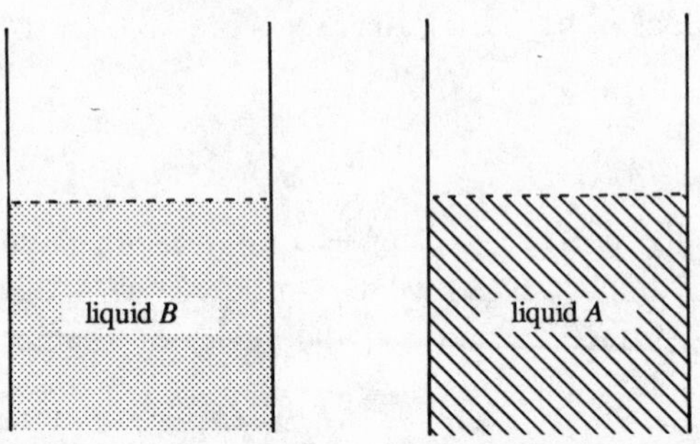

Fig. 4.7 The hypothetical separation of two immiscible liquids held in a vertical column.

and ΔQ_B respectively. Note, however, that some energy is recovered in the form of heat by the action of destroying the interface between A and B; let us call this quantity of heat ΔQ_{AB}. The equation that describes the conservation of energy for the hypothetical process of separation is

$$E_{AB} + W_{AB} + \Delta Q_{\mathrm{A}} + \Delta Q_B - \Delta Q_{AB} = E_A + E_B, \qquad (4.27)$$

where E_A and E_B are the total surface energies of the separated liquids in contact with their own vapours; the corresponding surface tensions are denoted γ_A and γ_B. However, eqn (4.7) gives a general relationship between total surface energy, surface tension and heat, namely

$$E = \gamma + \Delta Q$$

so that

$$E_A + E_B = \gamma_A + \Delta Q_A + \gamma_B + \Delta Q_B \qquad (4.28)$$

and

$$E_{AB} = \gamma_{AB} + \Delta Q_{AB}. \qquad (4.29)$$

Thus

$$W_{AB} = \gamma_A + \gamma_B - \gamma_{AB}. \qquad (4.30)$$

This important result is known as *Dupré's equation*. It has relevance to questions both of adhesion and the miscibility of liquids. The quantity W_{AB} is known as the *work of adhesion*. It is the work needed to separate two components by means of tensile forces applied normal to the interface between them. It is apparent that the work of adhesion increases as γ_{AB} decreases.

Since the interfacial tension, γ_{AB} and the interfacial energy E_{AB} are related in the same way as surface tension and surface energy, it is apparent that if γ_{AB} is positive the interfacial area tends to be minimized. Conversely, if γ_{AB} is negative, the area of the interface tends to be maximized, i.e. the two liquids tend to mix. From these arguments we deduce that for *complete miscibility* of two liquids, the interfacial tension should be either zero or negative. Further discussion of the contact between two different liquids appears in §4.18.

Dupré's equation (4.30) endows a new significance to the surface tension of a liquid. If the equation is applied to a (hypothetical) interface between two identical liquids, the equation gives the work needed to pull a liquid into two parts. This is usually called the *work of cohesion*. Although it would be consistent with our terminology to denote it by W, with L (for liquid) as a double suffix, we shall use the symbol W_C for the work of cohesion. From eqn (4.30)

$$W_C = \gamma_{LV} + \gamma_{LV} - \gamma_{LL}, \qquad (4.31)$$

where L refers to the single liquid phase and V to the vapour phase in contact with it, so that γ_{LV} is simply the normal surface tension of the liquid, previously denoted by γ. Clearly γ_{LL} must be zero, since a liquid is completely 'self miscible'. Therefore

$$W_C = 2\gamma_{LV}. \qquad (4.32)$$

In other words, surface tension is a measure of self adhesion or the *cohesion* of a liquid. This is hardly surprising in view of the connection established in §4.6 between surface tension and binding energy.

There are ways of measuring the surface energy, and hence making comparisons with $\frac{1}{2}W_C$ calculated from models of intermolecular potentials, the details of which do not concern us here. The results are found to give excellent agreement with measured values of the surface tension for various liquids.

4.10 The interfacial energy between a liquid and a solid

Similar to the situation of two liquids in contact, there is an interfacial energy for the surface of contact between a liquid and a solid. A consideration of this interfacial energy helps in the understanding of what is known as the *wetting* of solids by liquids. It is also important because the interfacial energies of liquid–solid systems are basic to such industrially-relevant topics as the science of adhesives, the solidification of alloys, and liquid-phase sintering (a method of forming dense bodies from powders without resorting to melting).

Everyday experience demonstrates that some liquids *wet* solid surfaces, while others do not. It is necessary, however, to decide precisely what is meant by wetting. Historically, wetting has often been considered in terms of a drop of liquid resting in equilibrium on a flat surface of a solid. We start with the same system and extend earlier conclusions (§4.5) concerning the equivalence of surface tension and surface energy to the interface between a liquid and a solid. Referring to the liquid drop on the flat, rigid solid shown in fig. 4.8(a), we assign interfacial tensions γ_{SV} and γ_{SL} to the solid–vapour and solid–liquid interfaces, respectively. We continue to use γ_{LV} for the surface tension of a liquid in contact with its own vapour. We now examine the origins of these forces.

The profile of the liquid drop depicted in fig. 4.8(a) is determined by the equilibrium between the forces acting at the periphery of the drop where all three phases, solid, liquid and gas are present. These determine the *contact angle* θ, which is defined as the angle between the tangent to the liquid–vapour surface and the tangent to the solid–vapour surface, both drawn normal to the curve defining the periphery of the drop in the solid–liquid surface. Thus we consider the forces acting to maintain static equilibrium of a small volume of liquid ABC adjacent to a unit length of the peripheral line of contact, as shown in fig. 4.8(b), making the assumption that BC somewhat exceeds the thickness of the surface layers of the liquid. In the light of the arguments applied in §4.4, only departures from the surrounding atmospheric pressure will affect the equilibrium.

Consider an imaginary surface through BC with unit length into the page. There are two contributions to the force across BC: one is the usual force of surface tension γ_{LV} originating from the layer below the liquid surface near B; the other comes from another layer in the liquid near C, whose properties differ from those of the interior of the drop to the left of BC. In the case illustrated, where the liquid wets the solid (i.e. is strongly attracted to it), we may make the reasonable assumption that there is a layer in the liquid close to the solid whose density exceeds that of the bulk. Thus for the volume ABC we can expect an increase in density towards C,

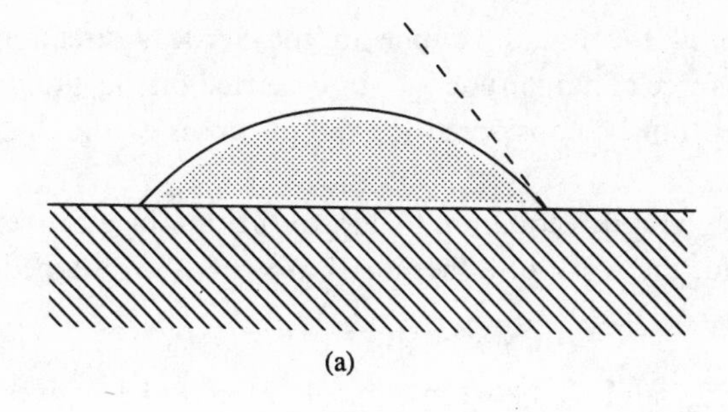

(a)

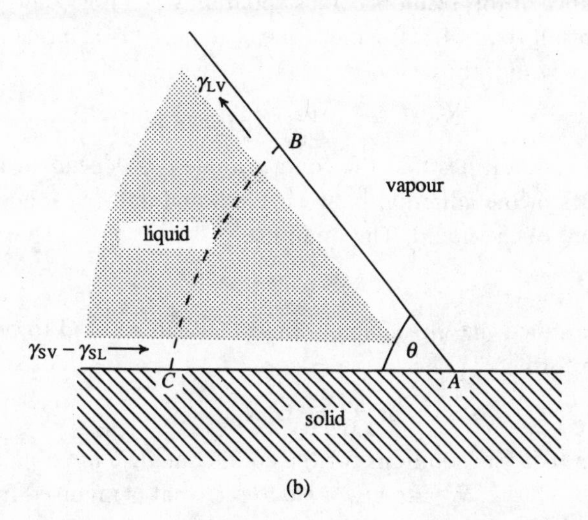

(b)

Fig. 4.8 (a) The equilibrium of a liquid drop on a flat solid surface; (b) enlarged view of the drop in the vicinity of A, indicating the layers of reduced and increased liquid density adjacent to the liquid–vapour and liquid–solid interfaces, respectively (after Berry, M.V. (1971). *Physics Education*, **6**, 79).

giving a pressure on BC near C, which we can associate with a negative surface tension force γ_{SL}. More accurately, consideration of the change in energy at the solid surface when vapour is replaced by liquid shows that the effective force acting towards the right is $\gamma_{SV} - \gamma_{SL}$. Equilibrium of the liquid volume shown in fig. 4.8(b) in the horizontal direction therefore requires that

$$\gamma_{LV} \cos \theta = \gamma_{SV} - \gamma_{SL}. \tag{4.33}$$

This is known as *Young's relation.* For solids and liquids whose relevant interfacial tensions are known, it enables determinations of the profile of the drop and the extent to which the liquid wets the surface of the solid.

Equilibrium of the liquid volume in the vertical direction requires a downwards force of magnitude $\gamma_{LV} \sin \theta$ exerted on the liquid by the solid, but this need not be considered for the purposes of the discussions that follow.

If we apply Dupré's equation (4.30), to the interface between the liquid and the solid, the work of adhesion W_A $(= W_{SL})$ is obtained as

$$W_A = \gamma_{SV} + \gamma_{LV} - \gamma_{SL}. \tag{4.34}$$

Eliminating γ_{SL} and γ_{SV} between eqns (4.33) and (4.34) gives

$$W_A = \gamma_{LV}(1 + \cos \theta). \tag{4.35}$$

The significance of this result becomes apparent when it is combined with the earlier result, eqn (4.32), which shows that γ_{LV} is a measure of the cohesion of a liquid, thus

$$\cos \theta = (2W_A/W_C) - 1. \tag{4.36}$$

Equation (4.36) indicates that the contact angle θ is dependent upon the relative values of the adhesion between the liquid and the solid, and the 'self adhesion' of the liquid. The quantity

$$C_s = 2(W_A - W_C)/W_C \tag{4.37}$$

is sometimes called the *spreading coefficient*. It can be used to recast eqn (4.36) in the form

$$\cos \theta = C_s + 1 \tag{4.38}$$

and it is seen that real solutions for the contact angle θ are obtained only when $0 \leqslant C_s \leqslant -2$. We are now able to see what is involved in wetting and to compare it with *spreading*.

Wetting is said to occur when, on lowering a solid into a liquid, the existing solid–vapour interface is replaced by a solid–liquid interface, figs. 4.9(a) and 4.9(b). In this case the work of cohesion of the liquid is greater than the effective work of adhesion between liquid and solid in the presence of the displaced vapour, that is

$$0 \leqslant W_A \leqslant W_C \quad \text{or} \quad -\gamma_{LV} \leqslant \gamma_{SV} - \gamma_{SL} \leqslant \gamma_{LV} \quad \text{or} \quad -2 \leqslant C_s \leqslant 0. \tag{4.39}$$

Spreading is a term applied when the liquid forms a film over the solid so that both solid–liquid and liquid–vapour interfaces are formed, as illustrated in fig. 4.9(c). Spreading only occurs spontaneously when the effective work of adhesion is greater than the work of cohesion of the liquid in the presence of the vapour, that is

$$W_A \geqslant W_C \quad \text{or} \quad \gamma_{SV} - \gamma_{SL} \geqslant \gamma_{LV} \quad \text{or} \quad C_s \geqslant 0. \tag{4.40}$$

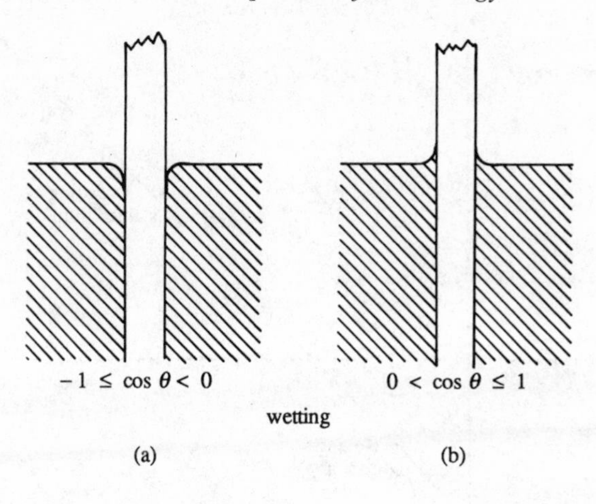

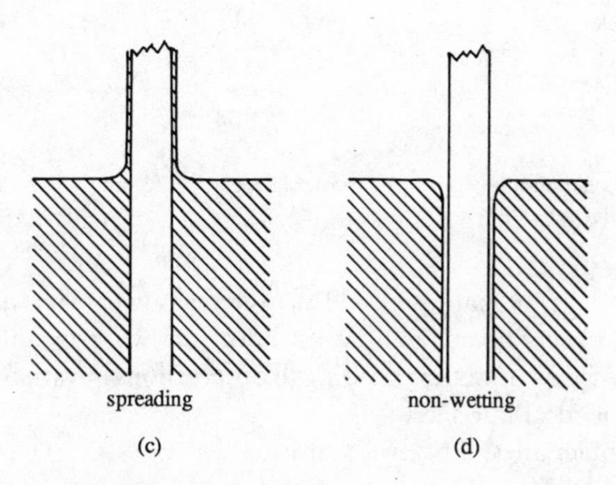

Fig. 4.9 Wetting, spreading and non-wetting systems in terms of the behaviour of different liquids when solid plates are lowered vertically into them.

Non-wetting is a relative term, that may be used to describe the converse behaviour to spreading, which is illustrated diagrammatically in fig. 4.9(d). Non-wetting implies that the effective work of adhesion between the solid and the liquid is negative, that is

$$W_A < 0 \quad \text{or} \quad \gamma_{SV} - \gamma_{SL} < -\gamma_{LV} \quad \text{or} \quad C_s < -2. \tag{4.41}$$

The energy required to immerse the solid is therefore minimized if a vapour film is maintained between the liquid and the solid.

In all the foregoing definitions the wording 'effective work of adhesion' has been used. This terminology is intended to recognize the limitations of the particular physical model of a droplet on a flat surface. In addition,

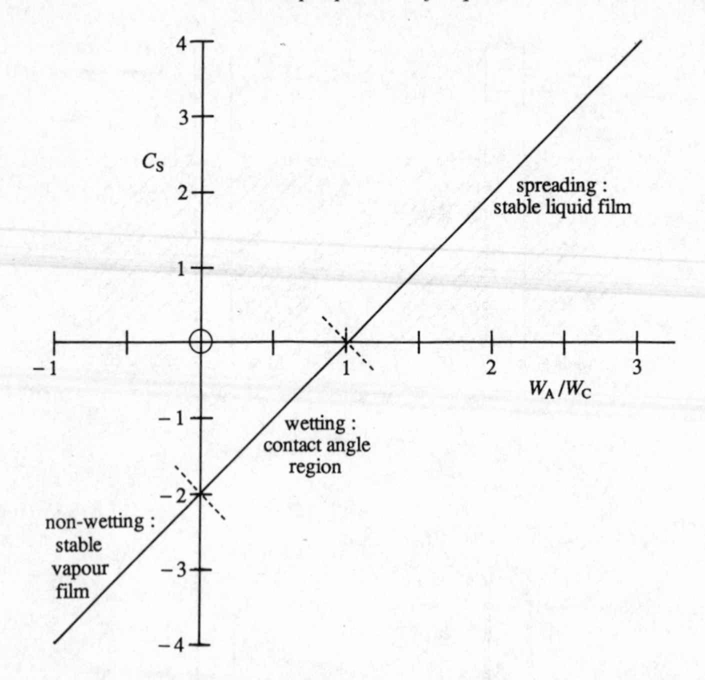

Fig. 4.10 The relationship between the spreading coefficient, C_s and the work of adhesion W_A (after Padday J.F. (1978)).

there is the reality of many practical measurements where it is not possible to avoid surface contaminants which affect the work of adhesion and thereby the degree of wetting, leading to observations at variance with the theory of the ideal interface.

If the contact angle θ is zero, that is

$$W_A = W_C \quad \text{or} \quad \gamma_{SV} - \gamma_{SL} = \gamma_{LV} \quad \text{or} \quad C_s = 0, \tag{4.42}$$

a liquid spreads spontaneously over a solid surface and thus wets it completely. Both water and mercury wet clean glass, but to different extents. For water on very clean glass, $\theta = 0°$ and spontaneous spreading occurs. However, mercury only partially wets glass, since $\theta \simeq 140°$, a result of particularly strong cohesive forces in the metal.

The three regimes described above can be illustrated by plotting the spreading coefficient as a function of the work of adhesion. From eqn (4.37) this must yield a straight line, as shown in fig. 4.10. The diagram indicates the changeover from wetting to spreading by the intercept of the straight line with the horizontal axis. Spreading behaviour is discussed further in the next section. For a particular system there may exist a temperature at which there is an abrupt transition from one regime to another. The science of surface tension effects includes the study of such *wetting transitions* but they are beyond the scope of this book. We do, however, consider some related effects that occur as the chemical composition of the liquid is changed.

4.11 Critical effects and surface tension

4.11.1 Critical surface tension for spreading

The transition from wetting to spreading behaviour can be clearly illustrated by experiments in which a series of chemically-related liquids are placed on the same solid surface, and the variation of contact angle with the surface tensions of the liquids is measured. Figure 4.11 shows the measured dependence of cos θ on γ_{LV} for a series of n-alkanes, with the chemical formula C_nH_{2n+2}, in contact with polytetrafluoroethylene

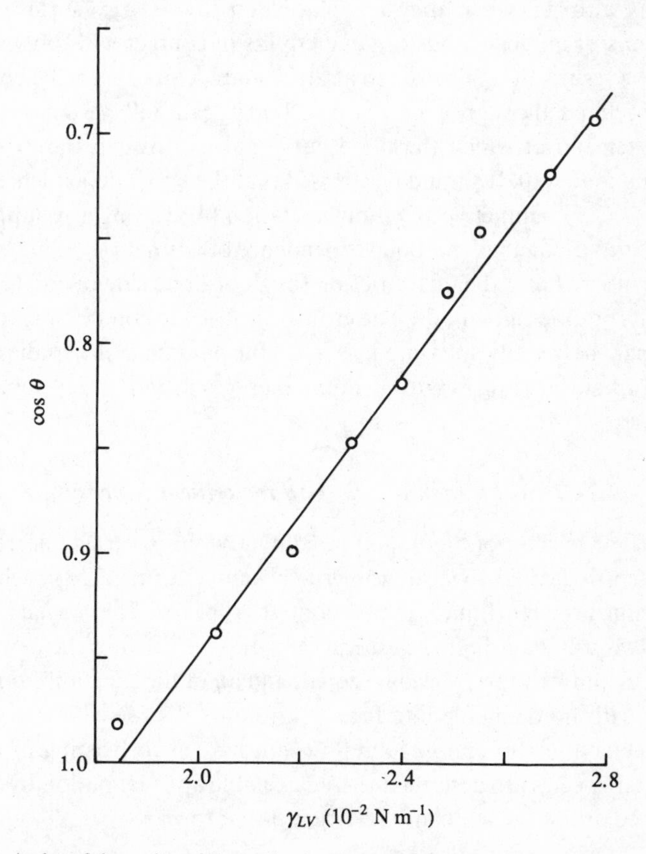

Fig. 4.11 A plot of the cosine of the angle of contact with polytetrafluoroethylene versus the surface tension for the homologous series of n-alkanes, C_nH_{2n+2} (data from Fox, H.W. & Zisman, W.A. (1950). *Journal of Colloid Science*, **5**, 514).

(PTFE). These and other polymeric and organic compounds have low surface energies. For the liquids considered, results are available for *n*-alkanes with 10 different values of *n*. Nine of the liquids exhibit finite contact angles, and these correspond to the points plotted in fig. 4.11. The tenth liquid, pentane, spreads over the PTFE surface. It is apparent that the measured points lie approximately on a straight line of the form

$$1 - \cos\theta = K(\gamma_{LV} - \gamma_c), \tag{4.43}$$

where K is a constant. The intercept with the horizontal axis, at which the contact angle falls to zero and the liquid just, and only just, spreads spontaneously over the given surface, is called the *critical surface tension for spreading*, denoted by γ_c.

Other series of chemically-related liquids show similar linear variations of $\cos\theta$ with γ_{LV} when they are placed on low energy surfaces. Even collections of miscellaneous organic liquids in contact with polymer surfaces give results that fall close to straight lines. More generally, $\cos\theta$ and γ_{LV} are related through a power law, but γ_c can still be defined as the surface tension at which the fitted curve passes through the axis corresponding to $\theta = 0$. It should be stressed that the linear dependence of eqn (4.43) is not in conflict with Young's relation (4.33); the latter appears at first sight to predict a hyperbolic dependence of $\cos\theta$ on γ_{LV}, but it should be remembered that the quantities on the right-hand side of eqn (4.33) are different for different liquids. The critical surface tension for spreading, γ_c, which can be readily measured, is a useful parameter in studies of the wetting of surfaces of low-to-medium energy and in the development of adhesives.

4.11.2 Surface tension close to the critical temperature

The surface tension of any liquid in contact with its vapour decreases as the system is heated towards the critical temperature T_c, for which the distinction between liquid and vapour disappears. The surface tension must obviously vanish at the critical temperature because the densities of the liquid and its vapour become equal, and all of the substance turns into vapour with no definable interface.

The behaviours of simple liquid–vapour systems consisting of a single molecular species are described to a reasonable approximation by van der Waals' equation, which can be written in the form

$$(P_r + 3/V_r^2)(3V_r - 1) = 8T_r, \tag{4.44}$$

where P_r, V_r and T_r are *reduced pressure*, *reduced volume* and *reduced temperature*, respectively, defined by

$$P_r = P/P_c, \quad V_r = V/V_c \quad \text{and} \quad T_r = T/T_c \qquad (4.45)$$

with P_c, V_c and T_c being respectively the pressure, volume and temperature at the critical point of the substance. This form of van der Waals' equation states that the P–V–T surfaces for all substances are the same if the measured pressure, volume and temperature are scaled by their critical values. Thus any two substances with the same values of P_r, V_r and T_r are said to be in corresponding states, and eqn (4.44) expresses the *law of corresponding states.*

Most physical properties have a temperature variation in the vicinity of the critical temperature that can be expressed as some power of $|T_c - T|$, where the power is the *critical exponent* for the property concerned. Thus most of the temperature-dependent variables tend towards zero or infinity at $T = T_c$, depending on whether the critical exponent is positive or negative. Some of the critical exponents can be calculated by making an expansion of the variables in van der Waals' equation around their values at the critical point, but others are more difficult to calculate. Experimental determinations of critical point exponents provide data against which the predictions of quite sophisticated theories of phase transitions can be tested.

Figure 4.12 shows the measured values of the surface tension γ_{LV} for CO_2 at temperatures close to T_c. The measurements were made by the technique of surface light-scattering spectroscopy described in § 5.11. The measured points lie on a straight line in a log–log plot, and the variation can be closely represented by

$$\gamma_{LV} = \gamma_0\{1 - (T/T_c)\}^{\mu} \quad (T \leqslant T_c) \qquad (4.46)$$

where γ_0 is a constant and

$$\mu = 1.29 \quad \text{for} \quad CO_2 \qquad (4.47)$$

is the *critical exponent for the surface tension.* Experimental results for measurements on a variety of substances in which the critical temperature is approached from below at constant volume give critical exponents in the range 1.28 ± 0.06. Differentiation of eqn (4.46) gives

$$d\gamma_{LV}/dT = -\mu\gamma_{LV}/(T_c - T), \qquad (4.48)$$

and this confirms the positive sign of the thermal contribution to the total surface energy in eqn (4.9).

Contact angles are also temperature dependent, although it was some years before this was recognized, because lack of attention to scrupulous

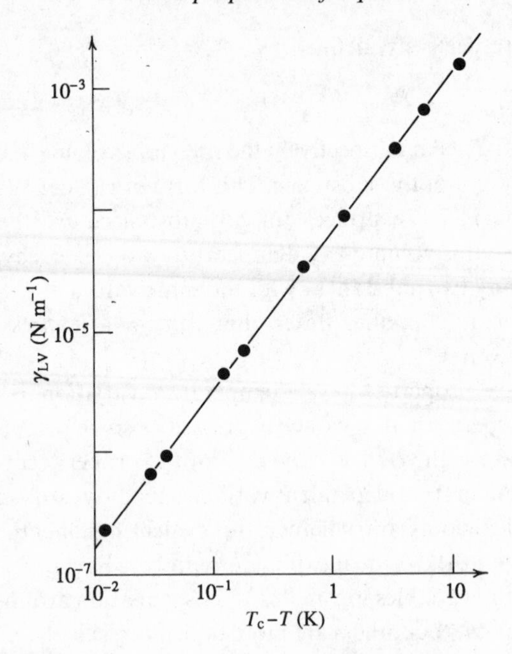

Fig. 4.12 Measured surface tension of CO_2 in the vicinity of its critical point, $T_c = 304$ K (from Herpin, J.C. & Meunier, J. (1974). *Journal de Physique*, **35**, 847).

cleanliness and to very careful experimentation can easily give erroneous values for θ. It is now acknowledged that contact angles are greatly affected both by the cleanliness and the finish (roughness) of the solid surface. Measurements of contact angles frequently exhibit spurious hysteresis effects which are a consequence of poor experimental conditions (e.g. contamination, surface roughness). Methods of measuring surface tension and contact angles are discussed in §4.15. If the dependences upon temperature of both properties are known, one may use the data to determine the *heat of wetting*. This is the energy released on replacing a solid–vapour interface by a solid–liquid one, where the liquid wets and forms a very thin continuous film. The heat of wetting ΔQ_W is defined as

$$\Delta Q_W = (\gamma_{SL} - T\mathrm{d}\gamma_{SL}/\mathrm{d}T) - (\gamma_{SV} - T\mathrm{d}\gamma_{SV}/\mathrm{d}T). \qquad (4.49)$$

Combining this equation with Young's relation (4.33) gives

$$\Delta Q_W = -\{\gamma_{LV} \cos\theta - T\mathrm{d}/\mathrm{d}T(\gamma_{LV}\cos\theta)\}. \qquad (4.50)$$

It is therefore possible to find ΔQ_W if both $\mathrm{d}\theta/\mathrm{d}T$ and $\mathrm{d}\gamma_{LV}/\mathrm{d}T$ can be measured.

4.12 Static drops and bubbles

In this and the following section we consider some simple systems where the effects of surface tension are important. The present section is concerned with the shapes of fluid drops and bubbles in contact with flat horizontal solid surfaces. The shapes of very small drops and bubbles approximate to spheres because this form minimizes their surface areas and gravity has little effect. Larger drops of liquid or bubbles of vapour are significantly distorted by the effects of gravity and some typical shapes are shown in fig. 4.13. Drops and bubbles that are prevented by the surface from moving under the influence of gravity are called *sessile* and *captive*, respectively (figs. 4.13(a) and (b)), while drops and bubbles that cling to the surface against the pull of gravity are called *pendant* (figs. 4.13(a) and (d)). Sessile drops are flattened by the effects of gravity, and the flattening becomes very pronounced for large volumes of liquid, as with a pool of mercury on a sheet of glass.

Let us now consider the physics of a moderately large sessile drop, intermediate in shape between a sphere and a film, which partially wets a supporting flat surface ($90° < \theta < 180°$). The liquid–vapour interface is a surface of revolution about a vertical axis of symmetry, and fig. 4.14 shows a meridional section through the drop. Cartesian axes with their origin at the top of the drop are shown in the diagram. Applying eqn (4.22) for the pressure difference across a curved surface to closely adjacent points O, O' on the vertical axis, just inside and just outside the liquid surface, gives

$$p_O - p_{O'} = 2\gamma/R_O, \tag{4.51}$$

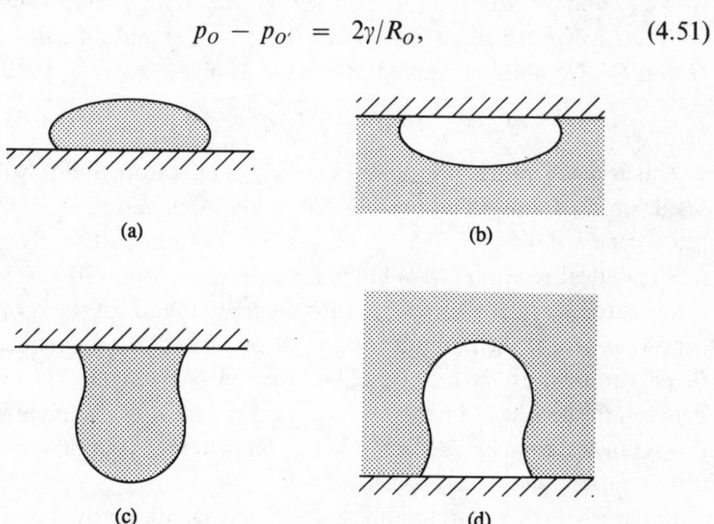

Fig. 4.13 Typical shapes of: (a) sessile drop; (b) captive bubble; (c) pendant drop; (d) pendant bubble.

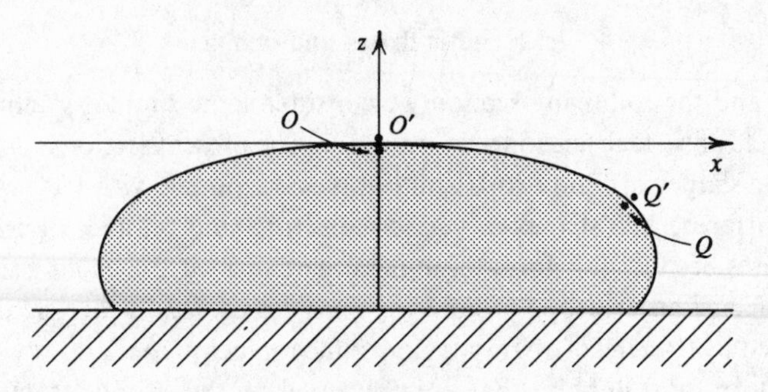

Fig. 4.14 A liquid drop resting on a flat solid surface, with angle of contact, θ such that $\pi/2 < \theta < \pi$.

where p_O and $p_{O'}$ are the pressures at O and O' respectively, R_O is the radius of curvature of the drop at the point O, and the LV suffix on γ has been omitted. The difference in pressure between a similar pair of points Q and Q' at a general position on the surface is

$$p_Q - p_{Q'} = \gamma(1/R_Q + 1/r_Q), \tag{4.52}$$

where r_Q and R_Q are the radii of curvature of the drop at Q in the plane of the figure and in the perpendicular plane, respectively. Considerations of hydrostatic pressure also give

$$\begin{aligned} p_Q - p_O &= -\varrho_L g z \\ p_{Q'} - p_{O'} &= -\varrho_V g z, \end{aligned} \tag{4.53}$$

where ϱ_L and ϱ_V are the liquid and vapour densities, respectively, and z is the (negative) vertical coordinate of the points Q and Q' with respect to O and O'. Combining eqns (4.51) to (4.53) gives

$$-(\varrho_L - \varrho_V)gz = \gamma(1/R_Q + 1/r_Q - 2/R_0). \tag{4.54}$$

This is a general relation between radii of curvature and position for a sessile drop. It shows that the reduced radius of curvature $R_Q r_Q/(R_Q + r_Q)$ must decrease as the point Q moves away from O. Exactly the same mathematical relation (4.54) holds for a pendant drop if the z-coordinate is measured vertically upwards from the lowest point of the drop. Thus z is now *positive* at the point Q and eqn (4.54) shows that the reduced radius of curvature must *increase* as Q moves away from O. This difference between the sessile and pendant cases is reflected in the different shapes of drops illustrated in figs. 4.13(a) and (c). The shapes of bubbles mimic those of the corresponding drops.

Large sessile drops and bubbles are flattened sufficiently that the curvature on the z-axis may be put equal to zero, so that $R_O \simeq \infty$. It may also

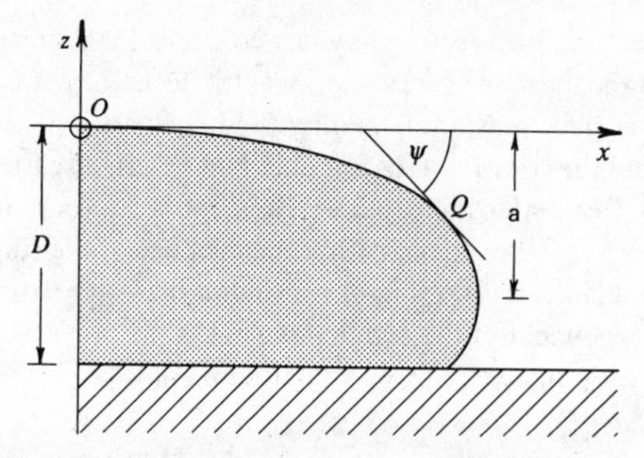

Fig. 4.15 The method by which the result (4.55) for the liquid drop shown in fig. 4.14 may be expressed in terms of the height of the drop.

be assumed that $R_Q \gg r_Q$, so that the result (4.54) becomes

$$-(\varrho_L - \varrho_V)gz = \gamma/r_Q. \tag{4.55}$$

The curvature of the sides of the drop is not easy to measure accurately, so it is better to express the results given above in different terms, by relating r_Q to the height of the drop. Referring now to fig. 4.15, let O and Q be in the liquid–vapour interface, let the arc OQ be of length s and let us extend the tangent to the surface at Q to cut the x-axis at an angle ψ. Then from eqns (4.55) and (A2.5)

$$\begin{aligned} g(\varrho_L - \varrho_V)z &= -\gamma/r_Q = -\gamma\,\mathrm{d}\psi/\mathrm{d}s \\ &= -\gamma\frac{\mathrm{d}\psi}{\mathrm{d}z}\frac{\mathrm{d}z}{\mathrm{d}s} \\ &= \gamma(\mathrm{d}\psi/\mathrm{d}z)\sin\psi. \end{aligned} \tag{4.56}$$

Hence

$$g(\varrho_L - \varrho_V)\int z\,\mathrm{d}z = \gamma\int \sin\psi\,\mathrm{d}\psi \tag{4.57}$$

and this can be integrated to obtain

$$\tfrac{1}{2}g(\varrho_L - \varrho_V)z^2 = \gamma(1 - \cos\psi), \tag{4.58}$$

where we have used the fact that $\psi = 0$ for $z = 0$. The position $z = -D$ of the base of the drop is obtained by putting ψ equal to the contact angle θ, whence

$$D^2 = 2\gamma(1 - \cos\theta)/g(\varrho_L - \varrho_V), \tag{4.59}$$

and the position $z = -\mathrm{a}$ of the point where the tangent is vertical is obtained by putting $\psi = 90°$, giving

$$\mathrm{a}^2 = 2\gamma/g(\varrho_L - \varrho_V). \tag{4.60}$$

The distances D and a are marked in fig. 4.15.

The distance a defined in this way is a constant for a given liquid and it is known as the *capillary constant*. It sets the scale for many effects that depend upon surface tension, including the heights of sessile drops considered here and the rise of liquids up tubes and plates, discussed in §§4.13, 4.14 and 4.16. For water the capillary constant is 3.93 mm at 0° C.

It is seen from the above results that the distance of the ring of vertical tangents, or equivalently the ring of maximum diameter, from the top of the drop is independent of the contact angle. On the other hand, the total height of the drop increases from a minimum value

$$D_{\min} = \mathrm{a} \quad \text{at} \quad \theta = \tfrac{1}{2}\pi \tag{4.61}$$

to a maximum value

$$D_{\max} = \mathrm{a}\sqrt{2} \quad \text{at} \quad \theta = \pi, \tag{4.62}$$

as the contact angle is increased. Of course, the ring of vertical tangents disappears for contact angles smaller than $\frac{1}{2}\pi$ and the distance a loses its physical significance, but the height of the drop is still given by eqn (4.59).

Separate measurements of D and a provide values for both the surface tension and the contact angle of a liquid, through the use of eqns (4.59) and (4.60). Reasonably satisfactory measurements of surface tension can be made rapidly from a sessile drop by using eqn (4.60) and determining the vertical distance between the top of the drop and the position at which the side of the drop is vertical. This can be achieved optically by finding when an image, formed by reflection off the side of the drop, is in the same horizontal plane as the object. Even when this approach is adopted the results obtained for γ and θ from sessile drops still do not give the accuracy which can be achieved with other techniques, partly because the probability of sessile drops becoming contaminated is high.

4.13 Rise of a liquid in a tube with a fine bore – capillarity

The ascent of a liquid in an open tube of fine bore when the bottom is dipped into a reservoir of the liquid is a well known effect. The amount of *capillary rise*, as it is called, is affected by both the radius R of the bore and the surface tension of the liquid. Historically, this effect has formed a basis of measurements of γ.

It is assumed that the shape of the surface of the liquid in the tube is concave towards the overlying vapour, as shown in fig. 4.16. Comparing the region of the meniscus in fig. 4.16 with fig. 4.8, we see that the concave surface is in a state of tension because a layer beneath it is at a pressure

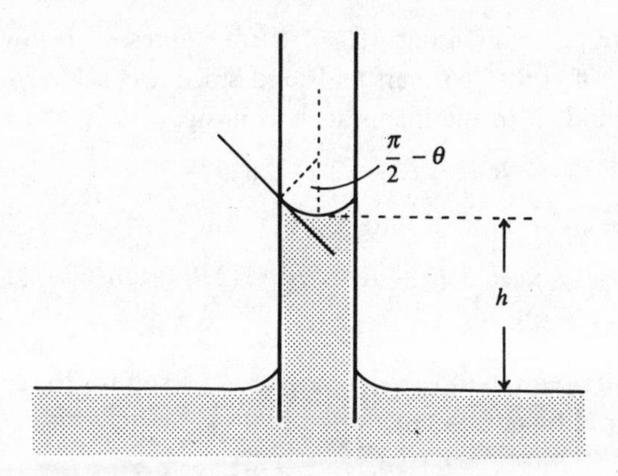

Fig. 4.16 The rise of a liquid in a vertical tube with a very fine bore (capillary rise).

which is lower than the atmospheric pressure, p_0. However, a cylindrical layer in the liquid adjacent to the capillary walls is at a pressure greater than p_0 for the reasons argued in §4.10. It was shown there that the effective force in the liquid parallel to the liquid–solid interface is equal to $\gamma_{SV} - \gamma_{SL}$, the difference between the solid–vapour and solid–liquid interfacial tensions (see fig. 4.8). The periphery of the meniscus may be viewed not as pulled upwards, as sometimes is argued, but as supported from below by the thin cylindrical layer at overpressure. Consider then the equilibrium of the liquid column. Its weight (we ignore the volume of the meniscus 'lens') is supported by the force from the overpressure, so that

$$2\pi R(\gamma_{SV} - \gamma_{SL}) = \pi R^2(\varrho_L - \varrho_V)gh. \tag{4.63}$$

Resolving pressures vertically at the periphery of the meniscus gives Young's relation (4.33),

$$\gamma_{SV} - \gamma_{SL} = \gamma_{LV}\cos\theta, \tag{4.64}$$

where θ is the angle of contact, so that

$$(\varrho_L - \varrho_V)Rgh = 2\gamma_{LV}\cos\theta. \tag{4.65}$$

For a liquid with a zero contact angle this becomes

$$(\varrho_L - \varrho_V)gh = 2\gamma/R, \tag{4.66}$$

where the LV suffix has been dropped.

The result, eqn (4.65), is only approximate, because the weight of the liquid in the meniscus lens has been neglected. However, eqn (4.65) shows that the capillary rise depends on the bore, as observed, being inversely proportional to the radius. It is apparent that the liquid in the tube rises only if $\theta < \frac{1}{2}\pi$, otherwise the column in the tube is either level with the surface of the reservoir liquid ($\theta = \frac{1}{2}\pi$) or depressed below it. A more exact version of eqn (4.65) can be found since it can be shown that the volume of liquid, V in the meniscus lens in fig. 4.16 is

$$V = \pi R^3\{\sec\theta + (2/3)\tan^3\theta - (2/3)\sec^3\theta\}, \tag{4.67}$$

which may be incorporated into eqn (4.63) to give

$$\gamma = \tfrac{1}{2}gR(\varrho_L - \varrho_V)\sec\theta\{h + R\sec\theta + (2R/3)\tan^3\theta - (2R/3)\sec^3\theta\}. \tag{4.68}$$

Putting $\theta = 0$ in eqn (4.68) gives a more exact result for the case of perfect wetting covered by eqn (4.66), which is

$$\gamma = \tfrac{1}{2}gR(\varrho_L - \varrho_V)(h + R/3), \tag{4.69}$$

and this can be rearranged to yield a simple relation between the height of the liquid column and the radius of the tube,

$$h = \mathrm{a}^2/R - R/3, \tag{4.70}$$

where a is the capillary constant defined in eqn (4.60) and the relation is valid only for $R \ll \mathrm{a}$.

The rise of a liquid in a capillary tube can yield reasonably good measurements of the surface tension of a liquid which has a zero contact angle, but a practical difficulty is that of adequately cleaning the fine bore before making measurements.

4.14 Other simple cases of capillarity

Results can easily be deduced for the capillary rise in some other systems with simple geometries. Two examples follow:

4.14.1 Two closely-spaced parallel, vertical plates

Consider again the arrangement depicted in fig. 4.16, but this time we assume that the verticals represent a section through two parallel plates which are separated by a distance d and stand in a tank of liquid. If the liquid wets the plates and has a contact angle, $\theta = 0$, a simple calculation similar to that of the previous section gives

$$2\gamma/d = (\varrho_L - \varrho_V)gh, \tag{4.71}$$

where h is the height to which the liquid rises between the plates.

4.14.2 Vertical plates with a small included angle

Now consider two plates which stand vertically in a liquid, but are non-parallel, so that they make a small angle ϕ, measured in a horizontal plane

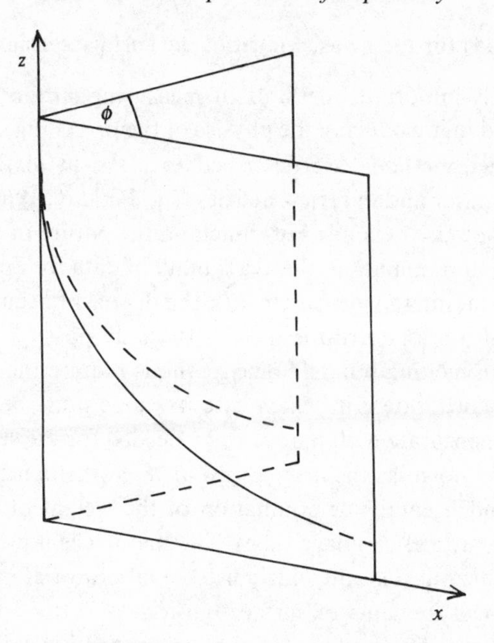

Fig. 4.17 The rise of a liquid between a pair of vertical plates inclined at a small angle, ϕ.

through the plates. The plates meet along a vertical line, which is taken as the z-axis. The x-axis is then taken to be a horizontal line lying in one of the plates, as shown in fig. 4.17, level with the surface of the liquid in the tank. The distance d between the plates at any x-coordinate is given approximately by

$$d = x\phi. \tag{4.72}$$

Combining this expression with eqn (4.71) gives

$$g(\varrho_L - \varrho_V)z = 2\gamma/x\phi, \tag{4.73}$$

where z is the height of the liquid surface between the plates corresponding to the horizontal x-coordinate. Rearranging eqn (4.73) gives

$$xz = 2\gamma/(\varrho_L - \varrho_V)g\phi = \mathrm{a}^2/\phi, \tag{4.74}$$

where a is the capillary constant defined in eqn (4.60). The right-hand side of eqn (4.74) is clearly constant for a particular liquid, so that the surface of the liquid between the plates traces a hyperbolic curve. Equation (4.74) is only an approximate result, valid for very small ϕ.

4.15 Methods for the measurement of the surface tension of liquids

Two historically-important methods of measuring surface tension have been mentioned in introducing the physics of the preceding sections, § 4.12 and 4.13. Other methods are described in classical textbooks on the properties of matter and in review articles (e.g. Padday, 1969). The choice of method in practice depends very much on the nature of the liquid, the possibilities of contamination, the availability of data for applying corrections, ease of making the measurements, the degree of accuracy required, etc. The use of a sessile drop is usually the only possible approach for solids with high melting points. Some methods require the density of the liquid to be incorporated, in which case its value must be known to an accuracy commensurate with the accuracy needed for the surface tension. One method is known as the *drop weight method* (the name is almost self explanatory and it entails determination of the weight of drops as they detach from a vertical capillary tube). The method is now considered to give poor results, but it is still widely used in laboratories making routine comparative measurements of surface tension. A better commonly-used method of investigating surface tension, originated by Wilhelmy around the middle of the nineteenth century, will be described in the following section. It is especially appropriate for the determination of contact angles and is also very easy and rapid in operation. A more recent method (Padday, 1979), which involves measurement of the maximum force exerted on a cone suspended in the free surface of a liquid, is capable of great accuracy. Another recent approach which utilizes the scattering of light from a laser by surface ripples (or capillary waves) is described in § 5.11. This is a very sensitive technique for determining small values of surface tension. This section concludes with a list of values of surface tension for common substances, Table 4.1.

4.16 The Wilhelmy plate

The measurement of contact angle is best made using a geometry where the liquid rises up an easily-accessed, external surface. The immersed object should have a simple shape, and so it is normally either a cylinder or a plate suspended vertically. Wilhelmy originally suggested that surface tension should be measured by finding the maximum force needed to pull a small plate vertically from a liquid surface, and variations of the method are now also called after him. Nowadays it is not usual practice to remove the plate from contact with the liquid, but merely to measure the downward force on the plate when its bottom surface lies in the same plane as the free surface of the liquid, as depicted in fig. 4.18. This avoids the need

Table 4.1. *Measured surface tension, γ of various liquids for given temperatures in the presence of their vapours or gas plus vapour.*

Substance	(Vapour)	Temperature, K	γ, $10^{-3}\,\mathrm{N\,m^{-1}}$
He	He	4	0.12
Ar	Ar	85	13.12
CO_2	CO_2	248	9.13
N_2	N_2	90	5.99
Methyl alcohol, CH_4O	Air	293	22.50
Carbon tetrachloride, CCl_4	CCl_4	293	27.0
Benzene, C_6H_6	Air	293	28.88
Glycerol, $C_3H_8O_3$	Air	293	63.4
Water	Water	293	72.0
Sodium, Na	Argon	373	209.9
Lead, Pb	Vacuum	633	446
Mercury, Hg	Hg	298	485.5
Tin, Sn	Air	553	523

These values mostly come from Kaye, G.W.C. & Laby, T.H. (1986). *Tables of Physical and Chemical Constants*, 15th edn. London: Longmans.

for buoyancy corrections. The method is suitable where high precision is required as, for example, in the measurement of the temperature dependence of contact angle.

At a small distance from one end of the plate the profile of the meniscus becomes constant and does not change until the other end is approached, i.e. the meniscus is cylindrical in form. If z is the height of a point on the meniscus surface above the free liquid surface, and ψ is the angle between the tangent at the point and the horizontal, it is not difficult to show, following the methods of §4.12, that these variables are related by an equation identical to eqn (4.58). This can be written

$$z^2 = \mathrm{a}^2(1 - \cos\psi), \tag{4.75}$$

where from eqn (4.60)

$$\mathrm{a}^2 = 2\gamma/g(\varrho_\mathrm{L} - \varrho_\mathrm{V}), \tag{4.76}$$

and a is the capillary constant. Figure 4.19 shows the profile of the meniscus, which is completely defined by the capillary constant. The angle

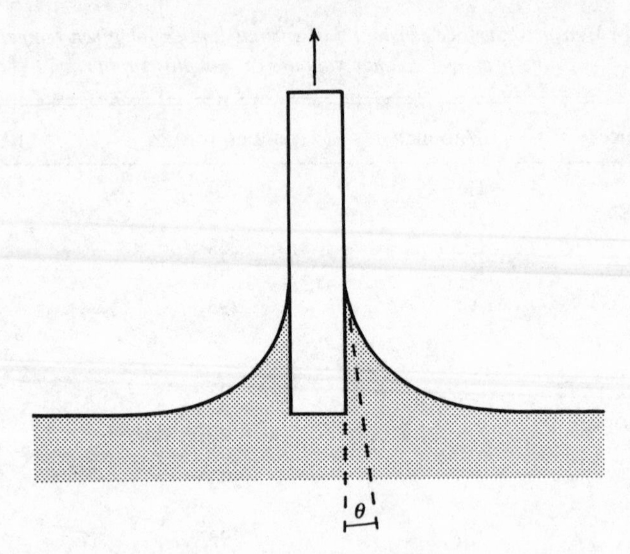

Fig. 4.18 The Wilhelmy plate method of determining surface tension.

of contact θ with the vertical plate determines the point at which the meniscus touches the plate, but it does not affect the meniscus shape. The height h of capillary rise at the plate is determined from eqn (4.75) by setting ψ equal to $\frac{1}{2}\pi - \theta$, giving

$$h^2 = a^2(1 - \sin\theta). \tag{4.77}$$

When the plate is suspended from the arm of a balance, the upwards

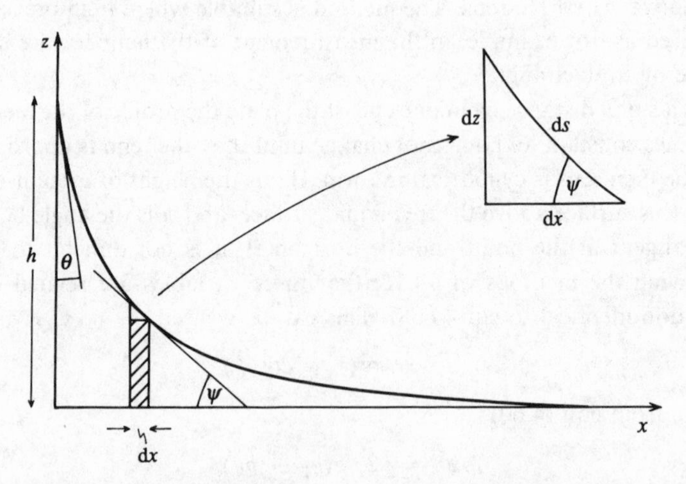

Fig. 4.19 Cross-section of cylindrical meniscus meeting a vertical Wilhelmy plate, showing capillary rise h, contact angle θ, tangential angle ψ, and an enlargement of the elementary triangle.

force on the plate exactly equals the weight of liquid held above the level of the free liquid surface, as we now show. Consider a unit length of cylindrical meniscus, whose cross-section is represented by fig. 4.19, where the insert shows an enlarged view of the small surface element of length $\mathrm{d}s$, at the top of a liquid column of width $\mathrm{d}x$ and height z. The volume of liquid in the column, $\mathrm{d}V$ is approximately given by

$$\mathrm{d}V = z\,\mathrm{d}x. \tag{4.78}$$

It follows from the properties of the triangle in fig. 4.19 that

$$\mathrm{d}z/\mathrm{d}x = \tan\psi, \tag{4.79}$$

and differentiation of eqn (4.75) gives

$$2z\,\mathrm{d}z = \mathrm{a}^2 \sin\psi\,\mathrm{d}\psi. \tag{4.80}$$

These results enable us to put eqn (4.78) in the form

$$\mathrm{d}V = \tfrac{1}{2}\mathrm{a}^2 \cos\psi\,\mathrm{d}\psi, \tag{4.81}$$

and integration over ψ between 0 and $\frac{1}{2}\pi - \theta$ gives the volume V per unit length of meniscus on one side of the plate as

$$V = \tfrac{1}{2}\mathrm{a}^2 \cos\theta. \tag{4.82}$$

This can be rearranged with the use of eqn (4.76) in the form

$$V(\varrho_\mathrm{L} - \varrho_\mathrm{V})g = \gamma\cos\theta, \tag{4.83}$$

showing that the resolved force resulting from surface tension acting vertically on the plate is indeed equal to the weight of the meniscus.

The above derivation ignores effects on the height h of the meniscus near the ends of the plate, but this is not important because when using the Wilhelmy method it is usual to employ a plate some 20 mm in width and to make shape measurements only over the central one-third of the plate. A more sophisticated version of the method has been developed where end corrections have been measured and incorporated into a calibration procedure so that the plate can be suspended from a direct-reading high-precision micro-balance. For a liquid which does not completely wet the plate, the Wilhelmy method is capable of accurate measurement of the angle of contact, assuming that the surface tension can be found by another means. The relationship between the contact angle and the capillary rise, h is obtained from eqns (4.76) and (4.77) as

$$\sin\theta = 1 - \{gh^2/2\gamma\}(\varrho_\mathrm{L} - \varrho_\mathrm{V}). \tag{4.84}$$

4.17 The energy for the rise of a capillary

The liquid supported in a capillary above the free liquid surface has potential energy. This energy is acquired from energy released as the liquid advances over the solid surface, replaces vapour and wets the solid. Consider as an example, the rise of liquid to a height h in a capillary tube of radius R. Since the centre of gravity of the column of liquid is at height $\frac{1}{2}h$ above the free surface, the potential energy of the liquid in the tube is

$$U = \tfrac{1}{2}h \times \pi R^2 \varrho g h = \tfrac{1}{2}\pi \varrho g R^2 h^2, \tag{4.85}$$

where

$$\varrho = \varrho_L - \varrho_V. \tag{4.86}$$

Let us now check to see whether this much energy is available from the process of wetting. On wetting the solid, unit area of interface with energy γ_{SV} is replaced by unit area of interface with energy γ_{SL}, which for perfect wetting (i.e. spontaneous spreading with $\cos\theta = 1$) gives, from eqns (4.34) and (4.35), the energy of wetting E_W as

$$E_W = \gamma_{SV} - \gamma_{SL} = \gamma_{LV}. \tag{4.87}$$

Thus for a column of height h, the total energy E released by wetting the bore, radius R, is

$$E = \gamma_{LV} 2\pi R h. \tag{4.88}$$

The condition for energy conservation, obtained by setting U from eqn (4.85) equal to E from eqn (4.88), gives a height

$$h = 4\gamma_{LV}/\varrho g R. \tag{4.89}$$

Now the height of the liquid column in static equilibrium, which we here denote as h_0, was obtained previously in eqn (4.66) as

$$h_0 = 2\gamma_{LV}/\varrho g R. \tag{4.90}$$

It is therefore seen that energy conservation gives

$$h = 2h_0, \tag{4.91}$$

suggesting that the liquid column should rise to the height $2h_0$ rather than the height h_0 derived previously.

The apparent discrepancy between the two results arises because we have neglected to take account of the viscosity of the liquid, a subject dealt with in the following chapter. Although viscosity can be ignored in static conditions, it is important when a liquid flows, as in the present case of establishing a capillary. The interpretation of the above results is that, but for viscous effects, the liquid in the tube would initially rise to a height $2h_0$, then fall to zero again, and proceed to oscillate about a mean position h_0. Simple observations of the behaviour of liquids show that flow is heavily damped (a stirred cup of tea, for example), so one should not expect to see large oscillations in a capillary tube. Nonetheless, the 'missing' energy of wetting has been used in overcoming viscosity and it is manifested as a slight rise in the temperature of the liquid. For a liquid that does not spread spontaneously, one must use a more general result for the energy of wetting, instead of eqn (4.87).

4.18 Films on liquid surfaces

4.18.1 Spreading conditions

Consider what happens when the surface of a liquid supports a drop of some less dense liquid. The second liquid may retain its drop-like character, disturbing the initially plane surface of the first liquid in the manner shown in fig. 4.20(a). The new configuration is determined by the equilibrium of the forces of surface tension at the circle of contact of the vapour and the two liquids. Figure 4.20(a) shows the cross-section in a plane normal to the circle of contact, and fig. 4.20(b) shows the triangle of balanced interfacial

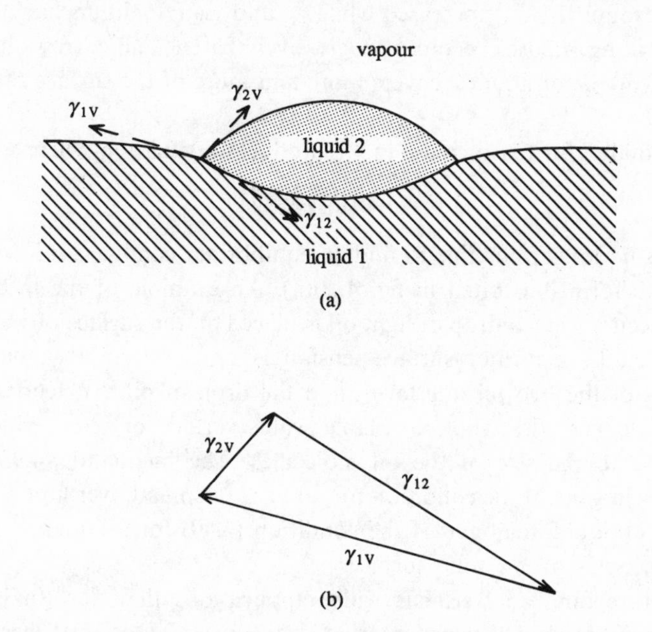

Fig. 4.20 The equilibrium of a drop of liquid, floating on the surface of another liquid (the liquids are immiscible): (a) meridional section through the floating drop, showing interfacial tensions acting at the line of triple contact; (b) equilibrium triangle of forces.

tension forces required for the stability of the floating drop. The system illustrated here should be compared with the liquid–solid system of fig. 4.8(a), where the equilibrium of the liquid drop arises from the upper subsurface layer of reduced pressure and the lower subsurface layer of overpressure shown in fig. 4.8(b). The interfacial tension forces shown in fig. 4.20(b) are associated with analogous pressure changes from the bulk values that occur also in the liquid–liquid system.

The various angles of the floating drop can be obtained from the familiar geometrical properties of the triangle in fig. 4.20(b), if the three surface tensions are known. The resulting conditions generalize Young's relation (eqn (4.33)) for the drop on a solid surface. However, our main interest here is the question of whether the drop can exist at all, i.e. its immiscibility. If

$$\gamma_{1V} < \gamma_{12} + \gamma_{2V}, \tag{4.92}$$

where the suffixes refer to the surface tensions between liquid 1 and vapour, liquid 1 and liquid 2, etc., then it is possible to draw the equilibrium triangle, which means that the configuration of drop shown in fig. 4.20(a) can in principle exist. A culinary example where eqn (4.92) holds is provided by the globule of fat often found floating on the surface of chicken soup! If γ_{1V} is increased while γ_{12} and γ_{2V} remain unchanged, the triangle in fig. 4.20(b) becomes progressively flatter, and correspondingly a fixed volume of liquid 2 covers more and more of the surface of liquid 1.

Eventually, further increase in γ_{1V} produces a situation where

$$\gamma_{1V} \geqslant \gamma_{12} + \gamma_{2V} \tag{4.93}$$

and it is no longer possible to find an equilibrium configuration with the drop-like form illustrated in fig. 4.20(a). An example of the inequality (4.93) occurs when a drop of light oil is placed on the surface of water. In this case, the air/water surface tension is greater than the combined tensions of the two oil interfaces, and the drop of oil is extended until either it covers the whole available water surface or it is reduced in thickness to the size of the oil molecules. The inequality (4.93) thus provides in general the condition for liquid 2 to spread over liquid 1. It is the liquid/liquid analogue of the condition (4.40) for a liquid to spread spontaneously over a solid surface.

The foregoing analysis has wider applications than just to liquids. Because of the equivalence of interfacial tension and interfacial energy, the equations resulting from the treatment above can, for example, be extended to problems of *heterogeneous nucleation* in solids (i.e. where second phases nucleate on crystal defects in preference to forming a uniform distribution throughout the volume). The shape of embryo precipitates of second phase nucleating on grain boundaries of solids is determined by the need to minimize total interface free energy. Thus the angles between the surfaces of the phases at the common line of contact reflect the balancing of the local interfacial forces. Embyro precipitates on internal surfaces in solids therefore normally have shapes resembling that shown in fig. 4.20(a). Where other internal surfaces that have characteristic energies (e.g. twin boundaries and grain boundaries) meet in thermally-equilibrated solids, it is also possible to use their geometries at the line of contact to assess relative interfacial energies.

4.18.2 Molecular monolayers

Quite surprisingly large areas of water can be covered by a small amount of oil that has spread out to form a film of single-molecule thickness, or *molecular monolayer*. We find that if the molecular diameter is about 10^{-9} m, a 5 ml medicine spoonful of oil forms a monolayer of area $\sim 5000\,\mathrm{m}^2$ (about $1\frac{1}{4}$ acres!). Conversely, the observation of the area covered by a known amount of the second liquid enables the molecular size to be estimated from measurements of macroscopic properties.

Many large molecules have an elongated structure with a *hydrophobic* (water insoluble) end and a *hydrophilic* (water soluble) end. These molecules take up preferred orientations with their hydrophilic ends in the water surface and their hydrophobic ends pointing upwards. The floating, oriented monolayer is known as a *Langmuir film*. In many cases, the monolayer can be removed from the water surface by slowly pulling a piece of a suitable substrate material vertically out of the liquid. The monolayer is deposited on to the substrate, and successive monolayers can be deposited onto the same substrate by repeated withdrawal through the film-covered liquid surface. Such multiple monolayer structures, known as *Langmuir–Blodgett* films are made in a *Langmuir trough* and they have interesting optical, magnetic, and electronic properties. We consider here, however, only the effects of films in modifying the properties of the liquid surface.

The main effect of a film on a liquid surface arises from the repulsive forces between the oriented molecules, which resist any compression of the monolayer in the plane of the surface. The repulsion corresponds to a surface pressure that acts to reduce the surface tension γ_{free} of the free surface to some smaller γ characteristic of the film-covered surface. If ϱ_s is the surface molecular density (i.e. the number of molecules per square metre of surface), the magnitude of these effects is described by the *elasticity* of the film

$$\mathscr{E} = -\varrho_s \partial\gamma/\partial\varrho_s. \tag{4.94}$$

Thus the surface tension at the equilibrium density ϱ_{s_0} is

$$\gamma = \gamma_{\text{free}} - \int_0^{\varrho_{s0}} (\mathscr{E}/\varrho_s)\,d\varrho_s. \tag{4.95}$$

The action of soaps and detergents relies on their ability to reduce the surface tension of water from its free-surface value, so that the soapy water can more effectively wet solid surfaces in accordance with eqn (4.33) or (4.40).

4.19 Suggestions for further reading

Origin of surface tension; solid–liquid interfaces

Brown, R.C. (1947). The fundamental concepts concerning surface tension and capillarity. *Proceedings of the Physical Society, London*, **59**, 429–48.

Berry, M.V. (1971). The molecular mechanism of surface tension. *Physics Education*, **6**, 79–84.

Woodruff, D.P. (1973). *The solid–liquid interface*. Cambridge: Cambridge University Press.

Jaycock, M.J. & Parfitt, G.D. (1981). *Chemistry of interfaces*. Chichester: Ellis Horwood.

Wetting and spreading

Padday, J.F. (1978). In *Wetting, spreading and adhesion* (ed. J.F. Padday). London: Academic Press.

de Gennes, P.-G. (1985). Wetting: statics and dynamics. *Reviews of Modern Physics*, **57**, 827–63.

Cazabat, A.-M. (1987). How does a droplet spread? *Contemporary Physics*, **28**, 347–64.

Thermodynamics of surface tension

Adkins, C.J. (1983). *Equilibrium thermodynamics*, 3rd edn. London: McGraw–Hill.

Guggenheim, E.A. (1967). *Thermodynamics*. 5th edn. Amsterdam: North Holland Publishing Co., p. 163.

Measurement of surface tension

Padday, J.F. (1969). Surface Tension, Parts 1 & 2. In *Surface and Colloid Sciences* (ed. E. Matejevic & F. Eirich), Vol. 1. New York: Wiley–Interscience, p. 40 and p. 101.

Padday, J.F. (1979). Menisci formed at a cone by a free liquid surface. *Journal of the Chemical Society (Faraday Transactions)*, **75**, 2827–38.
Jaycock, M.J. & Parfitt, G.D. (1981). *Chemistry of interfaces*. Chichester: Ellis Horwood.

Problems

1. When a saturated vapour (a gas at $T < T_c$) is condensed to yield a liquid, the volume of resultant liquid is typically 10^{-3} times that of the vapour at the same temperature and pressure. What is the ratio of the mean separation between molecules in the vapour and the mean separation between molecules in the liquid? Estimate the volume of 1 mole of the vapour at standard temperature and pressure, taking a reasonable value for the average separation of molecules in a liquid.

2. Two soap bubbles of radii a and b, and surface tension γ are brought into contact and then coalesce to form a single bubble of radius R. If the external pressure is p, show that the surface tension is given by the expression

$$\gamma = \frac{p(R^3 - a^3 - b^3)}{4(a^2 + b^2 - R^2)}.$$

3. The capillaries of the xylem (the tiny tubes carrying fluid) in the trunk of a tree have diameters of approximately 0.02 mm. Assuming that sap has a similar value of surface tension to that of water, investigate whether capillarity can explain the transport of sap to the branches of a mature tree of average height. What other explanation can you propose?

4. A liquid drop will just break away from the lower flat end of a capillary tube when the sides of the drop are vertical (this is an experimental observation). Assume that the mass of the drop at separation is solely a function of the surface tension γ, the density of the liquid ϱ, the radius of the capillary r, and the acceleration due to gravity g. Show with the help of dimensional analysis that, in fact, the mass of the separating drop is only a function of γ, r and g.

5. The angle of contact between mercury and glass is 140°. What is the minimum internal diameter for the glass tubing to be used in the construction of a barometric column if the height correction for capillarity effects is to be less than 0.1 mm of mercury?

6. The surface tension of a particular liquid varies linearly with temperature and the temperature coefficient is $-2 \times 10^{-3}\,\mathrm{N\,m^{-1}\,^\circ C^{-1}}$. The liquid is used to fill a U-tube in which the narrow vertical arms have different diameters. Find the value of the surface tension of the liquid at 25 °C if the difference in the levels between the liquid in the two arms at 50 °C is one fifth of the difference in the levels at 25 °C. (Assume that the density of the liquid is constant over this temperature range.)

7. The radius of a soap bubble, at the end of a narrow capillary tube connected to a frictionless syringe, is initially 20 mm. Calculate the mechanical work done in using the syringe to increase the bubble radius to 40 mm, assuming that the temperature of the soap film remains constant and neglecting the viscosity of air. (Take the surface tension of the soap solution to be $2.5 \times 10^{-2}\,\mathrm{N\,m^{-1}}$.)

8. When mercury is poured onto a clean flat glass surface, the glass is partially wetted, the contact angle being 140°. It is observed that the mercury spreads out into a pool of uniform thickness, independent of the diameter of the pool. Find the thickness of the mercury pool.

9. Surface tension enables a steel sewing needle to float on water when the needle is first held horizontally and then lowered carefully until in contact with the water. Assuming that the angle of contact is 180° and treating the needle as a cylinder, show that a needle of radius 0.2 mm floats with approximately half its volume above the free surface. You should assume that the slope of the water surface everywhere in the vicinity of the needle is sufficiently small that the approximation (A1.8) for the inverse radius of curvature is valid. (The surface tension of water is given in Table 4.1 and the density of steel is $8.6 \times 10^3\,\mathrm{kg\,m^{-3}}$.)

10. Show that the profile of the meniscus at a vertical plate is indeed given by eqn (4.75), i.e.

$$z^2 = \mathrm{a}^2(1 - \cos\psi),$$

where a^2 is equal to $2\gamma/g(\varrho_L - \varrho_V)$ and the coordinates and angle are defined in fig. 4.19.

11. A thin, single crystal wire is maintained at a temperature slightly below its melting point and made to support a small weight. The mechanical strength of the wire is negligible at the given temperature and the weight is almost entirely supported by the solid–vapour tension of the wire's surface. (a) Derive the condition for the equilibrium of the system; (b) consider why it is that, if the wire was initially not a single crystal, but instead was polycrystalline, the grain boundaries become transverse to the length of the weighted wire, when it is close to its melting point ('bamboo structure'); (c) derive an expression for the equilibrium of the system in situation (b).
(N.B. This method is used to obtain approximate values for the surface energies of metals close to their melting points, using the fact that the interface tension of a grain boundary is equal to about $\frac{1}{3}\gamma_{SV}$ at elevated temperatures.)

12. The interfacial tension between benzene and clean water is initially $3.5 \times 10^{-2}\,\mathrm{N\,m^{-1}}$ (they are slightly miscible and it therefore increases with time). Use the information given in Table 4.1 to find whether a drop of benzene placed on water immediately spreads, or remains as a drop. Say what you think will happen after some time.

5

Dynamic properties of liquids

Liquids in motion can display very complicated dynamical behaviour, as in the familiar examples of water at high pressure emerging from a hosepipe or sea waves breaking on a beach. Such everyday phenomena often involve high liquid velocities and/or large amplitude surface waves, where *nonlinear* terms in the equations of motion become important. These nonlinear terms introduce a dependence of the liquid acceleration on the *square* of its velocity, and the equations of motion become very difficult to solve except in some simple special cases. The studies of the nonlinear and turbulent motions of fluids (i.e. gases and liquids) are active research areas and some aspects of fluid behaviour are not completely understood.

Many interesting and important examples of fluid motion involve only small velocities or small disturbances from a static equilibrium state of the fluid, and such motions form the main topic of the present chapter. The simplest model is that of an *ideal* incompressible liquid, where it is assumed that the effects of the internal (i.e. intermolecular) forces are so small that they can be neglected. These forces are of course important in the formation of the liquid state itself, but their influence on liquid flow is small in many situations.

This most basic model of a liquid can be developed in various ways to provide a better description of the motion of real liquids. Thus, it is not difficult to allow for the effects of liquid compressibility, which are important for understanding the propagation of sound waves. Also, real liquids, as distinct from ideal ones, exhibit a resistance to flow – a viscosity – which is clearly a manifestation of the interatomic and intermolecular forces, and which must be allowed for in many flow processes. A further manifestation of the internal forces is surface tension, discussed in Chapter 4, which controls the behaviour of short-wavelength ripples on a liquid surface. These various generalizations to embrace non-ideal liquids are treated in the present chapter, together with some discussion of the breakdown of any kind of simple model description as the flow conditions approach the onset of turbulence (see § 5.16). Although the main concern of this chapter is the motion of liquids, many of the formal results apply equally well to gases, and some attention is given to specifically gaseous properties.

5.1 Steady flow of ideal liquids: Bernoulli's equation

The flow of a fluid is said to be *steady* if all three components of the velocity vector $\mathbf{v}$ of a fluid element are independent of the time, that is

$$\partial \mathbf{v}/\partial t = 0, \tag{5.1}$$

at all points in the fluid. The points in the fluid are fixed with respect to a coordinate system independent of the fluid. The coordinate system is usually at rest, but the conditions for steady flow are also satisfied if eqn (5.1) holds at points fixed with respect to a coordinate system moving in a straight line with constant velocity. For example, the flow of air around an aircraft wing may be steady with respect to a set of coordinate axes attached to the wing. Any accelerated motion of the coordinate axes is however excluded. A familiar example of steady flow with respect to stationary coordinates occurs in the gliding of water over a weir with a smooth glass-like surface.

Any flow of a fluid at a particular instant can be represented by a field of arrows, each of which indicates the instantaneous local velocity at a particular point, as shown in fig. 5.1(a). With sufficient arrows, the velocities of all the elements of fluid can be represented. The curves formed by sequences of infinitesimally small arrows, head to tail, are called *streamlines*. The streamlines thus show the simultaneous instantaneous motions of the different fluid elements; they differ in general from the paths traced out by the motions of individual fluid elements over the course of time. However, in the special case of steady flow, the streamlines do coincide with the paths of motion. Thus in steady *streamline flow*, each element of a fluid moves on a continuous curve whose tangent points in the direction of flow. Streamlines do not intersect by definition, so that fluid elements cannot move between them, and there can be no flow transverse to the streamlines.

Consider now an imaginary streamtube in an ideal liquid, consisting of a tube bounded by streamlines as in fig. 5.1(b). This behaves like a frictionless pipe, with cross section A_1 at one end and A_2 at the other, these

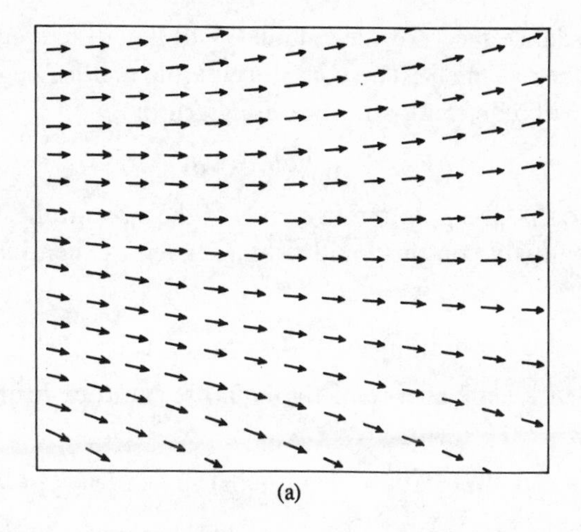

(a)

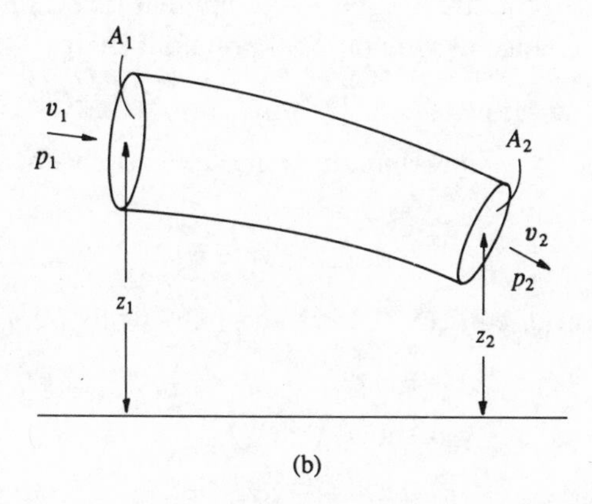

(b)

Fig. 5.1 (a) Streamline flow of a liquid, each arrow indicating the direction of flow at the coordinate point where the arrow head appears; (b) a streamtube in an ideal incompressible fluid.

areas being measured perpendicular to the streamline directions. We take the fluid velocities into and out of the streamtube to be respectively v_1 and v_2. The flow behaviour of an ideal liquid is simply determined by the conditions for conservation of mass and energy. With a fixed value ϱ for the density of the incompressible liquid, the condition for equal rates of mass flow into and out of the streamtube is

$$\varrho A_1 v_1 = \varrho A_2 v_2. \tag{5.2}$$

This is equivalent to the equation of continuity for the flow of an incompressible liquid. It shows that the streamlines are close together where the liquid flows quickly and far apart where it flows slowly.

In order to obtain the condition for energy conservation, we consider the energy balance sheet for the volume δV of liquid that enters or leaves the streamtube in time δt. The volumes entering and leaving are equal in accordance with eqn (5.2), and they are given by

$$\delta V = A_1 v_1 \delta t = A_2 v_2 \delta t. \quad (5.3)$$

With pressures p_1 and p_2 at the two ends of the streamtube, the net work done by the pressure on the liquid volume, given by the usual formula of force × distance moved is

$$\delta W = p_1 A_1 v_1 \delta t - p_2 A_2 v_2 \delta t = (p_1 - p_2)\delta V. \quad (5.4)$$

The change in kinetic energy of the liquid in the tube in the same time interval δt is

$$\delta K = \tfrac{1}{2}(\varrho A_2 v_2 \delta t) v_2^2 - \tfrac{1}{2}(\varrho A_1 v_1 \delta t) v_1^2 = \tfrac{1}{2}\varrho(v_2^2 - v_1^2)\delta V. \quad (5.5)$$

If the two ends of the streamtube are at different vertical heights z_1 and z_2 there is also a change in gravitational potential energy

$$\delta P = (\varrho A_2 v_2 \delta t) g z_2 - (\varrho A_1 v_1 \delta t) g z_1 = \varrho g(z_2 - z_1)\delta V. \quad (5.6)$$

The work done by the pressure difference must equal the change in total energy,

$$\delta W = \delta K + \delta P \quad (5.7)$$

and substitution of eqns (5.4), (5.5) and (5.6) yields *Bernoulli's equation*

$$p_1 + \tfrac{1}{2}\varrho v_1^2 + \varrho g z_1 = p_2 + \tfrac{1}{2}\varrho v_2^2 + \varrho g z_2. \quad (5.8)$$

Alternatively, the analysis shows that

$$p + \tfrac{1}{2}\varrho v^2 + \varrho g z = B_s, \quad (5.9)$$

where B_s is a constant, for all points on the *same* streamline s.

The conserved quantity B_s includes the usual kinetic and potential energy contributions, together with an additional term for the 'pressure energy'. Looked at another way, Bernoulli's equation is the familiar equation of hydrostatics modified to include the kinetic energy of the moving liquid. The most immediate consequence of the equation, assuming a constant gravitational potential, is the increase in pressure p with decrease in velocity v and vice versa. This is exemplified by the wind force caused by the increase in pressure as air impinges on and is brought to rest by an object.

5.2 Applications of Bernoulli's equation

A simple application of Bernoulli's equation is the *Venturi meter*, a device for measuring the velocity of flow through a pipe. As is illustrated in fig.

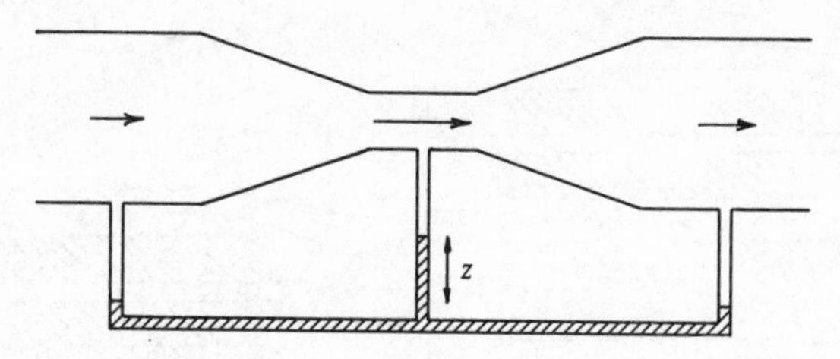

Fig. 5.2 Venturi meter for measuring fluid flow.

5.2, a section of the pipe is replaced by a tube of narrower diameter. The increase in velocity in the constricted section leads to a reduction in pressure. A reading of the difference in pressure between the two manometers shown in the figure and knowledge of the two cross-sectional areas of pipe lead, with the use of Bernoulli's equation, to a determination of the flow velocity.

On a larger scale, the effect utilized by Venturi can cause problems on both land and sea. Tall buildings placed close together cause the acceleration of winds blowing between them, so that windows and doors tend to be sucked outwards. In one practical example, a group of large water-cooling towers with smooth aerodynamic lines built in close proximity to one another were unable to withstand the lateral forces created by the reduced pressures between the towers in high winds. The Venturi effect also causes two ships passing close together on parallel courses to be pulled towards one another by the suction effect from the water rushing through the intervening channel at a higher velocity than that of the water on the distant sides of the ships.

The interaction of a steadily-flowing fluid with a smooth obstacle can be understood in terms of the configuration of streamlines illustrated in fig. 5.3. The streamlines divide around the obstacle but the one that meets the object at right angles terminates at its surface. This central streamline defines the *stagnation point* at the nose of the obstacle. Here the fluid comes to rest, and the pressure p exceeds that in the undeflected stream p_0

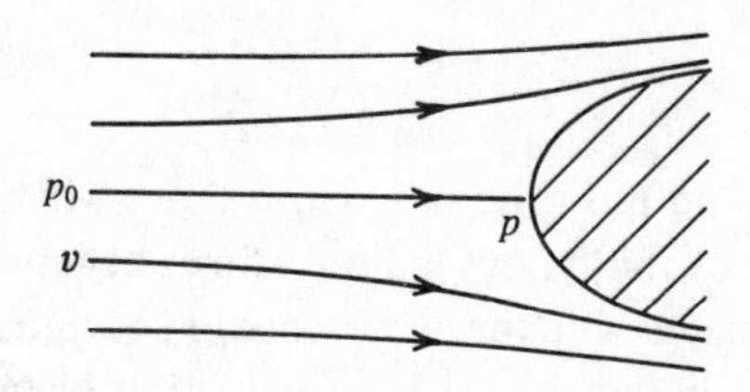

Fig. 5.3 Streamlines around an obstacle showing the stagnation point.

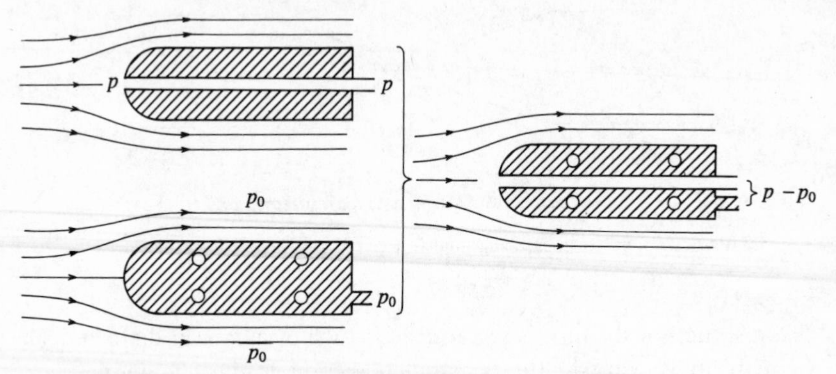

Fig. 5.4 Complete assembly and component parts of a Pitot-static tube for measuring air speed, indicating the physical principles involved.

by an amount

$$p - p_0 = \tfrac{1}{2}\varrho v^2, \tag{5.10}$$

where v is the velocity in the main flow. This pressure difference is known as the *dynamic pressure*.

The expression (5.10) provides the principle of operation of a device for measuring aircraft speed known as the *Pitot tube*, illustrated in fig. 5.4. A narrow tube has its open end at the centre of the nose of a larger tube, whose forward end is closed. The diagram also shows the device broken down into its basic components so as to illustrate their particular functions. The narrow tube is connected to a gauge, which registers the pressure p at the stagnation point. Holes along the sides of the larger tube allow its internal pressure to take up the value p_0, which is also measured. The device thus provides a reading of the dynamic pressure, which can be calibrated in terms of the air speed v.

The relation (5.10) between the pressure p at a point of zero velocity and the pressure p_0 at a point of velocity v on the same streamline can also be applied to an experiment where p is held fixed as the velocity v is increased. It is seen that p_0 must become negative when v exceeds a maximum velocity

$$v_{\max} = (2p/\varrho)^{1/2}. \tag{5.11}$$

Negative pressures do not arise in gaseous flows when proper account is taken of the compressibility, but in liquid flows negative pressures lead to an instability associated with the occurrence of *cavitation*. Cavitation is the vapourization of a liquid caused by a reduction of its internal pressure below the liquid vapour pressure. Effectively the liquid starts to boil,

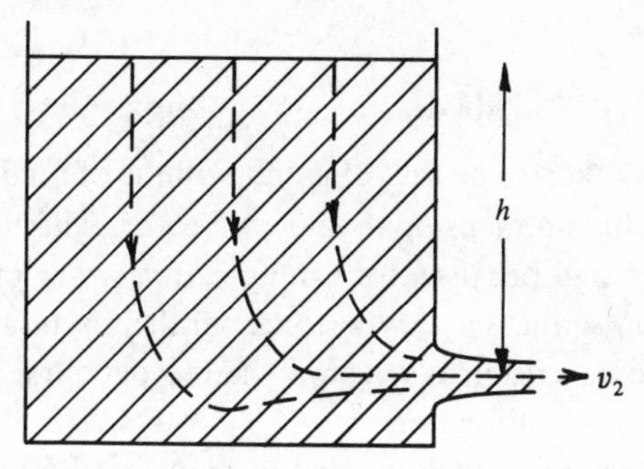

Fig. 5.5 Flow of liquid through an orifice.

forming vapour pockets distributed through the region of reduced pressure. Cavitation can be a serious problem in practical systems since it destroys steady flow, giving rise to turbulence and hence accelerated corrosion of pipes and components. It can be avoided in principle by designing systems so that the liquid flow does not reach the velocity [eqn (5.11)] above which the internal pressure becomes negative.

The applications of Bernoulli's theorem discussed above all refer to horizontal flow, where the gravitational potential plays no role. It does play a role in our final example, of liquid flow through an orifice at the bottom of a reservoir, illustrated in fig. 5.5. The diameter of the orifice is assumed to be small compared with the horizontal dimensions of the tank so that the liquid level falls only slowly and the liquid exits as a steady jet. We apply Bernoulli's theorem, eqn (5.8), with subscript 1 referring to the upper surface and subscript 2 to the orifice. The appropriate parameter values are

$$p_1 = p_2 = p_0, \quad v_1 \simeq 0 \quad \text{and} \quad z_1 - z_2 = h,$$

where p_0 is atmospheric pressure. Thus

$$\varrho g h = \tfrac{1}{2}\varrho v_2^2 \quad \text{or} \quad v_2 = (2gh)^{1/2}. \tag{5.12}$$

This result shows that the liquid emerges with the velocity that it would have acquired in free fall under gravity from a height h, a conclusion first reached by Torricelli by analogy with Galileo's law of falling bodies.

5.3 Dynamical equations of fluid mechanics

For more detailed studies of fluid motion, it is convenient to generalize slightly the equations used in the derivation of Bernoulli's equation in § 5.1, and to write them in differential form.

The Bernoulli equation itself is closely related to Newton's law of motion

$$\varrho\, \mathrm{d}\mathbf{v}/\mathrm{d}t \;=\; -\nabla p + \varrho\mathbf{g}, \tag{5.13}$$

where the mass is taken to be that of a unit volume of fluid (i.e. the density ϱ), and $\mathbf{v}$ is the three-dimensional velocity vector. The first force on the right has a minus sign because, for example, a *negative* pressure gradient in the x-direction produces a *positive* force parallel to the x-axis. The usual gravitational force is written as a vector, where $\mathbf{g}$ of course points vertically downwards, in the $-z$-direction.

To see the connection between Newton's law and Bernoulli's equation, we must allow for a general dependence of the velocity vector on both the time t and the position $\mathbf{r}$, expressed in terms of Cartesian coordinates x, y and z. The total time derivative on the left of eqn (5.13) can be expressed in terms of partial derivatives in the usual way,

$$\frac{\mathrm{d}\mathbf{v}}{\mathrm{d}t} \;=\; \frac{\partial \mathbf{v}}{\partial t} + \frac{\mathrm{d}x}{\mathrm{d}t}\frac{\partial \mathbf{v}}{\partial x} + \frac{\mathrm{d}y}{\mathrm{d}t}\frac{\partial \mathbf{v}}{\partial y} + \frac{\mathrm{d}z}{\mathrm{d}t}\frac{\partial \mathbf{v}}{\partial z} \;=\; \frac{\partial \mathbf{v}}{\partial t} + (\mathbf{v}\cdot\nabla)\mathbf{v} \tag{5.14}$$

since

$$\mathbf{v} \;=\; (\mathrm{d}x/\mathrm{d}t,\ \mathrm{d}y/\mathrm{d}t,\ \mathrm{d}z/\mathrm{d}t). \tag{5.15}$$

The total and partial time derivatives represent distinct physical characteristics of the fluid flow. The *partial* derivative, $\partial\mathbf{v}/\partial t$, represents the rate of change of velocity at a point fixed with respect to the coordinate axes, and it is this quantity that vanishes in steady flow as in § 5.1. Newton's law of motion describes the acceleration of a fluid element, and the *total* derivative, $\mathrm{d}\mathbf{v}/\mathrm{d}t$, represents the rate of change of velocity of the moving element. It need not vanish in steady flow because the velocities at different spatial positions are generally different, and the second term on the far right of eqn (5.14) accordingly makes a non-zero contribution.

Substitution of eqn (5.14) into (5.13) gives

$$\varrho(\partial\mathbf{v}/\partial t) + \varrho(\mathbf{v}\cdot\nabla)\mathbf{v} \;=\; -\nabla p + \varrho\mathbf{g}. \tag{5.16}$$

This is known as *Euler's equation*. It can in principle be solved for the fluid velocity $\mathbf{v}$, although it should be noted that the equation is *nonlinear* in $\mathbf{v}$ on account of the second term on its left-hand side. It is sometimes useful to re-express the nonlinear term with the use of a standard vector operator identity,

$$(\mathbf{v}\cdot\nabla)\mathbf{v} \equiv \tfrac{1}{2}\nabla(v^2) - \mathbf{v}\times(\nabla\times\mathbf{v}), \tag{5.17}$$

but Euler's equation in any form is generally difficult to solve. Many problems of practical interest involve, however, only low fluid velocities, where the second-order nonlinear term may be negligibly small.

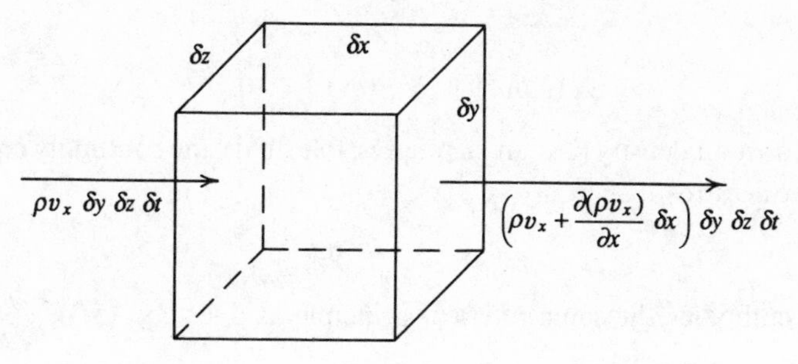

Fig. 5.6 Elementary volume showing mass transfers through opposite faces.

The nonlinear term also causes no trouble in conditions of steady streamline flow where eqn (5.1) is satisfied. We take the components of the remaining terms in eqn (5.16) along a streamline. Let **s** be a unit vector parallel to the streamline and remember that **v** is everywhere parallel to the streamline. Thus the gradient operators in eqn (5.16) become simply $\partial/\partial s$, and the direction cosine of the streamline relative to the gravitational force in the $-z$-direction is $-\partial z/\partial s$. Euler's equation therefore reduces to

$$(\partial/\partial s)\{\tfrac{1}{2}\varrho v^2 + p + \varrho g z\} = 0. \tag{5.18}$$

The quantity in brackets is constant along a given streamline, and we recover Bernoulli's equation (5.9).

The condition (5.2) or (5.3) for continuity of fluid flow can also be expressed in a useful, more general form. Consider the elementary rectangular volume of fluid shown in fig. 5.6. We abandon for the moment the restriction of our discussion to incompressible fluids and allow the density ϱ to vary in space and time. As far as the faces perpendicular to the x-axis are concerned, the net flow of mass *into* the volume in a short time δt is

$$-\frac{\partial(\varrho v_x)}{\partial x}\,\delta V \delta t \quad \text{where} \quad \delta V = \delta x \delta y \delta z.$$

Similar expressions hold for the other pairs of faces, and the total mass transfer into the rectangular volume is

$$-\nabla\cdot(\varrho\mathbf{v})\,\delta V \delta t. \tag{5.19a}$$

The mass transfer causes a change $\delta\varrho$ in fluid density such that the mass increase

$$\delta\varrho\,\delta V \tag{5.19b}$$

of the volume is equal to the quantity in the expression (5.19a). The condition for equality, written in differential form, is the *equation of continuity*:

$$(\partial \varrho / \partial t) + \boldsymbol{\nabla} \cdot (\varrho \mathbf{v}) = 0. \tag{5.20}$$

For a constant density (i.e., an incompressible fluid), the continuity equation thus reduces to

$$\boldsymbol{\nabla} \cdot \mathbf{v} = 0, \tag{5.21}$$

which embodies the same physical principle as does eqn (5.3).

5.4 Irrotational flow

The angular velocity of any rotating body can be represented by a vector $\boldsymbol{\omega}$ whose direction lies parallel to the axis of rotation. For a rotation axis that passes through the origin of coordinates, the linear velocity $\mathbf{v}$ at the point with vector position $\mathbf{r}$ is given by the vector product

$$\mathbf{v} = \boldsymbol{\omega} \times \mathbf{r}. \tag{5.22}$$

It is not difficult to show from the standard definition of the curl of a vector in Cartesian coordinates that

$$\boldsymbol{\nabla} \times \mathbf{v} = 2\boldsymbol{\omega}. \tag{5.23}$$

Motion for which there is zero angular velocity, and hence for which

$$\boldsymbol{\nabla} \times \mathbf{v} = 0 \tag{5.24}$$

is said to be *irrotational*.

Varieties of steady streamline flow of a fluid, for which this condition for irrotational motion is satisfied, correspond to a special case of Bernoulli's theorem. Thus with the nonlinear term in Euler's equation (5.16) expressed in the form (5.17), insertion of the conditions (5.1) and (5.24) enables the equation to be written

$$\boldsymbol{\nabla}\{\tfrac{1}{2}\varrho v^2 + p + \varrho g z\} = 0. \tag{5.25}$$

This is very similar to the form of Euler's equation (5.18) but no reference has been made to differentiation along a particular streamline, and the $\boldsymbol{\nabla}$ operator in eqn (5.25) includes differentiation with respect to all three Cartesian coordinates. Thus we can write

$$\tfrac{1}{2}\varrho v^2 + p + \varrho g z = B, \tag{5.26}$$

which is similar to eqn (5.9), but the same constant value of B applies throughout the fluid. This condition on Bernoulli's theorem can be used as an alternative specification of irrotational flow.

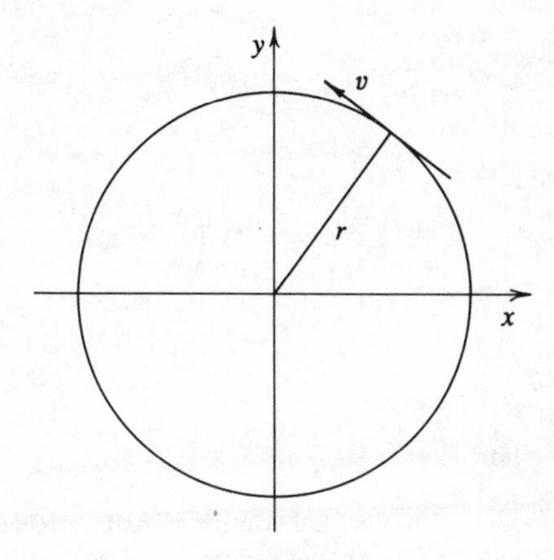

Fig. 5.7 Circular motion of a fluid element in the horizontal plane of a whirlpool.

The physical significance of the designation 'irrotational' can be understood in terms of the *circulation* of a fluid. Consider a closed curve l with length element $\mathbf{dl}$ in the fluid. The integral of the component of fluid velocity parallel to $\mathbf{dl}$ around the curve is defined to be the circulation

$$C = \int_l \mathbf{v} \cdot \mathbf{dl}. \tag{5.27}$$

This quantity can be evaluated for a small element of fluid when the curve l is chosen to encircle the element, and it clearly provides a measure of the rate of rotation of the element caused by the fluid's motion. According to Stokes theorem of vector analysis, the line integral in eqn (5.27) can be re-expressed as a surface integral to give

$$C = \int_S \nabla \times \mathbf{v} \cdot \mathbf{dS}, \tag{5.28}$$

where S is any surface bordered by the curve l, and $\mathbf{dS}$ is an element of area with direction perpendicular to the surface. The circulation obviously vanishes in regions of the fluid where the condition (5.24) for irrotational flow is satisfied.

A familiar example of irrotational motion is provided by the whirlpool. That this demonstrably rotating object should be an example of irrotational motions seems highly paradoxical, but we proceed to show that the motion is indeed irrotational at all points away from the centre of the whirlpool. Consider the motion of a fluid element in a circle centred on the whirlpool in the horizontal xy-plane, shown in fig. 5.7. The fluid velocity v is assumed to be constant around the circle but it varies with the radius r. At a point on the circle with coordinates (x, y) the velocity vector is

$$\mathbf{v} = (-vy/r, vx/r). \tag{5.29}$$

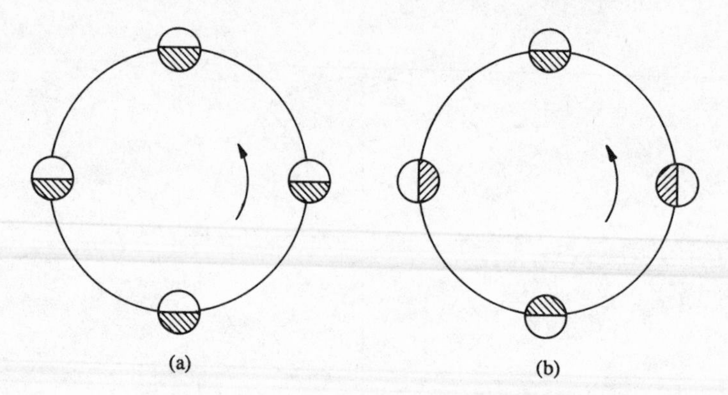

Fig. 5.8 Examples of the motion of a fluid element around a circle with the condition for irrotational motion (a) satisfied and (b) not satisfied.

The z-component of the irrotational condition (5.24) gives

$$(\nabla \times \mathbf{v})_z = (\partial v_y/\partial x) - (\partial v_x/\partial y) = 0. \tag{5.30}$$

Thus inserting the velocity components from eqn (5.29),

$$\begin{aligned} \frac{\partial}{\partial x}\left(\frac{vx}{r}\right) + \frac{\partial}{\partial y}\left(\frac{vy}{r}\right) &= \left(x\frac{\partial}{\partial x} + y\frac{\partial}{\partial y}\right)\frac{v}{r} + \left(\frac{2v}{r}\right) \\ &= r\frac{\partial}{\partial r}\left(\frac{v}{r}\right) + \frac{2v}{r} \\ &= \frac{\partial v}{\partial r} + \frac{v}{r} \\ &= \frac{1}{r}\frac{\partial}{\partial r}(rv) = 0. \end{aligned} \tag{5.31}$$

The solution is therefore

$$rv = \text{constant}, \tag{5.32}$$

showing that fluid elements at smaller radii move faster than those at larger radii. This is the characteristic feature of whirlpool motion.

Figure 5.8(a) attempts to resolve the paradox of irrotational motion in a rotating fluid. A fluid element, with partial shading to indicate its orientation, is shown at four positions in its motion around the circle. Despite the circular path of the element as a whole, the initial orientation is maintained and the element itself suffers no rotation. The circulation in eqn (5.27) vanishes for a path around the element, and the motion is irrotational in accordance with eqn (5.24). Figure 5.8(b) shows the contrasting behaviour of an element of fluid in a rotating bucket for conditions in which all of the fluid has the same angular velocity. The element turns through 360° on a single circuit around its circle, and the conditions for irrotational motion are *not* satisfied. For irrotational circulation the fluid elements must slip past each other without inducing rotations, which is possible because there is no friction between the elements in an ideal fluid.

To return to the whirlpool, the fluid behaviour at its centre is clearly highly singular since, according to eqn (5.32), the fluid velocity must tend to infinity as the radius tends to zero. This singularity causes exceptions to the general vanishing of the circulation in the cases of all closed curves that encircle the origin. Indeed the circulation of eqn (5.27) around a circle of radius r centred on the origin is simply

$$C_0 = 2\pi r v = \text{constant}. \tag{5.33}$$

The centre of the whirlpool is known as a *vortex*, and it does not belong to the region of irrotational motion. The whirlpool as a whole is an example of a *free vortex*. The quantity C_0 in eqn (5.33) provides a measure of the strength of the vortex.

The profile of the liquid surface in the vicinity of a whirlpool can be found with the help of Bernoulli's theorem. Thus with p taken equal to the constant atmospheric pressure p_0, and the velocity and fluid surface height set equal to zero for large values of the radius, eqn (5.26) gives

$$\tfrac{1}{2}\varrho v^2 + p_0 + \varrho g z = p_0. \tag{5.34}$$

Elimination of the velocity with the use of eqn (5.33) gives

$$z = -C_0^2/8\pi^2 g r^2. \tag{5.35}$$

The calculated surface shape is shown in fig. 5.9. It is seen that the problem

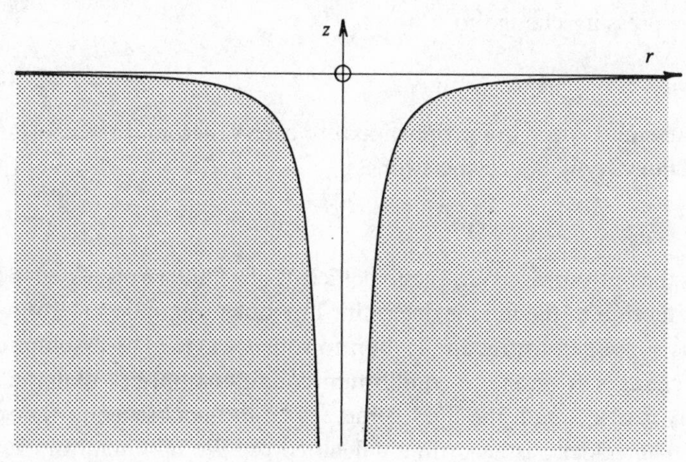

Fig. 5.9 Cross-section of a whirlpool showing the surface profile.

of infinite fluid velocity only arises at infinite depths, and the whirlpool has a hole at its centre in practice. The above derivations do not distinguish clockwise and anticlockwise rotations of the whirlpool. The sense of rotation of a terrestrial whirlpool is determined in principle by the *Coriolis force* caused by the Earth's rotation. Ideally the circulations of vortices should be in opposite senses in the northern and southern hemispheres but, in practice, even in laboratory experiments, accidentally-introduced circulatory effects tend to be the determining factors.

5.5 Sound waves in fluids

Most of the discussion so far has assumed that the fluid density is a fixed quantity. However, real liquids are somewhat compressible and gases highly so, and one well-known consequence is their ability to transmit sound waves, otherwise known as *acoustic waves* or *pressure waves*.

Let ϱ_0 and p_0 be the static equilibrium density and pressure in a fluid and let $\delta\varrho$ and δp be the small changes in density and pressure associated with a sound wave, so that

$$\varrho = \varrho_0 + \delta\varrho \tag{5.36}$$

and

$$p = p_0 + \delta p. \tag{5.37}$$

The small changes are related to the variables of elasticity theory discussed in §§ 2.1 and 2.2. Thus the density change corresponds to a strain

$$\varepsilon = -\delta\varrho/\varrho_0 \tag{5.38}$$

and the pressure change to a stress

$$\sigma = -\delta p. \tag{5.39}$$

The stress and strain are proportional to each other, and the constant of proportionality is the elastic stiffness

$$c_{11} = \sigma/\varepsilon = \varrho_0 \delta p/\delta\varrho. \tag{5.40}$$

This is the only non-zero stiffness coefficient of a fluid, as discussed in § 3.5.

The motion of the fluid is determined by Euler's equation (5.16), which may be simplified for its application to sound waves. The assumption of small changes from static equilibrium conditions implies that the fluid velocity is also small, the fluid being at rest in the absence of the sound wave. The velocity is accordingly denoted $\delta\boldsymbol{v}$. We now retain in Euler's equation only terms of first order in the small quantities and the nonlinear term therefore makes no contribution. Thus for waves propagated horizontally in the x-direction, where the gravitational term also makes no contribution, we find

$$\varrho_0(\partial/\partial t)\,\delta v + (\partial/\partial x)\,\delta p = 0. \tag{5.41}$$

This equation relates the pressure gradient to the rate of change of velocity. Euler's equation (5.16) was itself derived for an incompressible fluid, but the generalization to a compressible fluid makes no difference to eqn (5.41) since there is no contribution from the small density change. The latter does, however, contribute to the equation of continuity, eqn (5.20), where a similar first-order approximation gives

$$(\partial/\partial t)\,\delta\varrho + \varrho_0(\partial/\partial x)\,\delta v = 0. \tag{5.42}$$

The velocity can be removed by differentiation of eqn (5.41) with respect to x and differentiation of eqn (5.42) with respect to t, whence

$$(\partial^2/\partial t^2)\,\delta\varrho = (\partial^2/\partial x^2)\,\delta p. \tag{5.43}$$

However δp and $\delta\varrho$ are proportional to each other in accordance with eqn (5.40), and either can be removed to obtain

$$(\partial^2/\partial t^2)\,\delta\varrho = v_s^2(\partial^2/\partial x^2)\,\delta\varrho, \tag{5.44}$$

or the similar equation for δp, where

$$v_s^2 = c_{11}/\varrho_0 = \delta p/\delta\varrho \rightarrow (\partial p/\partial\varrho)_s \tag{5.45}$$

in agreement with the longitudinal velocity in a liquid given in eqn (3.71). The final step in this equation recognizes the normal conditions in sound wave propagation through fluids, where the changes in density and pressure are adiabatic, i.e., they are too rapid for the temperature to become uniform. Thus v_s is known as the adiabatic sound velocity and eqn (5.44) is the one-dimensional equation for waves of this velocity. Typical sound velocities are

$$H_2O:\quad v_s = 1400\,\mathrm{m\,s^{-1}} \qquad \mathrm{Hg:}\quad v_s = 1500\,\mathrm{m\,s^{-1}}.$$

The solutions of the wave equation can be written in the usual forms, for example

$$\delta\varrho = \delta\varrho_0 \cos(\omega t - qx)$$

or

$$\delta\varrho = \delta\varrho_0 \exp(-\mathrm{i}\omega t + \mathrm{i}qx), \tag{5.46}$$

with similar forms for δp and δv. The frequency ω and wavevector $\mathbf{q}$

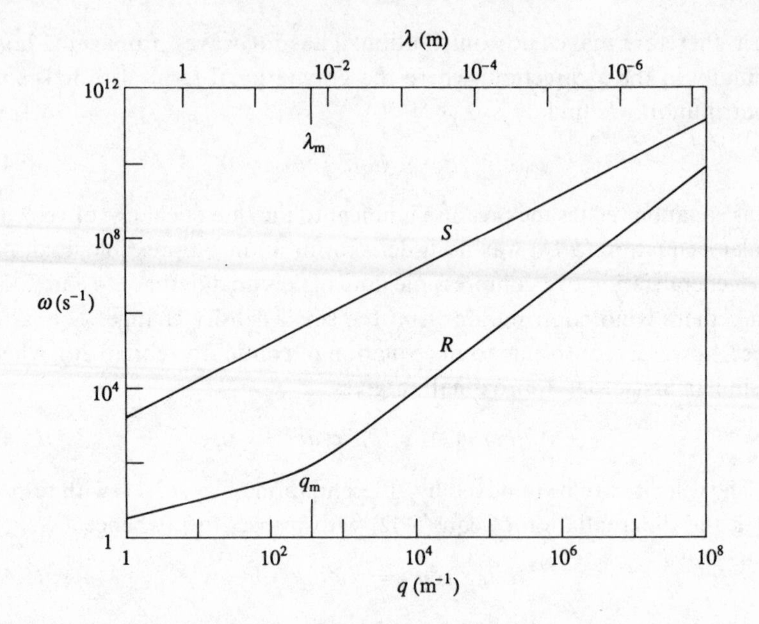

Fig. 5.10 Dispersion relations for sound waves (S) and surface ripples (R) in water. The wavevector q_m and the wavelength λ_m are defined in eqn (5.76).

($|\mathbf{q}| = 2\pi/\lambda$) are related by the sound wave dispersion relation

$$\omega = v_s|\mathbf{q}|, \tag{5.47}$$

and this is plotted for water in fig. 5.10. It is seen that for a given wavevector, the sound wave frequency is generally substantially higher than the surface ripple frequency, to be derived in § 5.7. The elastic stiffness provides a much stronger restoring force than do either surface tension or gravity. The sound waves are longitudinally polarized since the particle motion is parallel to the direction of wave propagation.

5.6 Surface waves on a liquid

Consider waves of small amplitude on a deep incompressible liquid whose undisturbed surface is the $z = 0$ plane. The fluid velocity $\mathbf{v}$ is assumed to be everywhere so small that

$$(\mathbf{v} \cdot \nabla)\,\mathbf{v} \ll \partial\mathbf{v}/\partial t. \tag{5.48}$$

Terms in v^2 can then be neglected and Euler's equation (5.16) can be approximated by

$$\varrho\partial\mathbf{v}/\partial t = -\nabla p + \varrho\mathbf{g}. \tag{5.49}$$

The continuity equation (5.21) for an incompressible fluid is

$$\nabla \cdot \mathbf{v} = 0. \tag{5.50}$$

We seek solutions of these equations that represent travelling waves in the *zx*-plane, where the fluid velocity is given by the real part of the complex vector

$$\mathbf{v} = \mathbf{v}_0 \exp(\mathrm{i}q_x x + \mathrm{i}q_z z - \mathrm{i}\omega t). \tag{5.51}$$

The corresponding fluid displacement has the form

$$\mathbf{u} = \mathbf{u}_0 \exp(\mathrm{i}q_x x + \mathrm{i}q_z z - \mathrm{i}\omega t), \tag{5.52}$$

where the velocity and displacement are related in the usual way by

$$\mathbf{v} = \mathrm{d}\mathbf{u}/\mathrm{d}t = -\mathrm{i}\omega\mathbf{u}. \tag{5.53}$$

Substitution of the assumed solution (5.51) into eqn (5.49) gives

$$-\mathrm{i}\omega\varrho v_x = -\partial p/\partial x \tag{5.54}$$

$$-\mathrm{i}\omega\varrho v_z = -(\partial p/\partial z) - \varrho g, \tag{5.55}$$

while substitution of the solution (5.51) into eqn (5.50) gives

$$q_x v_x + q_z v_z = 0. \tag{5.56}$$

The pressure is eliminated from eqns (5.54) and (5.55) by differentiating them respectively with respect to z and x, whence

$$q_z v_x = q_x v_z. \tag{5.57}$$

The waves thus provide another example of irrotational motion, defined in accordance with eqn (5.24).

Elimination of the velocity components from the two previous equations gives

$$q_x^2 + q_z^2 = 0 \quad \text{or} \quad q_z = \pm \mathrm{i}q_x. \tag{5.58}$$

We expect on physical grounds that the amplitude of a surface wave must diminish, and certainly not grow, as the downwards distance $-z$ from the liquid surface increases. This requirement selects the minus sign in eqn (5.58), and it is convenient to simplify the notation by putting

$$q_x \equiv q \quad \text{and} \quad q_z = -\mathrm{i}q, \tag{5.59}$$

where q is chosen real and positive, corresponding to a wave that propagates parallel to the liquid surface. Thus q now denotes only the wavevector component parallel to the surface.

These properties of the wavevector components require that the velocity components must be related by

$$v_z = -\mathrm{i}v_x \tag{5.60}$$

in order to satisfy eqns (5.56) and (5.57). We take the velocity amplitude component v_{0x} to be real, and simplify the notation by setting

$$v_{0x} \equiv v_0 \quad \text{and} \quad v_{0z} = -\mathrm{i}v_0. \tag{5.61}$$

The real velocity components are then

$$\begin{aligned} v_x &= v_0\,\mathrm{e}^{qz}\cos(qx - \omega t) \\ v_z &= v_0\,\mathrm{e}^{qz}\sin(qx - \omega t). \end{aligned} \tag{5.62}$$

The velocity of wave propagation, the ripple velocity, is

$$v_\mathrm{R} = \omega/q. \tag{5.63}$$

The inequality (5.48) implies that

$$qv_0^2 \ll \omega v_0 \quad \text{or} \quad v_0 \ll v_\mathrm{R}, \tag{5.64}$$

and the velocity of a fluid element is thus much smaller than the wave propagation velocity.

We can now examine the displacement of liquid elements in a similar manner. The liquid displacement is related to its velocity by eqn (5.53). Thus the real fluid displacement components obtained with the use of eqns (5.52) and (5.61) are

$$\begin{aligned} u_x &= -u_0\,\mathrm{e}^{qz}\sin(qx - \omega t) \\ u_z &= u_0\,\mathrm{e}^{qz}\cos(qx - \omega t), \end{aligned} \tag{5.65}$$

where

$$u_0 = v_0/\omega \tag{5.66}$$

is the displacement amplitude at the fluid surface $z = 0$. The amplitude falls to $1/\mathrm{e}$ of its surface value at a depth

$$-z = 1/q = \lambda/2\pi, \tag{5.67}$$

where λ is the wavelength. The inequality (5.64) shows that

$$u_0 \ll 1/q, \tag{5.68}$$

and the wave amplitude is therefore much smaller than its wavelength.

The solution obtained above represents a circular but irrotational motion of the liquid elements. The general features of the motion are represented in fig. 5.11. The velocity vector rotates in a clockwise sense, keeping a fixed

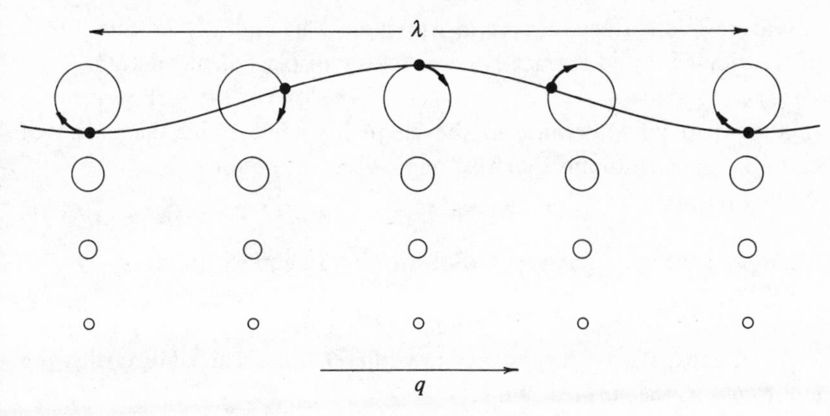

Fig. 5.11 Motion of liquid elements as a function of lateral distance and depth in the presence of a surface ripple.

magnitude at a given depth. The motion is somewhat similar to that induced by a Rayleigh wave on a solid surface, illustrated in fig. 3.16. However, the sense of the rotation is opposite for the two kinds of surface wave, and the Rayleigh motion is elliptical rather than circular.

5.7 Surface ripple dispersion relation

The relation between the surface wavevector component q and frequency ω is obtained by consideration of the pressure conditions at the liquid surface. The variation of pressure with position in the liquid is determined by eqns (5.54) and (5.55). It is not difficult to verify that these are satisfied by

$$p = p_0 + (\varrho\omega/q)\,v_x - \varrho g z, \tag{5.69}$$

where the relations (5.59) and (5.61) between the wavevector and velocity components in the complex velocity, eqn (5.51), are used. It is clear that p_0 is the pressure at the liquid surface in the absence of any disturbance, and it can be set equal to the atmospheric pressure.

The presence of a ripple causes the liquid surface to curve slightly, bringing into play a surface-tension-induced difference between atmospheric pressure and the pressure p_l just inside the surface. A surface wave propagating parallel to the x-axis produces a curvature only in the x-direction and if R is the corresponding radius of curvature, the pressure difference obtained from eqn (4.26) is

$$p_l - p_0 = -\gamma/R, \tag{5.70}$$

where γ is the surface tension. A general expression for the radius of curvature of a surface is given in Appendix 1 by eqn (A1.7). The profile of the rippled liquid surface is given by the displacement relations, eqns (5.52) or (5.65) with z set equal to zero. The slope of the surface is much smaller than unity because of the inequality (5.68), and the radius of curvature is accordingly given by eqn (A1.8) in the form

$$1/R = \mathrm{d}^2u_z/\mathrm{d}x^2 = -q^2u_z, \tag{5.71}$$

where u_z is evaluated at $z = 0$. Substitution in eqn (5.70) gives

$$p_1 - p_0 = \gamma q^2 u_z. \tag{5.72}$$

Another expression for the pressure difference is obtained by evaluating eqn (5.69) at the surface. We set

$$p = p_1, \quad v_x = \mathrm{i}v_z = \omega u_z \quad \text{and} \quad z = u_z, \tag{5.73}$$

where eqns (5.53) and (5.61) have been used. Thus eqn (5.69) gives

$$p_1 - p_0 = (\varrho\omega^2/q)\,u_z - \varrho g u_z. \tag{5.74}$$

The right-hand sides of eqns (5.72) and (5.74) must be equal, and rearrangement of the resulting equation gives the surface-ripple dispersion relation

$$\omega^2 = gq + (\gamma q^3/\varrho). \tag{5.75}$$

Figure 5.10 shows the variation of angular frequency with wavelength and wavevector for ripples on water. The frequency is mainly determined by gravitation for small q or large λ and by surface tension for large q or small λ. The two forces have equal influence for waves on water at the wavevector and wavelength

$$q_\mathrm{m} = (\varrho g/\gamma)^{1/2} = 2^{1/2}/\mathrm{a} = 370\,\mathrm{m}^{-1}$$

and

$$\lambda_\mathrm{m} = 2\pi/q_\mathrm{m} = 17\,\mathrm{mm}, \tag{5.76}$$

where a is the capillary constant defined in eqn (4.60). These values are marked on the horizontal axes of fig. 5.10. Surface waves that have wavelengths much greater than λ_m are commonly called *gravity waves*, while those with wavelengths smaller than λ_m are called *ripples* or *capillary waves*. The frequency and period of the wave on water corresponding to the wavevector given by eqn (5.76) are

$$\omega_\mathrm{m} = 86\,\mathrm{s}^{-1}$$

and

$$T_\mathrm{m} = 2\pi/\omega_\mathrm{m} = 0.073\,\mathrm{s} \tag{5.77}$$

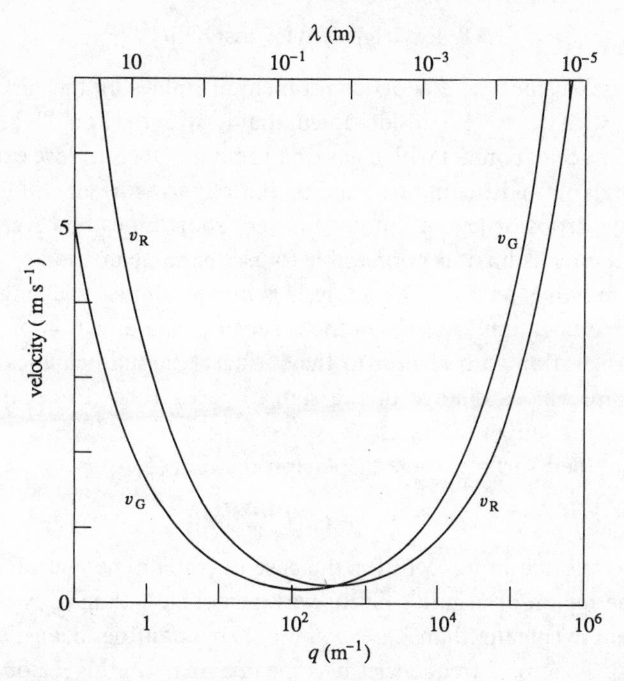

Fig. 5.12 Variations of the surface ripple phase velocity v_R and group velocity v_G with wavevector and wavelength for water.

The surface-wave phase velocity is

$$v_R = \omega/q = \{(g/q) + (\gamma q/\varrho)\}^{1/2}, \tag{5.78}$$

and the variation of this velocity for water is shown in fig. 5.12. The velocity has a minimum value of about 0.23 m s^{-1} at the wavevector and wavelength given in eqn (5.76). The flow of energy occurs, as usual, at the group velocity derived from eqn (5.75) as

$$v_G = \frac{d\omega}{dq} = \frac{\varrho g + 3\gamma q^2}{2\varrho^{1/2}(\varrho g q + \gamma q^3)^{1/2}}. \tag{5.79}$$

This rather complicated relation is best appreciated by reference to fig. 5.12 where its wavelength dependence is plotted. The group velocity also has a minimum value, whose parameters for water are

$$q = 140\,\mathrm{m}^{-1}, \quad \lambda = 44\,\mathrm{mm} \quad \text{and} \quad v_G = 0.18\,\mathrm{m\,s}^{-1}. \tag{5.80}$$

Waves of about this wavelength are typically seen on the surface of a pond disturbed by a thrown stone, after the more rapidly-moving waves have dispersed. The group and phase velocities are equal at the wavevector q_m given in eqn (5.76).

5.8 Rayleigh–Taylor instability

Now let us consider the analogous problem of ripples on the surface of a liquid but with the system upside-down, that is, with the liquid lying above the free surface of contact with a gas or a vacuum. Of course we expect the liquid to fall out of its container, and an initially flat surface will break up into falling drops or jets of liquid. For very short times however, before the surface breaks up, it is permissible to use the same method of analysis as in the previous section. The analysis is indeed almost unchanged, and the only modification arises from the reversal of the direction of gravitational acceleration with respect to the surface. The mathematical results thus require only a change in the sign of g for them to apply to the upside-down configuration.

The modified surface ripple dispersion relation (5.75) is

$$\omega^2 = -gq + (\gamma q^3/\varrho), \tag{5.81}$$

and this is plotted in fig. 5.13 for the case of water. The main feature of note is the region of negative ω^2 for wavevectors less than $q_m = 370\,m^{-1}$ or wavelengths greater than $\lambda_m = 17$ mm, these quantities being defined in eqn (5.76). The ripple frequencies become imaginary in this region. Corresponding to the travelling wave solutions (5.65) of the fluid equations of motion, there are also standing wave solutions, for real ω, of the form

$$u_z = u_0\,e^{qz} \cos(qx) \cos(\omega t). \tag{5.82}$$

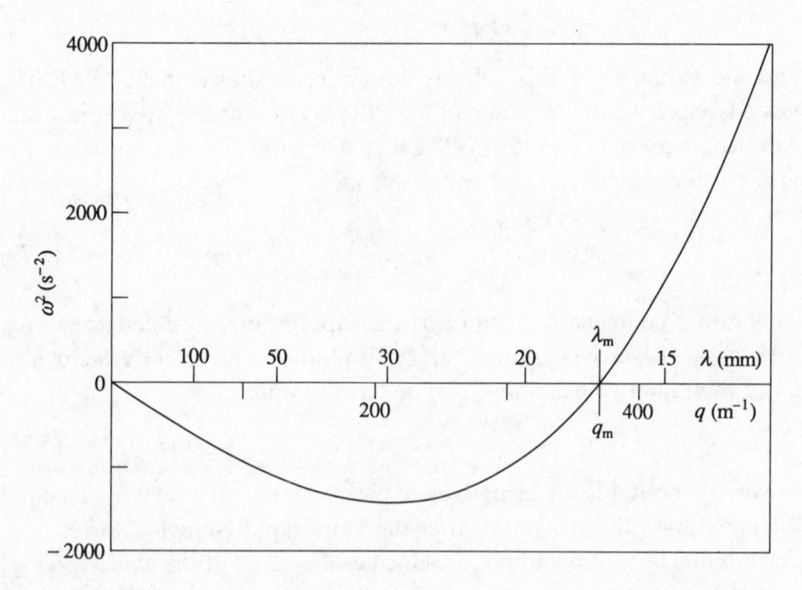

Fig. 5.13 Ripple dispersion relation for an inverted water surface.

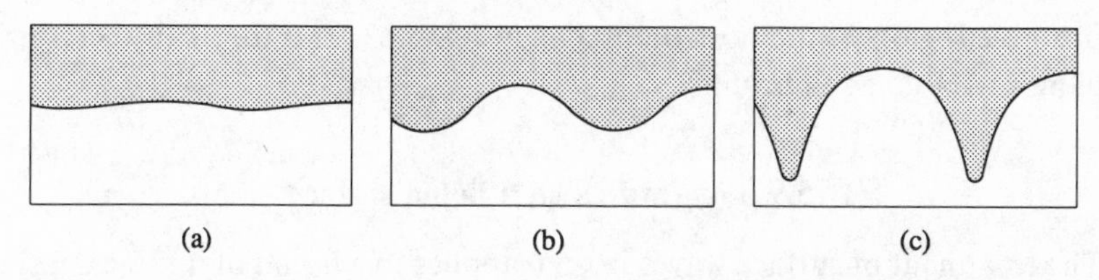

Fig. 5.14 Development in time of a Rayleigh–Taylor instability showing: (a) small distortions from the horizontal inverted surface, (b) linear growth, and (c) nonlinear growth of the distortion (from Evans, R.G. (1986). *Canadian Journal of Physics*, **64**, 893).

When ω is imaginary with $\omega = \mathrm{i}|\omega|$, this solution is converted to

$$u_z = u_0 \, \mathrm{e}^{qz} \cos(qx) \cosh(|\omega| t). \tag{5.83}$$

The stable oscillatory solutions are thus replaced by unstable displacements that grow exponentially with increasing time, and it is these solutions that represent the breakup of the liquid surface.

It is seen from fig. 5.13 that the exponential growth is most rapid for excitations of wavelength close to 30 mm, and it is these that dominate the unstable behaviour. The oscillatory character of the waves is maintained by the predominance of surface tension effects for wavelengths shorter than 17 mm. The phenomenon of exponential growth is known as the *Rayleigh–Taylor instability*. The solution (5.83) applies only in the initial stages of development of the instability when the liquid displacements are small and Euler's equation can be approximated by the linear form, eqn (5.49). The nonlinear term in Euler's equation (5.16) cannot be neglected when the displacement amplitude u_0 no longer satisfies eqn (5.68). Figure 5.14 shows the schematic form of the growth of the Rayleigh–Taylor instability, including the effects of the nonlinear term.

The initial configurations of a liquid with a flat surface in a container that is either upright or inverted are special cases of a range of possibilities in which the container is upright but subjected to an arbitrary vertical acceleration. With an acceleration of magnitude g directed downwards, the liquid becomes effectively weightless, while a downwards acceleration of $2g$ is mathematically equivalent to turning the container upside-down. Intermediate accelerations give less rapid developments of the Rayleigh–Taylor instability.

The above theory strictly applies only to initially flat liquid surfaces of large area, but it nevertheless also gives some indication of the behaviour of suspended drops. Thus a water drop of diameter larger than 17 mm hanging from the underside of a horizontal surface can support unstable surface waves, and it will accordingly break up and fall. Drops of diameter smaller than 17 mm are too small to support surface wavelengths in the unstable region of the horizontal axis in fig. 5.13, and they can therefore hang in stable equilibrium.

5.9 Long waves on a liquid surface

The treatment of surface waves in § 5.6 applies to a liquid of infinite depth. The depth h is effectively infinite if the wave displacement in eqn (5.65) is negligible at $z = -h$, i.e. if

$$qh \gg 1 \quad \text{or} \quad \lambda \ll 2\pi h. \tag{5.84}$$

This condition is normally satisfied by capillary waves, where the wavelength for water is smaller than about 17 mm and the liquid depth only needs to be larger than 10 mm or so. However, the inequality (5.84) is often not satisfied for surface waves in the gravity wave regime where, in many practical situations, the wavelength is greater than the depth of liquid. We must therefore consider the effects of the finite fluid depth on these long waves. In view of the above remarks, it follows that their wavelength is normally sufficiently large for the effects of surface tension to be negligible.

The long waves can be treated by the same method that we have already used, and the general forms of solutions (5.51) and (5.52) remain valid. However, q_z can now take both positive and negative values since $-z$ has a maximum value of h. Thus with the notation

$$q_x \equiv q, \tag{5.85}$$

as in eqn (5.59), the general solution for the liquid velocity is

$$\begin{aligned} v_x &= (Ae^{qz} + Be^{-qz}) \exp(iqx - i\omega t) \\ v_z &= (-iAe^{qz} + iBe^{-qz}) \exp(iqx - i\omega t), \end{aligned} \tag{5.86}$$

where the relative signs in the two components are determined in accordance with the equation of continuity (5.50).

The presence of a rigid bottom at $z = -h$ prevents any vertical motion there. The boundary condition

$$v_z = 0 \quad \text{at} \quad z = -h \tag{5.87}$$

provides the relation

$$A\,e^{-qh} = B\,e^{qh}. \tag{5.88}$$

It is convenient to express A and B in terms of the amplitude v_0 of the x-component of the velocity at the liquid surface, $z = 0$, given by eqn (5.86) as

$$v_0 = A + B. \tag{5.89}$$

The solutions of these last two equations are

$$A = \frac{v_0 \, e^{qh}}{2 \cosh (qh)}$$

and

$$B = \frac{v_0 \, e^{-qh}}{2 \cosh (qh)}, \tag{5.90}$$

and the real velocity components obtained from eqn (5.86) are

$$\begin{aligned} v_x &= v_0 \frac{\cosh [q(z + h)]}{\cosh (qh)} \cos (qx - \omega t) \\ v_z &= v_0 \frac{\sinh [q(z + h)]}{\cosh (qh)} \sin (qx - \omega t). \end{aligned} \tag{5.91}$$

These expressions reduce to eqn (5.62) for a deep liquid where $qh \gg 1$. The corresponding real displacement components are

$$\begin{aligned} u_x &= -u_0 \frac{\cosh [q(z + h)]}{\cosh (qh)} \sin (qx - \omega t) \\ u_z &= u_0 \frac{\sinh [q(z + h)]}{\cosh (qh)} \cos (qx - \omega t), \end{aligned} \tag{5.92}$$

where

$$u_0 = v_0/\omega. \tag{5.93}$$

The elements of fluid move in elliptical paths with axes in the ratio

$$\frac{\text{minor axis}}{\text{major axis}} = \tanh [q(z + h)], \tag{5.94}$$

where the various hyperbolic functions that occur in these solutions are illustrated in fig. 5.15. Figure 5.16 shows how the paths of the fluid elements vary with depth. The striking effect of the finite depth on the fluid motion is seen by comparing fig. 5.16 with the corresponding diagram, fig. 5.11, for a fluid of infinite depth.

The dispersion relation for long waves is obtained from eqn (5.69), which remains valid. In the absence of surface tension, the pressure p just inside the liquid surface equals the pressure p_0 just outside the surface. The velocity and displacement at the surface are given by eqns (5.91) and (5.92) as

$$\begin{aligned} v_x &= v_0 \cos (qx - \omega t) = u_0 \omega \cos (qx - \omega t) \\ z = u_z &= u_0 \tanh (qh) \cos (qx - \omega t). \end{aligned} \tag{5.95}$$

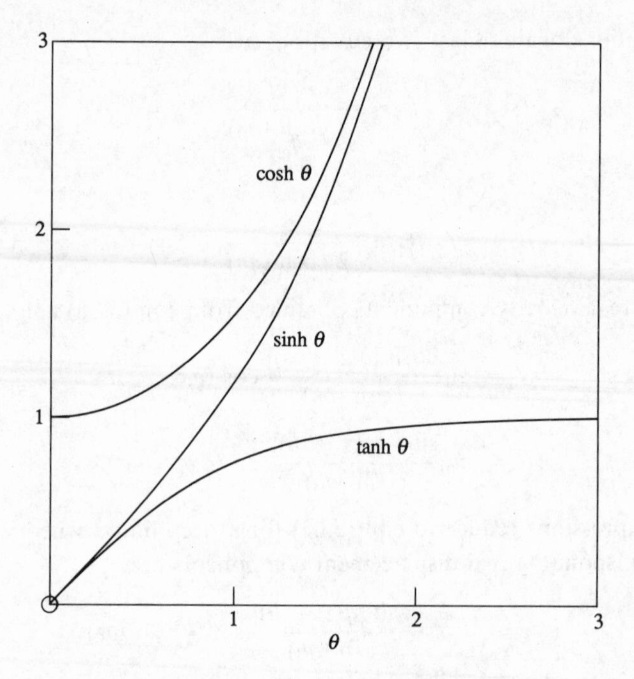

Fig. 5.15 Hyperbolic functions that occur in the theory of ripples on a shallow liquid.

Thus substitution in eqn (5.69) gives the dispersion relation

$$\omega^2 = gq \tanh (qh). \qquad (5.96)$$

In the limit of a very deep liquid where eqn (5.84) is satisfied, this dispersion relation reproduces the gravity wave result obtained from eqn (5.75) when the surface tension γ is neglected. With diminishing depth h, the surface wave frequency for a given wavevector or wavelength decreases on account of the tanh (qh) factor. This factor can be expanded in a Taylor series for

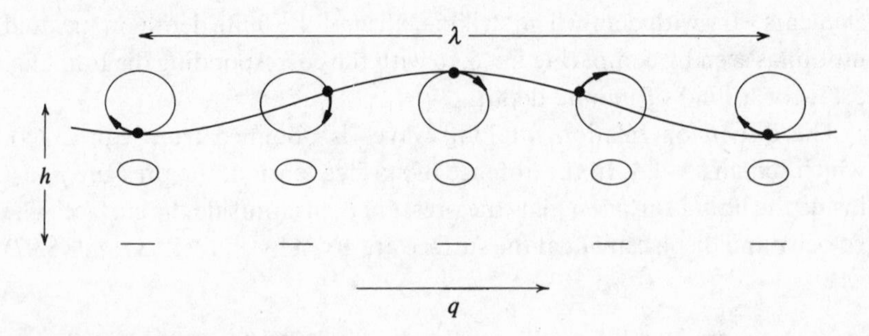

Fig. 5.16 Motion of liquid elements in the presence of a surface ripple for a sample of finite depth h.

small qh, and if only the first two terms are retained eqn (5.96) gives

$$\begin{aligned} \omega &= \{gq \tanh (qh)\}^{1/2} \simeq \{gq(qh - \tfrac{1}{3}q^3h^3)\}^{1/2} \\ &\simeq (gh)^{1/2}q\{1 - \tfrac{1}{6}q^2h^2\} \quad (\text{for } qh \ll 1). \end{aligned} \tag{5.97}$$

The waves are nondispersive, i.e. ω is proportional to q, when qh is so very small that the second, dispersive term on the right of eqn (5.97) can be neglected.

Important examples of long waves occur on the sea near a coastline and on rivers in estuaries. Waves incident at the coast that originate from deeper ocean typically have wavelengths of order 100 m, corresponding to angular frequencies of about $0.8\,\text{s}^{-1}$ and to periods of about 8 s. The period and angular frequency of the waves are unchanged as they encounter the shallower water near the shore. It follows from eqns (5.96) or (5.97) that the wavevector must increase and the wavelength must decrease in the shallower water.

5.10 Solitons

All of the liquid surface waves considered in the previous sections generate fluid velocities sufficiently small that the inequality (5.48) is satisfied and the nonlinear term in Euler's equation (5.16) can be neglected. The resulting linear equation (5.49) has sinusoidal solutions of the kind illustrated in figs. 5.11 and 5.16. The nonlinear term can no longer be neglected when large liquid velocities occur and/or surface waves have large amplitudes, in which cases the solution of the equation of motion presents a much more difficult problem. There is however a special case of the general problem, of great practical importance, for which a very simple solution exists.

Consider again the treatment of long waves given in the previous section, in the limit of shallow liquid where qh is very small. The solutions (5.91) for the velocity show that the liquid motion is mainly in the direction parallel to the surface,

$$v_x \gg v_z \quad \text{for} \quad qh \ll 1. \tag{5.98}$$

The x-component of the linearized Euler equation (5.49) gives

$$\varrho \partial v_x/\partial t = -\partial p/\partial x = -(\varrho\omega/q)\,\partial v_x/\partial x, \tag{5.99}$$

where the expression (5.69) for the pressure has been used. Substitution of the long wave dispersion relation from eqn (5.97) gives

$$\frac{\partial v_x}{\partial t} = -(gh)^{1/2}(1 - \tfrac{1}{6}q^2h^2)\frac{\partial v_x}{\partial x}$$

$$= -(gh)^{1/2}\left(\frac{\partial v_x}{\partial x} + \frac{1}{6}h^2\frac{\partial^3 v_x}{\partial x^3}\right), \tag{5.100}$$

where the second step recognizes that the second derivative of v_x with respect to x in eqn (5.91) generates a factor of $-q^2$.

The *linear* equation (5.100) is satisfied by the long waves of the previous section to an accuracy corresponding to the two terms retained in the expansion (5.97) of the dispersion relation in powers of qh. The first term on the right of eqn (5.100) generates the nondispersive part of ω, proportional to q, while the third derivative term generates the dispersive part of ω, proportional to q^3. The *nonlinear* Euler equation (5.16) can now be subjected to the same approximations based on the inequalities (5.98). The linear terms retain their forms shown in eqn (5.100), but the nonlinear term provides an additional contribution to give

$$\frac{\partial v_x}{\partial t} + v_x\frac{\partial v_x}{\partial x} + v_\text{p}\left(\frac{\partial v_x}{\partial x} + \frac{1}{6}h^2\frac{\partial^3 v_x}{\partial x^3}\right) = 0, \tag{5.101}$$

where v_p is the phase velocity for waves of very small q, given by

$$v_\text{p} = (gh)^{1/2}. \tag{5.102}$$

This is known as the *Korteweg–de Vries equation*, or more briefly, as the KdV equation. The derivation given here, particularly the insertion of the nonlinear term, is cavalier and not entirely correct; more careful analysis produces the same equation but shows that v_x is equal to $\frac{3}{2}$ of the physical fluid velocity.

The KdV equation has simple solutions in two special cases. One of these is the limit of small amplitudes, where eqns (5.48) and (5.64) are satisfied and the nonlinear term can be neglected. This is the case treated in the previous section, which has sinusoidal solutions for the velocity and displacement given in eqns (5.91) and (5.92). The KdV equation becomes more difficult to solve as the wave amplitude is increased and consequently the nonlinear term cannot be neglected. Numerical integration shows that as the amplitude grows, the wave peaks become sharper and the troughs between the peaks become flatter. The wave train tends towards a regular succession of propagating localized peaks separated by expanses of flat liquid.

Finally, for a velocity v_x of magnitude comparable with the phase velocity v_p, where the linear and nonlinear terms in eqn (5.101) have

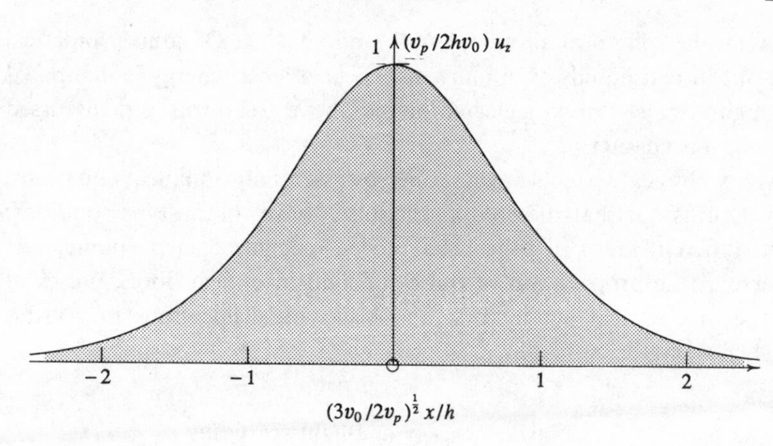

Fig. 5.17 Liquid surface profile associated with a soliton.

comparable importance, there exists a second kind of simple solution of the form

$$v_x = 3v_0 \operatorname{sech}^2 \left[\left(\frac{3v_0}{2v_p} \right)^{1/2} \frac{x - (v_0 + v_p)t}{h} \right]. \qquad (5.103)$$

It is not difficult to verify, by somewhat tedious differentiation, that this is indeed an exact solution of the nonlinear equation (5.101). It represents a single propagating pulse of disturbance and is known as a *solitary wave* or a *soliton*. As in the low-amplitude linear limit, the x-component of velocity has the same space and time variation as the z-component of the surface displacement, and the latter is plotted as a function of x at time $t = 0$ in fig. 5.17. Note that the velocity v_0 is arbitrary and that a pulse with a larger value of v_0 not only travels faster but also has a greater height and a reduced length in the propagation direction.

The soliton pulse has the remarkable property that it propagates alone along the liquid surface without suffering any change of shape. Its existence was first noted by Russell in the earlier half of the nineteenth century in observations of a single hump of water generated by the motion of a canal barge. The hump proceeded to travel some miles along the canal with little loss of its initial amplitude. On a smaller scale, solitons can be generated in a water trough in the laboratory. The trough is set up in a preliminary configuration where a short section at one end has its water level maintained a few centimetres higher than in the main part of the trough by a removable barrier. With suitable choices of the length and height of the short section, the sudden removal of the barrier can launch a soliton down the length of the main trough. Studies of such solitons confirm the calculated relation between the soliton amplitude, length and velocity. Of course, practical solitons in real liquids eventually disappear as their energy is dissipated through viscous forces neglected in the above treatment and discussed later in this chapter.

Many physical systems are described by essentially nonlinear equations, and it turns out that these equations often have soliton-type solutions. Thus different kinds of soliton have been recognized in phenomena as diverse as the propagation of pulses of light in optical fibres, pulses of charge in electrical conductors, and kinks on dislocations in crystals (§ 1.11).

5.11 Surface energy and light scattering

Consider the energy content of the surface wave described by eqn (5.65) over an area A of the xy-plane whose dimensions are large compared to the wavelength. The average energy is independent of the time so we set $t = 0$ for simplicity, and the coordinates (x, z) of a point in the surface satisfy

$$z = u_0 \cos(qx). \tag{5.104}$$

The displacement amplitude is again assumed to be much smaller than the wavelength so that

$$u_0 q \ll 1. \tag{5.105}$$

There are contributions to the surface energy from the surface tension and gravitational forces. The increase in surface area caused by the wave is determined by the element δs of surface length in the zx-plane

$$\begin{aligned} \delta s &= (\delta x^2 + \delta z^2)^{1/2} = \{1 + (\mathrm{d}z/\mathrm{d}x)^2\}^{1/2}\delta x \\ &= \{1 + u_0^2 q^2 \sin^2(qx)\}^{1/2}\delta x. \end{aligned} \tag{5.106}$$

Thus retaining only the first two terms in the Taylor expansion of the square root and averaging the sinusoidal term, the increase in surface area is

$$\Delta A = \tfrac{1}{4} u_0^2 q^2 A. \tag{5.107}$$

The contribution to the surface energy of the wave resulting from the opposition of surface tension forces to an adiabatic increase in the area [eqn (4.7)] is therefore

$$\Delta E_{\mathrm{S}} = \tfrac{1}{4} \gamma u_0^2 q^2 A. \tag{5.108}$$

The amount of gravitational energy is that needed to lift liquid from the wave troughs to the wave peaks, and is therefore given by

$$\Delta E_G = \varrho g A(q/\pi) \int_0^{\pi/q} u_0^2 \cos^2 qx \, dx = \tfrac{1}{4}\varrho g u_0^2 A. \quad (5.109)$$

The *total* surface energy is therefore

$$\Delta E_G + \Delta E_S = \tfrac{1}{4}u_0^2 A(\varrho g + \gamma q^2). \quad (5.110)$$

The thermal excitation of the surface wave of wavevector **q** at temperature T is obtained from the Boltzmann distribution by a calculation that is mathematically identical to that given in § 3.7 with ε replaced by u_0, and with ΔE in eqn (3.79) replaced by the expression (5.110). The thermal equilibrium mean-square amplitude of the surface wave is accordingly

$$\langle u_0^2 \rangle_{\mathbf{q}} = 2k_B T / A(\varrho g + \gamma q^2). \quad (5.111)$$

This quantity refers to a surface wave of given wavevector **q** parallel to the x-axis, but waves of all wavevector magnitudes and directions in the xy-plane are thermally excited. The inverse proportionality of the mean-square amplitude to the arbitrary area A is removed when eqn (5.111) is summed over all wavevectors to obtain the total mean-square displacement

$$\langle u_0^2 \rangle = \sum_{\mathbf{q}} \langle u_0^2 \rangle_{\mathbf{q}}. \quad (5.112)$$

A calculation based on eqns (5.111) and (5.112) gives a root-mean-square surface displacement for water at room temperature of

$$\langle u_0^2 \rangle^{1/2} \simeq 4 \times 10^{-10} \text{ m}, \quad (5.113)$$

equal to about 8 Bohr radii.

This displacement produced by the thermally-excited surface waves is of course much too small to be observed directly, but it is sufficiently large to produce observable light scattering, similar to the light scattering by Rayleigh waves on solids described in § 3.9. It can be shown that the strength of the scattered light is proportional to the mean-square displacement, so that according to eqn (5.111) waves of small q, or long wavelength, scatter light more strongly than do waves of large q, or short wavelength. Thus, as for Rayleigh waves, the strongest scattering occurs in directions close to that of the reflected beam. It is seen by reference to fig. 5.10 that the range of wavevectors **q** accessible to light scattering measurements, given by eqn (3.76), is such that surface tension is much more important than the gravitational force in determining the surface ripple frequency, and eqn (5.75) can be approximated by

$$\omega = (\gamma q^3/\varrho)^{1/2}. \quad (5.114)$$

Measurements of the scattered light spectrum yield a value for ω at a known q, so that the surface tension γ can be determined if the liquid density ϱ is known.

The light scattering method turns out to be more sensitive than other techniques for determining very small surface tensions. Such small surface tensions occur when a liquid in contact with its vapour is heated up to the critical temperature T_c for the liquid–vapour phase transition. The critical behaviour of the surface tension has been discussed in §4.11.2, and fig. 4.12 shows the data obtained by light scattering measurements on CO_2.

It is seen from eqn (5.111) that a vanishing γ produces an enhanced amplitude for the surface fluctuations. The intensity of light scattering thus increases as T approaches T_c, an effect known as *critical surface opalescence*. This expected increase is somewhat lessened by the disappearance of the liquid surface, from which the scattering takes place, at the transition temperature! However, some critical opalescence remains when all the relevant effects are taken into account.

5.12 Flow in real liquids: viscosity

The interatomic and intermolecular forces in real liquids produce a resistance to flow in the form of a liquid viscosity, which we have so far ignored. One way of looking at the behaviour of a viscous liquid is to draw an analogy with the behaviour of a rigid solid, since both depend on the same interatomic forces. Maxwell took the view that a liquid has a limited amount of rigidity, which breaks down continuously and strives to re-establish itself under the action of shearing forces. He called this effect *fugitive elasticity*. The viscosity can be defined by analogy with the rigidity modulus (§§ 1.9 and 2.3) of a solid according to

$$\text{solid: rigidity } G = \frac{\text{stress}}{\text{strain}} = \frac{\sigma_{xy}}{\mathrm{d}u_x/\mathrm{d}y} \tag{5.115}$$

$$\text{fluid: viscosity } \eta = \frac{\text{stress}}{\text{velocity gradient}} = \frac{\sigma}{\mathrm{d}v_x/\mathrm{d}y}. \tag{5.116}$$

This definition of viscosity is illustrated by the behaviour of a liquid flowing over a plane fixed surface S, shown in fig. 5.18. A velocity gradient is established in the liquid, with the liquid at rest immediately adjacent to the surface S and the velocity increasing with increasing distance y from S. In conditions of streamline flow, the velocity is constant over layers parallel to S, with different layers in relative motion. This is known as *laminar flow*. The viscosity or internal friction of the liquid is manifested by a drag force that the molecules in each layer exert on the molecules of the overlaying layer. This viscous drag force F for two layers of area A is

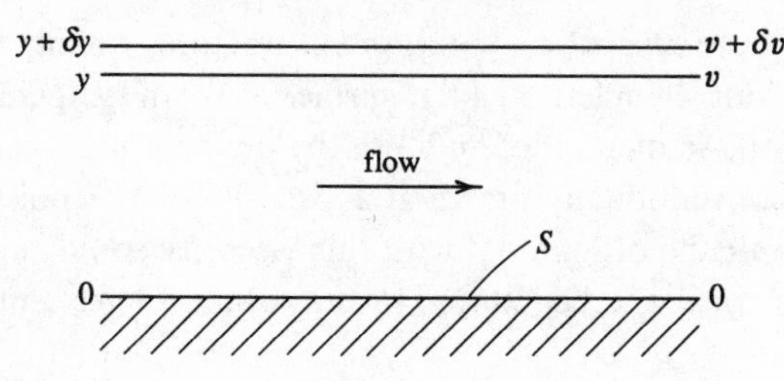

Fig. 5.18 Liquid flow adjacent to a plane surface.

given by Newton's law of viscous flow

$$F = \eta A \, \mathrm{d}v_x/\mathrm{d}y. \tag{5.117}$$

The quantity η, whose full name is the *dynamic coefficient of shear viscosity*, is thus given by

$$\eta = \frac{F/A}{\mathrm{d}v_x/\mathrm{d}y} = \frac{\sigma}{\mathrm{d}v_x/\mathrm{d}y}, \tag{5.118}$$

where σ is the shear stress, and we thus recover the definition of eqn (5.116).

Since the velocity gradient satisfies

$$\frac{\mathrm{d}v_x}{\mathrm{d}y} = \frac{\mathrm{d}}{\mathrm{d}y}\frac{\mathrm{d}u_x}{\mathrm{d}t} = \frac{\mathrm{d}}{\mathrm{d}t}\frac{\mathrm{d}u_x}{\mathrm{d}y} = \frac{\mathrm{d}\varepsilon}{\mathrm{d}t}, \tag{5.119}$$

where ε is the strain, eqn (5.118) can be re-expressed in the forms quoted in § 1.6.

$$\eta = \sigma/\dot{\varepsilon} \quad \text{or} \quad \dot{\varepsilon} = \mathscr{F}\sigma, \tag{5.120}$$

where

$$\mathscr{F} = 1/\eta \tag{5.121}$$

is the fluidity, first introduced in eqns (1.6) and (1.7). This form is similar to Hooke's law, eqn (1.32) except for the appearance of the time derivative on the left-hand side of eqn (5.120). Fluids for which the stress is accurately proportional to the velocity gradient or rate of change of the strain are called *Newtonian fluids*. This linear dependence is valid in practice over a wide range of velocity gradients for a variety of liquids.

The distinction between a liquid and a solid is somewhat arbitrary since a stressed medium may behave as either a liquid or as a solid depending on the length of time for which a stress is applied. For example silicone rubber ('silly putty') bounces like a rubber ball on sudden impact with the floor but it can also be poured like a liquid in slow motion from a container. A more ancient material, pitch, fractures cleanly when struck by a hammer, but when left on a flat surface at room temperature it flows out into a pool. As mentioned in § 1.6, the dividing line between liquids and solids is conventionally drawn at a viscosity of 10^{15} poise. In human terms, a 30 mm cube of material with this viscosity could support a man or woman for a year and be reduced by only about 10 per cent of its initial height.

A word of explanation is needed for the unit of viscosity, since the commonly-used poise (denoted by P), named after the French scientist Poiseuille, is a relic of the cgs system that has been taken over into SI units. Its exact equivalence in conventional SI units is

$$\text{dynamic viscosity } \eta\text{: 1 poise} = 10^{-1}\,\mathrm{N\,m^{-2}\,s}. \tag{5.122}$$

The poise is in fact inconveniently large for many liquids, and so the centipoise, cP, is still commonly used. The viscosity of water at room temperature is approximately 1 cP. Table 5.1 gives values of the dynamical coefficient of shear viscosity for some representative liquids and materials.

Both $\dot{\varepsilon}$ and σ in eqn (5.120) can be measured by instruments called *viscometers*, which thus provide values for η. Various designs of viscometer allow a wide range of liquid viscosities to be measured. The most common kind incorporates a cylinder, or bob, placed inside another concentric

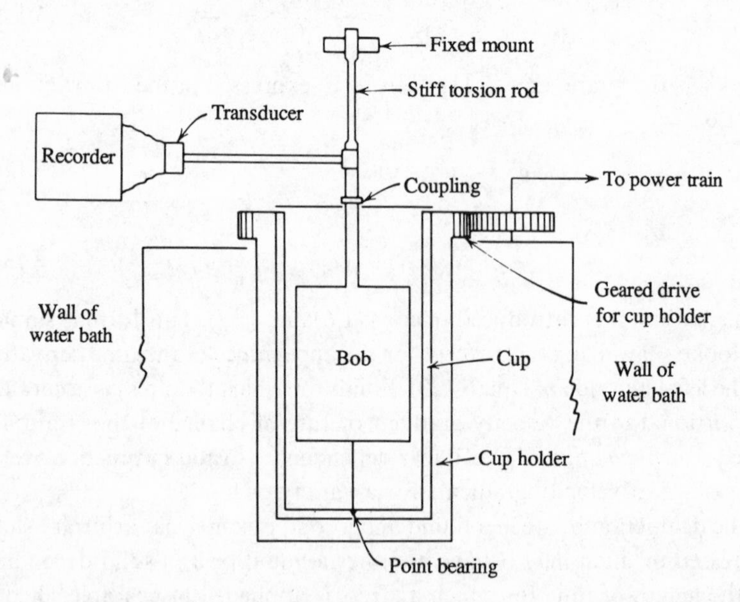

Fig. 5.19 Rotating cup viscometer (from Van Wazer, J.R., Lyons, J.W., Kim, K.Y. & Colwell, R.E. (1963). *Viscosity and Flow Measurement – a Laboratory Handbook of Rheology*. New York: John Wiley).

Table 5.1. *Some representative values of viscosity coefficients*

Substance	Temperature, °C	η, 10^{-3} P	ν, 10^{-3} St
Acetone	25	3.10	3.88
Benzene	25	6.03	7.54
Carbon tetrachloride, CCl_4	25	0.57	0.62
Castor oil	25	7×10^3	7.4×10^3
Methanol, CH_4O	25	5.43	6.79
Mercury	25	15.28	1.15
Olive oil	25	670	0.74×10^3
Pitch	25	2×10^{14}	—
	50	1.5×10^{10}	—
Sodium silicate glass	25	$\sim 10^{20}$	—
	1000	$\sim 10^3$	—
Water	25	8.92	8.94
Air	25	0.18	0.133
Aluminium	800	26.6	9.96
Argon	−172	0.197	—
Lead	400	23.2	2.06
Sodium chloride	816	15.0	6.95
Sodium	100	6.80	7.0
Basaltic lava	1000	$\sim 10^4$	$\sim 10^4$
Glacier ice	<0	$\sim 1 \times 10^{17}$	$\sim 1 \times 10^{17}$
Olivine, $(Mg, Fe)_2SiO_4$	1200	$\sim 3 \times 10^6$	$\sim 1 \times 10^6$
Silica, SiO_2	1200	$\sim 5 \times 10^{16}$	$\sim 2 \times 10^{16}$

These values mostly come from Kaye, G.W.C. & Laby, T.H. (1986). *Tables of Physical and Chemical Constants*, 15th edn. London: Longmans.

cylinder, or cup, containing the liquid. In the version shown schematically in fig. 5.19, the outer cylinder is rotated at a constant speed that sets the liquid into laminar flow with a known value of velocity gradient. The viscous drag on the inner cylinder applies a torque to the rod supporting it, which is measured by the transducer and thus provides a value for the applied stress. The whole system is enclosed in a temperature-regulated chamber so that the variation of η with temperature can be measured.

Other forms of viscometer are based upon the flow of liquid through a capillary tube and the fall of a small sphere (or the rising of a bubble) through the liquid. The dependences of both of these processes on the viscosity are discussed in the following section. Another measurement technique uses the draining of a small cup of the liquid through a standard orifice. The falling sphere method, widely used to characterize oils, provides a measurement of the ratio of dynamic viscosity η to the liquid density ϱ. This ratio is called the *kinematic viscosity* and denoted by

$$\nu = \eta/\varrho. \tag{5.123}$$

Its unit is the stokes (denoted by St), and the relation to standard SI units is

$$\text{kinematic viscosity } \nu\text{: 1 stokes} = 10^{-4}\,\text{m}^2\,\text{s}^{-1}. \tag{5.124}$$

Table 5.1 also gives values of ν for representative substances.

The only liquids whose measured viscosity values are encountered in everyday life are motor engine oils. These have kinematic viscosities expressed in terms of grades devised by the American Society of Automotive Engineers. Thus oil with grade SAE 20 (called 20W in the United Kingdom) has a kinematic viscosity of about 0.6 St at 38° C (100° F), while SAE 50 is equivalent to 2.0 St at the same temperature.

5.13 Examples of streamline viscous flow

5.13.1 Flow of liquid through a pipe: Poiseuille's equation

The nature of steady streamline flow of a liquid through a pipe of circular cross-section, illustrated in fig. 5.20, resembles the sliding of concentric tubes telescoping into one another. The outermost tube of liquid clings to the walls, and has zero velocity, while maximum velocity occurs at the centre of the pipe. The tubes of flowing liquid are analogous to the laminae in Newtonian flow over a flat surface.

Consider a cylinder of liquid of radius r and length L, as shown in fig. 5.21, with pressures p_1 and p_2 acting at the left- and right-hand ends respectively. In streamline flow the pressure only depends on the distance along the pipe. The cylinder of liquid does not accelerate once the flow is established, and the viscous drag force on its outer surface given by eqn (5.117) must therefore balance the force provided by the pressure difference,

$$-\eta 2\pi r L\,\mathrm{d}v/\mathrm{d}r = (p_1 - p_2)\,\pi r^2. \tag{5.125}$$

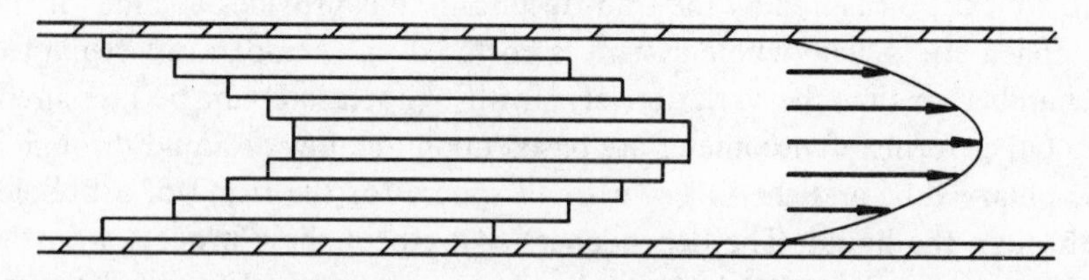

Fig. 5.20 Nested telescope representation of streamline flow through a circular tube, showing parabolic variation of velocity.

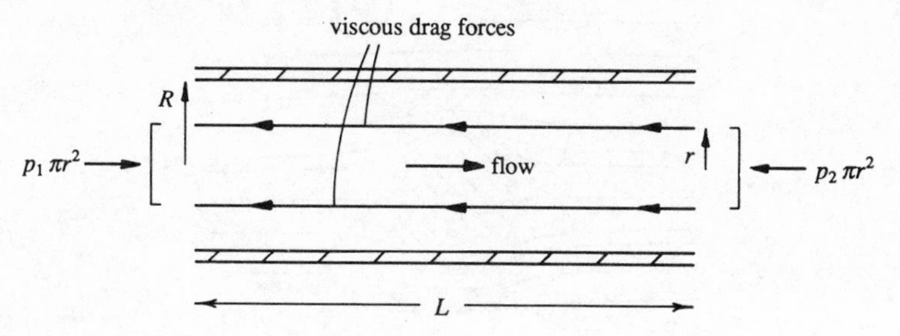

Fig. 5.21 Liquid flow through a tube showing the forces on a cylinder of radius r.

This can be rearranged and integrated from the wall at radius R where the velocity is zero to radius r where the velocity is v,

$$-\int_0^v \mathrm{d}v = \frac{p_1 - p_2}{2L\eta}\int_R^r r\,\mathrm{d}r,$$

giving

$$v = \frac{p_1 - p_2}{4L\eta}(R^2 - r^2). \tag{5.126}$$

This parabolic variation of velocity with radius is illustrated in fig. 5.20.

The rate Q of volume flow down the pipe can now be found by integrating the flow through an annular cross-section

$$Q = \int_0^R v2\pi r\,\mathrm{d}r = \frac{p_1 - p_2}{2L\eta}\pi\int_0^R (R^2 - r^2)r\,\mathrm{d}r$$

$$= \frac{(p_1 - p_2)\pi R^4}{8L\eta}. \tag{5.127}$$

This result is known as *Poiseuille's equation*, and it is obeyed experimentally for the flow of a Newtonian liquid through a long narrow pipe when the pressure difference is sufficiently small. From eqn (5.126) the maximum velocity at the centre of the pipe is

$$v_{\mathrm{max}} = (p_1 - p_2)R^2/4L\eta, \tag{5.128}$$

and Poiseuille's equation can be rewritten as

$$Q = \tfrac{1}{2}\pi R^2 v_{\mathrm{max}}. \tag{5.129}$$

The average flow velocity v, defined as the rate of volume flow divided by the cross-section, is thus

$$\bar{v} = Q/\pi R^2 = \tfrac{1}{2}v_{\mathrm{max}}. \tag{5.130}$$

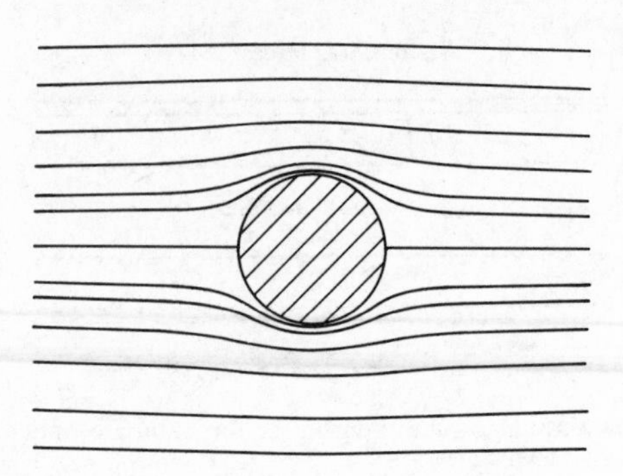

Fig. 5.22 Pattern of streamlines in the vicinity of a moving sphere.

The ratio of $\frac{1}{2}$ between the average and maximum flow velocities is characteristic of the steady streamline flow of a viscous liquid through a circular pipe.

Poiseuille's equation (5.127) presents a convenient technique for laboratory measurements of viscosity. For accurate results, the flow rate must be small, corresponding to a small pressure difference between the ends of the pipe. In practice the liquid must drip slowly from the end of a narrow pipe. A corrected form of Poiseuille's equation is used to extract the value of viscosity from the measured flow rate when this condition cannot be satisfied.

5.13.2 Viscous drag on a sphere: Stokes' law

When an ideal fluid with no viscosity flows past a sphere or, conversely, when a sphere moves through an inviscid stationary fluid, the streamlines form a symmetrical pattern, as shown in fig. 5.22. This implies that the pressure at any point on the upstream hemispherical surface is identical to that at the corresponding point on the downstream hemispherical surface, and the resultant force on the sphere is zero. With a real viscous fluid, the streamline flow pattern is not symmetrical, and there is a net force on the sphere because of viscous drag. The same is true for non-spherical bodies, but the size of the force is easily calculated only in the case of a sphere.

The form of the viscous drag force F can be obtained by the method of dimensions. Thus for a sphere of radius r in a fluid of viscosity η moving with velocity v well away from the sphere, we put

$$F = r^{\alpha}\eta^{\beta}v^{\delta}. \tag{5.131}$$

Thus expressing the variables in dimensions of mass M, length L and time

T, we have

$$\frac{ML}{T^2} = L^{\alpha}\left(\frac{M}{LT}\right)^{\beta}\left(\frac{L}{T}\right)^{\delta} \tag{5.132}$$

with solution

$$\alpha = \beta = \delta = 1. \tag{5.133}$$

A complete fluid-dynamical calculation, first carried out by Stokes, shows that the constant of proportionality in eqn (5.131) is 6π, and we arrive at *Stokes' law*

$$F = 6\pi r\eta v. \tag{5.134}$$

An important application of Stokes' law is the free fall of spherical particles under gravity, as occurs for example in Millikan's historic oil-drop experiments to determine the electronic charge. Let ϱ_S be the density of the material that forms the particles and ϱ_F be that of the surrounding fluid. The gravitational force is

$$F = m_{\text{eff}}g, \tag{5.135}$$

where the particle effective mass is

$$m_{\text{eff}} = (4/3)\pi r^3(\varrho_S - \varrho_F). \tag{5.136}$$

The falling particles accelerate until they achieve a steady terminal velocity v_T for which the viscous drag force given by eqn (5.134) exactly balances the gravitational force of eqn (5.135), so that

$$v_T = \frac{m_{\text{eff}}g}{6\pi r\eta} = \frac{2r^2(\varrho_S - \varrho_F)g}{9\eta}. \tag{5.137}$$

For spherical water droplets of radius 10^{-2} mm, such as found in a mist, falling through air ($\eta = 1.8 \times 10^{-5}\,\text{N}\,\text{m}^{-2}\,\text{s}$)

$$v_T \simeq 1.2 \times 10^{-2}\,\text{m}\,\text{s}^{-1}. \tag{5.138}$$

The falling sphere viscometer relies on the use of the formula (5.137) to obtain the viscosity η, or the kinematic viscosity in eqn (5.123), from a measurement of the terminal velocity of a standard sphere. In practice, corrections to eqn (5.137) are necessary because the viscous medium does not have an infinite extent, as assumed in the derivation of Stokes' law, and both the vertical walls and the bottom of the containment vessel influence the drag force. These and other empirical corrections are needed when the absolute viscosity is to be measured, but relative viscosities can be determined by writing eqn (5.137) in the form

$$\eta = \kappa(\varrho_S - \varrho_F)t, \tag{5.139}$$

where t is the time of fall between two fixed levels in the centre of the viscometer cylinder and κ is an instrumental constant that is taken to be the same for all liquids. The viscometer can thus be calibrated with a liquid of known density and viscosity.

5.13.3 Damping of torsional oscillations of a suspended disc

Consider a system consisting of a circular disc of radius R suspended from a torsion fibre, with the plane of the disc horizontal and lying at a height h from a flat horizontal surface. The system is immersed in a fluid of viscosity η, which dampens the rotational oscillations of the disc.

If conditions of laminar flow are assumed, the viscous drag force given by eqn (5.118) on an annulus of the disc defined by the radii r and $r + \mathrm{d}r$ takes the form

$$\mathrm{d}F = \eta 2\pi r\,\mathrm{d}r v/h, \tag{5.140}$$

where v is the linear velocity of the annulus, and the velocity gradient is taken to be constant across the space between disc and horizontal surface. Thus if v is set equal to $r\dot{\theta}$, where $\dot{\theta}$ is the angular velocity, the torque on the annulus is

$$\mathrm{d}\Gamma = r\,\mathrm{d}F = \eta 2\pi r^3\,\mathrm{d}r\dot{\theta}/h \tag{5.141}$$

and the total torque on the disc is

$$\Gamma = \frac{2\pi\eta\dot{\theta}}{h}\int_0^R r^3\,\mathrm{d}r = \frac{\pi\eta R^4}{2h}\,\dot{\theta}. \tag{5.142}$$

The equation of motion of the disc is therefore

$$\mathscr{I}\ddot{\theta} + (\pi\eta R^4/2h)\,\dot{\theta} + k\theta = 0, \tag{5.143}$$

where $\mathscr{I}$ is the moment of inertia of the disc and k is the torsional constant of the suspension. This is an equation for damped harmonic motion with an amplitude that decreases to 1/e of its initial value in a decay time

$$\tau = 4h\mathscr{I}/\pi\eta R^4. \tag{5.144}$$

This relation provides another method for measuring the viscosity η, by observing the decay time for torsional oscillations. It was first used by Maxwell for determining the viscosities of gases, where it has the advantage of avoiding complications associated with their large compressibilities. Gaseous viscosities are small compared to typical values for liquids, and Maxwell's experiments used a stack of circular glass discs suspended on a torsion fibre and set to rotate back and forth between a set of fixed glass plates in the gas container. With n discs whose upper and lower faces are both subjected to viscous drag, the decay time of eqn (5.144) is reduced by a factor of $2n$.

5.14 Settlement versus suspension: sedimentation

The analysis of § 5.13.2 suggests that any distribution of particles in a less-dense gas or liquid will eventually descend under gravity to settle and form a sediment on the bottom of the fluid container. This suggestion is correct at the absolute zero of temperature, but at finite temperatures the system tends to a steady state in which the particle density falls off exponentially with height above the bottom of the container, as we shall now show.

Let the z-axis point vertically upwards with its origin at the bottom of the container, assumed flat. The gravitational potential energy at height z is

$$m_{\text{eff}} g z,$$

where the effective mass is given by eqn (5.136), if the particles are spherical. The probability that thermal excitation at temperature T will bestow this amount of energy on a particle is given by the usual Boltzmann factor

$$\exp\,(-m_{\text{eff}} g z / k_{\text{B}} T).$$

The number of particles per unit volume is thus a function of the form

$$n(z) = n(0)\exp\,(-m_{\text{eff}} g z / k_{\text{B}} T). \tag{5.145}$$

The average particle height is calculated from

$$\langle z \rangle = \int_0^\infty z n(z)\,\mathrm{d}z \Big/ \int_0^\infty n(z)\,\mathrm{d}z \tag{5.146}$$

and a simple integration produces the result

$$\langle z \rangle = \frac{k_{\text{B}} T}{m_{\text{eff}} g} = \frac{3 k_{\text{B}} T}{4\pi r^3 (\varrho_{\text{S}} - \varrho_{\text{F}}) g}, \tag{5.147}$$

where the final form applies to spherical particles of radius r and density ϱ_{S} in a fluid of density ϱ_{F}.

These results show that, as stated above, all the particles settle on the bottom of the container at absolute zero, $T = 0\,\text{K}$, but otherwise the particles in principle remain in suspension up to a characteristic height $\langle z \rangle$. Figure 5.23 shows observed distributions of particles at four heights z, with number densities that fall off with z in accordance with eqn (5.145).

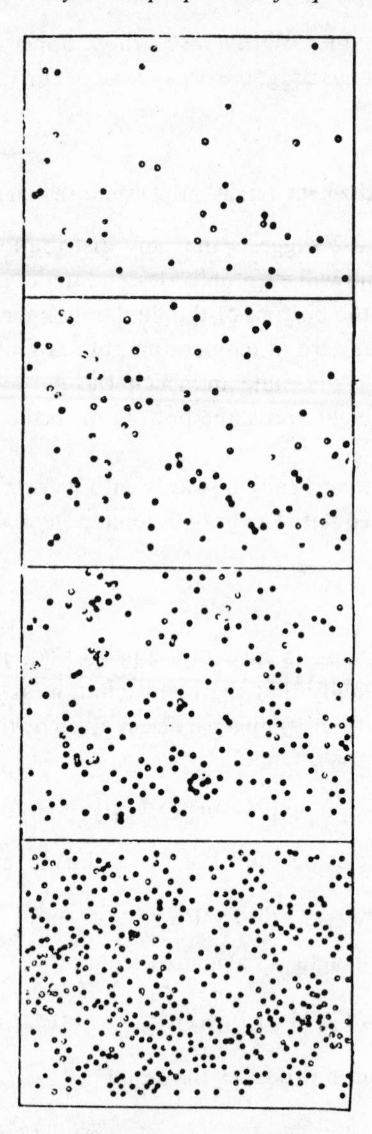

Fig. 5.23 Equilibrium distributions of gamboge particles of diameter 0.6 μm in water at heights successively separated by 10 μm (from Perrin, J. (1909). *Annales de Chimie et de Physique*, **18**, 1).

For the water droplets of radius 10^{-5} m in air, considered in § 5.13.2, the average height at room temperature is

$$\langle z \rangle = 10^{-10}\ \text{m}, \tag{5.148}$$

which is so small compared to the droplet radius that settlement effectively occurs over a time scale governed by the terminal velocity of eqn (5.138).

Consider, however, spherical particles of aluminium ($\varrho_S = 2700\,\mathrm{kg\,m^{-3}}$) of radius 0.1 μm immersed in water, where eqn (5.147) gives an average height at room temperature of

$$\langle z \rangle = 6 \times 10^{-5}\,\mathrm{m} = 60\,\mu\mathrm{m}. \tag{5.149}$$

This is again small but it greatly exceeds the particle radius and indicates an observable suspension. Furthermore, the terminal particle velocity of eqn (5.137) in this example is

$$v_T = 4 \times 10^{-8}\,\mathrm{m\,s^{-1}}. \tag{5.150}$$

This extremely small velocity corresponds to periods of the order of a week for an initially uniform distribution of aluminium particles in a typical glass beaker to settle to the average height of eqn (5.149).

Because of its role in controlling how quickly a suspension of particles in a fluid settles, the terminal velocity of eqn (5.137) is also called the *sedimentation velocity*. This is proportional to g, so that the sedimentation rate can be increased artificially by effectively increasing g. This is essentially the principle of operation of the *centrifuge*.

The analysis of the present section continues to apply for a distribution of particles with no supporting fluid, where $\varrho_F = 0$ and $m_{\mathrm{eff}} = m$, the ordinary particle mass. The number density of particles given by eqn (5.145) and the average height

$$\langle z \rangle = k_B T/mg \tag{5.151}$$

from eqn (5.147) are valid for any shape of particle. They can be applied to the important example of the earth's atmosphere, where we put $m = 4.65 \times 10^{-26}$ kg for air, assuming it to be mainly composed of nitrogen, giving

$$\langle z \rangle = 8.9\,\mathrm{km} \quad \text{at} \quad T = 293\,\mathrm{K}. \tag{5.152}$$

The number density of molecules is expected to fall off with height according to the decaying exponential in eqn (5.145). The air pressure is proportional to the density, in accordance with the discussion of § 5.5, and it falls off with the same exponential dependence. These conclusions are in fact only valid for an isothermal model of the atmosphere since T has been assumed to be constant. The variation of temperature with height in practice produces a more complicated dependence of density and pressure on z, although there are regions between 10 and 20 km and between 200 and 500 km where only small changes of temperature occur.

5.15 Films on liquid surfaces

Important modifications of liquid properties occur when the surface is covered by a thin film of a different substance. The elasticity $\mathscr{E}$ of such a film and its effect in reducing surface tension have been described in §4.18.2. As is mentioned there, the action of soaps and detergents relies on their reduction of the surface tension. The reduction also affects the propagation of the surface ripples, discussed in §§5.6 and 5.7. In the presence of a film, the surface tension γ in the surface ripple dispersion relation (5.75) is given by eqn (4.95), with a consequent reduction in the frequency ω of a wave of given surface wavevector q.

Waves on the surface of a real liquid do not have the purely sinusoidal forms represented by eqns (5.62) and (5.65), but they eventually die away because of the viscous damping of the liquid motion. The decay time, or half-life, of a surface wave in the absence of any film is given by

$$\tau_S = \varrho/2\eta q^2, \tag{5.153}$$

a formula originally derived by Stokes. The decay rate is thus directly proportional to the viscosity η. As an example, the decay time for water at the wavevector and wavelength given in eqn (5.76) is

$$\tau_S = 3.65\,\mathrm{s}\ (q = 370\,\mathrm{m}^{-1},\ \lambda = 17\,\mathrm{mm}). \tag{5.154}$$

In the presence of an oil monolayer, the film elasticity $\mathscr{E}$ defined in eqn (4.94) causes a resistance against the periodic surface expansions and contractions that accompany the wave motion. This changes the pattern of liquid flow underneath the surface in such a way that viscous forces in the bulk liquid dissipate the surface wave energy more rapidly. Thus the role of the film is to harness the intrinsic viscosity of its supporting liquid more effectively in the damping of the surface wave. The decay time is reduced in a way that depends on the magnitude of the film elasticity $\mathscr{E}$. However, in the limit of a stiff film where $\mathscr{E}$ is significantly larger than the surface tension γ, the decay time takes a relatively simple form derived by Reynolds,

$$\tau_R = (8\varrho/\eta\omega q^2)^{1/2} = 4(\tau_S/\omega)^{1/2}. \tag{5.155}$$

The numerical value to compare with eqn (5.154), assuming an unchanged ripple frequency, is

$$\tau_R = 0.82\,\mathrm{s}\ (q = 370\,\mathrm{m}^{-1},\ \lambda = 17\,\mathrm{mm}). \tag{5.156}$$

The half-life of these particular surface waves is thus reduced by a factor of about 4.5. The phenomenon provides a qualitative explanation of the calming effect of pouring oil upon troubled waters.

5.16 Turbulent flow versus streamline flow: Reynolds' number

The conditions of streamline flow assumed so far represent a rather special situation in the broad range of dynamic properties of liquids. Streamline and laminar flow tend to become unstable and break into turbulent motion as the liquid velocity is increased. The *turbulence* is usually associated with some geometrical restriction on the motion, as in the flow of liquid around a solid object placed in the flow stream, or in the flow of liquid down a pipe. The onset of turbulence is relatively easy to achieve for a liquid of low viscosity, as may be recognized by restricting the nozzle on the end of a garden hose to produce a turbulent jet of water. High viscosity produces low velocities of flow, and stable streamline behaviour is maintained, as in the example of the flow of ice in glaciers.

Reynolds performed some classical experiments on the effects of increasing velocities on the nature of liquid flow along a horizontal, circular, glass tube. As represented in fig. 5.24, he introduced a thread of coloured water into the centre of the main body of clear water flowing down the tube. The behaviour is streamline at low velocities, as shown by the straight thread in fig. 5.24(a), but at the high velocities represented in fig. 5.24(b) the motion becomes unstable with a thread that shows an eddying or turbulent behaviour. Further experiments by Reynolds showed that the flow velocity v corresponding to the onset of turbulence depends upon the viscosity η of the fluid and the radius r of the tube. He showed

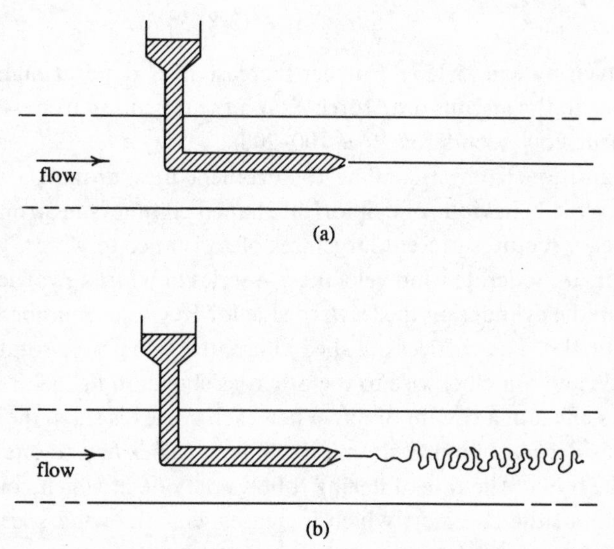

Fig. 5.24 Behaviour of coloured thread in flow of water along a tube under conditions of (a) streamline and (b) turbulent flow.

that the numerical value of the dimensionless parameter

$$R = \frac{\varrho v r}{\eta} = \frac{v r}{\nu}, \tag{5.157}$$

now known as *Reynolds' number*, determines whether streamline or turbulent flow occurs.

It is found empirically that the transition from streamline to turbulent flow in a circular tube occurs at a critical value $R \simeq 1000$ of Reynolds' number. Thus, for example, water with a kinematic viscosity $\nu \simeq 10^{-6}$ $\mathrm{m^2\,s^{-1}}$ flowing down a tube of diameter 20 mm cannot maintain a streamline flow at velocities in excess of about $0.1\,\mathrm{m\,s^{-1}}$. However, air with the much higher kinematic viscosity of $\nu = 1.33 \times 10^{-5}\,\mathrm{m^2\,s^{-1}}$ can pass through the same tube with streamline flow at velocities up to $1.33\,\mathrm{m\,s^{-1}}$. Indeed, with careful design to eliminate features that could initiate eddies, streamline flow can be maintained, even in curved pipes, to much higher values of Reynolds' number.

Similar considerations apply to the flow of liquid around an obstacle, for example a sphere, where the dimension r in Reynolds' number now characterizes the size of the obstacle. The treatment of flow around a sphere given in § 5.13.2 applies only for sufficiently low velocities when Reynolds' number is much smaller than unity. For larger values of Reynolds' number, but still less than unity, the first-order correction gives a generalization of Stokes' law (5.134) in the form

$$F = 6\pi r \eta v\{1 + (3/8)R\}, \tag{5.158}$$

with R given by eqn (5.157). Further increase in R requires higher order corrections to the viscous drag force F, and the transition from streamline to turbulent flow occurs for $R \simeq 100$–200.

The transition from streamline to turbulent flow around a sphere is relevant to the behaviour of cylindrical-shaped chimneys in strong winds. When such structures present large ares of resistance to winds, turbulent flow sets in at moderate wind velocities. A series of whirling vortices break away from the cylinder on the leeward side for Reynolds' numbers greater than about 100. The vortices are shed alternately, clockwise-rotating ones to one side and anticlockwise to the other, as shown in fig. 5.25(a) for the analogous effect in a stream of oil. When each vortex is shed, the chimney receives a sideways impulse, so that it will tend to flex first to one side and then to the other. The rate of flexing follows the rate at which vortices are spilled behind the chimney, which increases with the wind speed. If the natural vibrational frequency of the chimney is reached, resonance will occur and the ensuing larger amplitudes of motion may result in failure

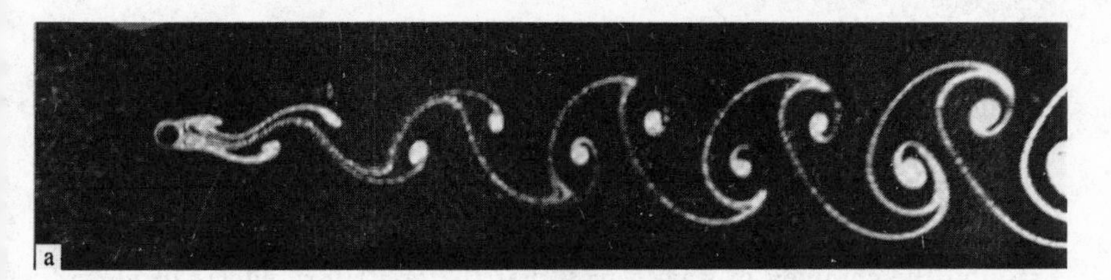

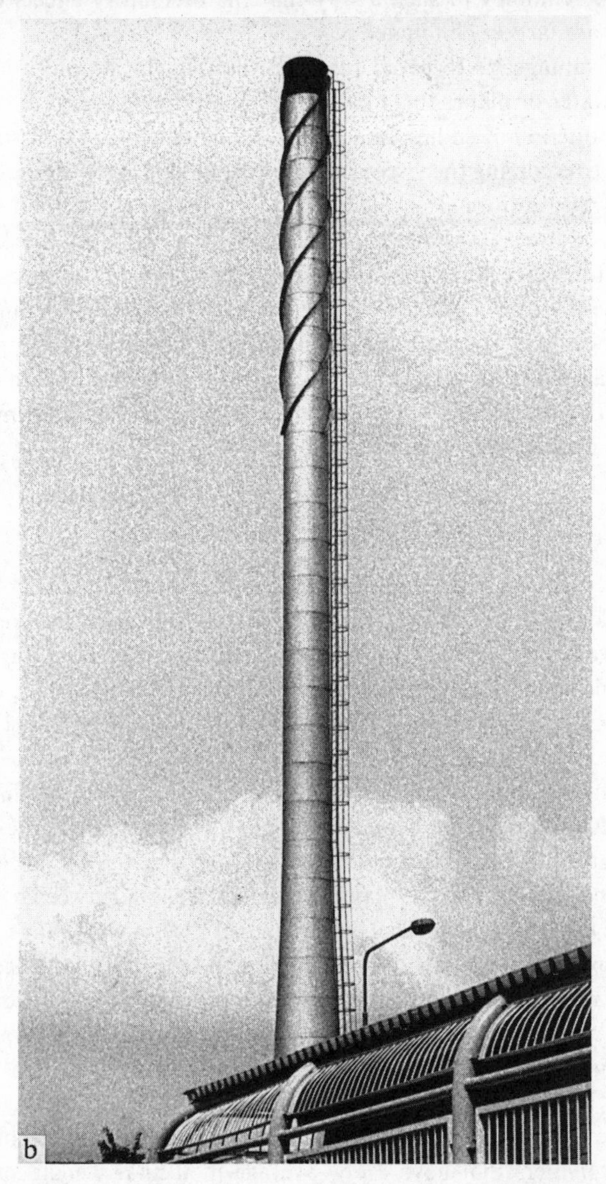

Fig. 5.25 (a) Formation of vortices in oil flowing past a vertical cylinder (from Homann, F. (1936). *Forschung auf dem Gebiete des Ingenieurwesens*, **7**, 1–10); (b) photograph of a tall, metal chimney fitted with a helical strake to reduce vortex production (at the University of Salford, courtesy R. Gerber).

of the structure. This could easily occur with lightweight chimneys, which are made from metal tubing. For this reason such industrial chimneys are often to be seen with helical metal strakes around them, as in the photograph of a tall chimney shown in fig. 5.25(b). The strakes modify the flow around the chimney in such a way that the oscillatory effects of vortex shedding are largely eliminated.

It is advantageous to avoid turbulent flow in the propulsion of ships through water or planes through the air, since kinetic energy is dissipated in the production of eddies, in addition to the unavoidable dissipation of energy in overcoming the viscous drag of the fluid. Careful design of ships' hulls and the additional use of hydrofoils gives access to high values of Reynolds' number, although the relatively low value of the kinematic viscosity for water limits the maximum speeds of water-borne craft (with streamline flow) to a fraction of the velocity of sound. Fortunately for man's aspirations with regard to flight, streamline motion through the air can be maintained to very large values of Reynolds' number, in excess of 10^6, and the achievement of high velocities is aided by the larger magnitude of the kinematic viscosity.

5.17 Non-Newtonian flow

We showed in Chapter 2 that elastic solids (i.e. materials that conform to Hooke's law) can be characterized by a set of elastic constants or moduli. These parameters were shown in § 2.4 to depend upon the strength of the interatomic bonds which produce restoring forces when atoms are displaced from their equilibrium positions. Resistance to local atomic displacements in liquids, which is a form of internal friction, is represented by the viscosity of the liquid. We saw from eqn (5.120) that η is somewhat similar to a rigidity modulus but, because of the nature of fluid flow, the equation involves strain *rate* and not merely strain.

Substances that do not display a linear relationship between σ and $\mathrm{d}\varepsilon/\mathrm{d}t$ are called *non-Newtonian*. This definition embraces a wide range of industrially-important materials. Pastes, slurries and gels have long been recognized for non-Newtonian behaviour. High polymers are a newer class of materials which are non-Newtonian. The structure of polymers depends upon their molecular weight, their composition and, of course, the temperature. Thus we find that some polymers are amorphous and some crystalline at room temperature as described in § 1.3, while others are liquids. Polymers that have a low average molecular weight have fairly short molecular chains. When such a polymer liquid is stressed, the short chains tend to straighten and become progressively disentangled, lying side by side in an orderly fashion as the stress is increased. Some bonds may rupture in achieving this disentangling but thereafter the molecules elongate and then glide past each other fairly easily. It is therefore frequently observed that flow is difficult to initiate but then the viscosity diminishes as the molecules align themselves with the flow direction. This type of behaviour also applies to many *gels* – these are concentrated colloidal solutions which 'set' when at rest at constant temperature. Gels consist of a continuous network of weakly-linked solute molecules with an interstitial liquid. Gelatine in water and rubber in benzene (rubber cement) are everyday examples.

The lowering of η with increased stress is called *thixotropy* (Greek: thixis – touch, trope – change). When allowed to rest for long times the molecules of thixotropic fluids gradually form more densely-packed networks, gelling to give a weak solid. When stirred vigorously, the cross-links are destroyed, e.g. in a traditional oil-based lead oxide paint. Thixotropic properties have been developed and exploited in many modern non-drip or gel-paints and adhesives. When worked, gel-paints break down and flow smoothly at a lower stress, facilitating 'brushing-out'. The resulting paint film starts to re-gel quickly, thus preventing 'runs' where the film is undesirably thick.

An increase of η as a consequence of increased stress (sometimes called *inverse thixotropy*) is a behaviour exhibited by some pastes, including wet sand. The flow properties of pastes, gravels, etc. depend upon the volume fraction, f of the solid particles and their shape, usually described by their aspect ratio l_r (the ratio of maximum dimension to minimum cross-sectional diameter). It is found that a dilute suspension of spheres in a fluid ($f \leqslant 0.2$) shows Newtonian behaviour and its viscosity is given by a relation attributed to Einstein

$$\eta = \eta_0\{1 + (5/2)f\}, \tag{5.159}$$

where η_0 is the viscosity of the pure fluid. However, for smooth rod-like or cigar-shaped particles eqn (5.159) is modified to

$$\eta = \eta_0(1 + bl_r^2 f), \tag{5.160}$$

where b is a constant ($\simeq 1$) which depends on whether we are considering macroscopic or microscopic particles (i.e. small shingle or fine grains of sand). In general, the effect of flow on a suspension of non-spherical particles is to rotate them so as to align their maximum dimension with the direction of flow. This effect is apparent with macroscopic particles, but with microscopic particles Brownian motion (§ 6.5) tends to oppose it by randomizing the particle orientations, thereby maintaining the friction between them, which is manifest as viscous drag.

For naturally-occurring particles, equation (5.160) gives values for η that are not normally more than ten times η_0. However, equations (5.159) and (5.160) no longer describe the viscous behaviour of a suspension when the concentration of particles becomes sufficiently high for the disturbed patterns of flow around adjacent particles to interact. The relationship between η and f then becomes nonlinear. Extreme cases are thick slurries, pastes, wet sand and wet shingle. A mixture of water and sand grains behaves as a stiff paste when $f > 0.65$ and as a Newtonian fluid when $f < 0.6$. Inverse thixotropy can be understood in the light of this finding. The increase of η caused by increased stress often results from the additional effect of the separation of two or more phases from a mixture. An applied stress squeezes out fluid from between the solid particles of a paste, so that the viscous fluid phase is much less influential than solid–solid friction. The pressure of one's foot on the wet sand at the seashore produces a dry patch as the sand grains are pressed into contact and water is displaced sideways to assist in compensatory flow of the surrounding sand.

The occurrence of stable sand-dunes on top of bends of dry adhesionless grains demonstrates that the packing density of grains is sufficient to imbue the beds with 'solid' rather than with 'fluid' properties. Settling of particles under the influence of gravity can be prevented either by immersing the bed in a fluid which has the same density as the constituent grains, or by forcing fluid up through the bed from the bottom. The fluid needs to be of low viscosity and is usually a gas. When the upwards force exerted by the gas balances the weight of the grains, the dry bed expands slightly as contacts between the grains are broken. The whole assembly – particles and gas – then assumes the properties not of a solid bed, but of a fluid of low viscosity. *Fluidized beds*, as they are called, are used in industry where it is necessary to anneal or heat-treat small components. A hot fluidized bed is advantageous over a simple furnace since it combines the heat capacity of the solid medium with the high fluidity of the gas, thus providing excellent heat transfer properties.

The use of fluids as vehicles for solid particles is common in industrial situations and it is still finding new applications. The fluid transport of particulate materials is sometimes used in the mining of coal and mineral ores, and is an important aspect of deep drilling operations, e.g. for oil. Water is pumped down through the drill tube so that a slurry (drilling mud) forms which carries the cuttings back up to the surface. Not only does the mud enable the drill to be extracted from the drill hole without damage, but mud remaining in the hole prevents the walls from collapse when drilling is suspended.

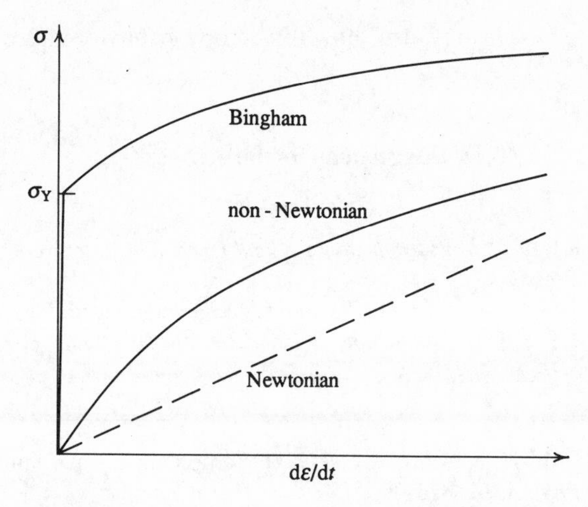

Fig. 5.26 Qualitative forms of relation between shear stress and strain rate for various kinds of flow.

Newtonian and non-Newtonian flows are represented on the stress versus strain rate diagram shown in fig. 5.26. The figure also illustrates another type of interesting behaviour called *Bingham flow*. Materials that demonstrate this are more like plastic solids than viscous fluids. A finite *yield stress* σ_Y is required to break down their internal structure and thereby to initiate flow, but above this stress they can deform with relative ease. This characteristic, which is crucial in the moulding of clay pottery, is approximately represented by the *Bingham equation*

$$\mathrm{d}\varepsilon/\mathrm{d}t = (\sigma - \sigma_Y)/\eta, \tag{5.161}$$

where σ is the applied stress and the viscosity η changes only slowly with stress. For clays σ_Y and η are typically $\simeq 200\,\mathrm{N\,m^{-2}}$ and $\simeq 20\,\mathrm{cP}$ respectively.

Clay particles form a colloidal suspension in water. When only a little water is added to dry clay, it becomes plastic (i.e. mouldable). More water increases the plasticity, the best 'modelling' properties being achieved typically at a water content of 30 vol %, corresponding to a water layer ~ 200 nm thick between the clay particles. Clays are silicates with 'layered' crystal structures and they consist of particles which are small, usually thin crystal platelets (often about 1–5 μm in diameter and $\sim 0.1\,\mu$m thick). Thin films of water molecules are bonded to the surfaces of the platelets electrostatically. At appropriate water contents, various clays respond to low stresses elastically because the weak bonds effectively linking the clay platelets are not broken. Above some threshold, the yield stress, σ_Y, they flow at a rate closely proportional to the excess stress, which is Bingham flow. However, clays may also show thixotropic behaviour above the yield stress.

5.18 Suggestions for further reading

Fluid dynamics

Batchelor, G.K. (1973). *An Introduction to Fluid Dynamics*. Cambridge: Cambridge University Press.

Landau, L.D. & Lifshitz, E.M. (1959). *Fluid Mechanics*. Oxford: Pergamon Press.

Sprackling, M.T. (1985). *Liquids and Solids*. London: Routledge & Kegan Paul.

Solitons

Olsen, M., Smith, H. & Scott, A.C. (1984). Solitons in a wave tank. *American Journal of Physics*, **52**, 826–30.

Viscosity and rheology

Cottrell, A.H. (1964). *The Mechanical Properties of Matter*. New York: Wiley.

Eirich, F.R. (1956). *Rheology*, Vol. 1. London: Academic Press.

Dinsdale, A. & Moore, F. (1962). *Viscosity and its Measurement*. London: Chapman & Hall.

Waves on liquids

Crapper, G.D. (1984). *Introduction to Water Waves*. Chichester: Ellis Horwood.

Lighthill, J. (1978). *Waves in Fluids*. Cambridge: Cambridge University Press.

Problems

1. Show that the velocity at which liquid flows from an orifice of area A in the side of a cylindrical tank of cross-sectional area B is equal to

$$\left(\frac{2gh}{1-(A/B)^2}\right)^{1/2},$$

where h is the height of the liquid surface above the orifice and g is the acceleration due to gravity. (This expression generalizes eqn (5.12), which is valid only for $A \ll B$.)

2. A tank is filled with liquid to a height H. A hole, small in comparison with the cross-sectional area of the tank, is made in one of the walls of the tank at a depth h below the surface of the liquid. Find: (a) the distance from the foot of the wall at which the stream of liquid strikes the floor on which the bottom of the tank rests; (b) whether there is another depth at which the hole could have been positioned so that the resulting stream of liquid would have had the same range and, if so, find this depth.

3. A channel has a horizontal bottom and vertical sides. In the neighbourhood

of point P, the channel is uniform and has a constant width. At some distance from P, the channel begins gradually to increase in width in a symmetrical fashion until near position Q the channel again becomes uniform and constant in breadth. Water flows steadily and smoothly along the channel from P towards Q. If v and h are respectively the velocity and depth of the water at P, and if the water has depth H at Q, prove that $H < h$ if

$$v^2 > \frac{2gH^2}{h + H}$$

where g is the acceleration due to gravity.

4. A submarine is travelling at a depth of 5 m and a velocity of 15 km h^{-1}. Assuming sea water to be incompressible and its density to be 1.0×10^3 kg m^{-3}, estimate the gauge pressure at the stagnation point at the front of the hull.

5. Prove by direct evaluation of the circulation integral in eqn (5.27), without using Stokes theorem, that the circulation C indeed vanishes for any circle that does not enclose the centre of the whirlpool, when the fluid velocity falls off inversely with radial distance, as in eqn (5.32).
(Hint: consider the contributions to the circulation of the two elements of a circle that lie on a common whirlpool radius.)

6. Prove the result embodied in eqns (5.22) and (5.23), that is

$$\nabla \times (\boldsymbol{\omega} \times \mathbf{r}) = 2\boldsymbol{\omega},$$

where $\boldsymbol{\omega}$ is the constant angular velocity of a body and $\mathbf{r}$ is the vector position of any point in the body.

7. Show that the minimum value of the liquid surface-ripple phase velocity is given by

$$(4g\gamma/\varrho)^{1/4},$$

and show that the group velocity has the same value at the corresponding wave-vector.

8. Show that the most rapid Rayleigh–Taylor instability occurs for the surface ripple of wavevector $q_m/\sqrt{3}$, where q_m is defined in eqn (5.76). Find the time taken for the initial amplitude u_0 of this ripple on an inverted water surface at $x = z = 0$ to increase by a factor of two, using the solution given in eqn (5.83).

9. A jar of salad oil is shaken vigorously, causing the oil to become turbid as a result of the formation of many tiny air bubbles. It is observed that it takes 10 minutes for the turbidity to disappear completely. Assuming that the smallest air bubbles were 0.5 mm in diameter, the depth of oil in the jar is 0.1 m, the densities of the salad oil and of air are 1.50×10^3 kg m^{-3} and 1.30 kg m^{-3} respectively, find the viscosity of the oil. (Neglect changes in hydrostatic pressure experienced by the bubbles, and interactions between the bubbles.)

10. When administering a blood transfusion, the plastic container of blood is set up so that the level of blood in it is 1.25 m above the needle inserted in the patient's arm. The needle is 30 mm long and has an internal diameter of 0.35 mm. Calculate the viscosity of blood if 4 ml is transfused into the patient in the first minute. (You may assume that the density of blood is equal to that of water.)

11. Two glass cylinders with identical forms are joined by a narrow capillary of length L and radius r, which passes horizontally between their bases. The capillary and the cylinders are filled with a liquid of viscosity η and density ϱ, but the cylinders are filled to different heights, h and $3h$. Show that the time taken to reduce the difference in the levels of the liquid to h is $4LA\eta \ln(2)/\pi r^4 \varrho g$, where A is the cross-section of each cylinder.

12. A cone is suspended, point downwards, and rotated at constant angular velocity ω, in a liquid of viscosity η. The cone has an apex half-angle θ and the apex almost touches a fixed circular plate of the same radius R as the base of the cone. The cone and plate together constitute a viscometer and the cone is subject to a torque Γ due to viscous drag. Show that (a) the rate of shear of the liquid is constant everywhere between the fixed and moving parts; (b) the shearing stress on the liquid has the value $3\Gamma/2\pi R^3$; (c) η is given by the expression

$$3\Gamma/2\pi\omega R^3 \tan\theta.$$

6
Diffusion

Systems that are not in equilibrium or in a steady state often move towards such conditions by means of diffusion processes. We consider particularly in this chapter non-uniform concentrations of different substances and non-equilibrium distributions of temperature in samples of condensed matter. The spatial and temporal changes in both mass concentration and temperature are governed by essentially the same theory, based on the diffusion equation. We first derive the diffusion equation and then apply it to various examples.

At the microscopic level, the interdiffusion of different substances involves the transport of atoms or molecules from regions where their concentration is high to regions where it is lower. The equilibration of temperature involves the transport of heat, in the form of atomic kinetic or vibrational energy, from regions of high temperature to regions of lower temperature. Our main concern is with the macroscopically observable variations in concentration and temperature to which these microscopic transport processes give rise. It is however helpful in deriving the diffusion equation to consider briefly the atomic or molecular motions involved.

6.1 Atomic picture of diffusion

Consider the diffusion of atoms through a crystal lattice. Of course, the atoms in a perfect crystal occupy all the sites on a regular lattice of points, and diffusion cannot easily occur. In order to remove an atom from one of the sites and thus create a vacancy, it costs an amount of energy, E_v say, and no such energy is available at absolute zero where all the sites should be occupied. However, at elevated temperatures T, where thermal energy

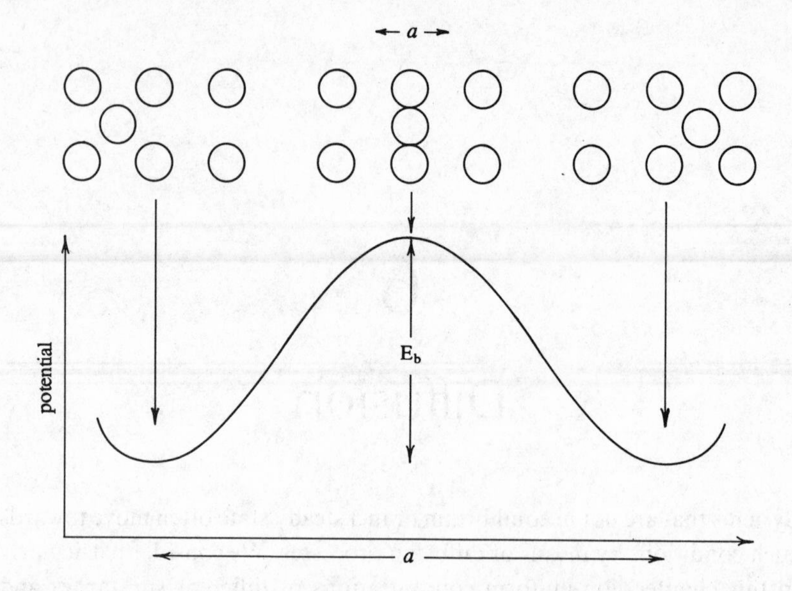

Fig. 6.1 Variation of atomic potential energy with position, showing potential minima at two equilibrium sites a distance a apart separated by a maximum of height E_b.

is available, Boltzmann's law gives a probability

$$\exp\left(-E_v/k_B T\right) \tag{6.1}$$

that a particular site may be unoccupied (see eqn (1.43)). With a finite number of vacancies, it becomes possible in principle for atoms to move about. For an adjacent atom to move onto a vacant site, further energy is in fact needed since the lattice sites are positions of minimum potential energy separated by potential barriers. Figure 6.1 shows the qualitative form of the atomic potential maximum that must be overcome for an atom to jump to a neighbouring vacant site. The requisite energy must again be provided thermally, and the rate at which jumps occur includes a further thermal factor

$$\exp\left(-E_b/k_B T\right). \tag{6.2}$$

The random jumping about that can occur in a crystal containing vacancies is the microscopic mechanism for the diffusion process. In most common crystals at room temperature, the thermal factors (6.1) and (6.2) are sufficiently small that no significant diffusion takes place, and it is generally necessary to heat solid materials towards their melting points before diffusion is easily observable. As a rule of thumb, it is usual to assume that diffusion in a solid does not become important until the temperature reaches $\sim 0.5T_{\text{M.P.}}$, where $T_{\text{M.P.}}$ is the melting point of the

(6.14) produces the three-dimensional diffusion equation

$$\partial n/\partial t = D\nabla^2 n. \tag{6.15}$$

6.4 Solution of the diffusion equation

The one-dimensional equation (6.12) has many solutions and we shall not attempt a complete treatment here. Instead we present in this section some particular solutions that are needed for the practical examples to be considered later in the chapter.

The simplest solution occurs in steady-state conditions when the concentration n does not depend on the time, so that

$$\partial n/\partial t = 0 \quad \text{and hence} \quad \partial^2 n/\partial x^2 = 0. \tag{6.16}$$

The most general steady-state spatial dependence of the concentration is thus a linear one, and it can be written in the same form,

$$n = n_0 - (J_0 x/D), \tag{6.17}$$

as eqn (6.8), where J_0 is the constant particle current and n_0 is the concentration at $x = 0$.

The simplest kind of time-dependent solution of the diffusion equation expresses the concentration as a product of a function of the time with a function of position. This is the technique of separation of the variables, and if the two functions are taken to be simple exponentials, we can put

$$n(x, t) = n_0 \, e^{\alpha t} \, e^{\beta x}, \tag{6.18}$$

where n_0, α and β are constants. Substitution in eqn (6.12) gives

$$\alpha = D\beta^2. \tag{6.19}$$

The particle density is not allowed to become infinite in real physical systems, even after an infinite time has elapsed, since the total number of particles is usually conserved. Thus α in eqn (6.18) must be negative, and since the diffusion coefficient D is always positive, eqn (6.19) implies that β must be purely imaginary. We therefore put

$$\beta = \pm ik, \tag{6.20}$$

and the solution (6.18) becomes

$$n(x, t) = n_0 \exp(-k^2 Dt) \exp(\pm ikx). \tag{6.21}$$

The real and imaginary parts of the right-hand side of this equation are also solutions, with the forms

$$\begin{aligned} n(x, t) &= n_0 \exp(-k^2 Dt) \cos kx \\ n(x, t) &= n_0 \exp(-k^2 Dt) \sin kx. \end{aligned} \tag{6.22}$$

It is easily verified that the solutions (6.21) and (6.22) satisfy the diffusion equation for any value of k, and this quantity is determined by the physical conditions of systems to which the theory is applied.

The above solutions have spatial oscillations in density that diminish with increasing time. It is also possible to find separated-variable solutions like eqns (6.18) in which the density has an exponential spatial dependence but oscillates in time. Accordingly we put

$$\alpha = -\mathrm{i}\omega \tag{6.23}$$

and the solution of eqn (6.19) for β is then

$$\beta = \pm(\omega/2D)^{1/2}(1 - \mathrm{i}). \tag{6.24}$$

Substitution back into the trial solution (6.18) gives

$$n(x, t) = n_0 \exp(-\mathrm{i}\omega t) \exp\{\pm(\omega/2D)^{1/2}(1 - \mathrm{i})x\}. \tag{6.25}$$

The complex conjugate of this function is also a solution of the diffusion equation, as are its real and imaginary parts. The latter can be written

$$\begin{aligned} n(x, t) &= n_0 \exp\{\pm(\omega/2D)^{1/2}x\} \cos\{\pm(\omega/2D)^{1/2}x + \omega t\} \\ n(x, t) &= n_0 \exp\{\pm(\omega/2D)^{1/2}x\} \sin\{\pm(\omega/2D)^{1/2}x + \omega t\}, \end{aligned} \tag{6.26}$$

and these are valid for any value of ω. The choices of cosine or sine solution, of n_0 and ω, and of the + and − signs are determined by the nature of a particular diffusing system.

More generally, the diffusion equation also has solutions in which the x and t variables cannot be separated. Consider for example the spatial variations of concentration illustrated in fig. 6.2, where the initial step-like distribution of density at $t = 0$ is replaced by the gradual fall-off at later times t. It is obvious that these two distributions cannot simply be the same function of x multiplied by the different values of a purely time-dependent function at the two times.

A simple kind of more general solution can be generated from the form of the separated solution given in eqn (6.21). Since the diffusion equation (6.12) is linear, the sum of two or more solutions like eqn (6.21) with different values of k is also a solution. An extreme form of a sum of solutions is the integral of eqn (6.21) over k, where all possible values of k are combined with equal weights to give

$$\int_{-\infty}^{\infty} \exp(-k^2Dt \pm \mathrm{i}kx)\,\mathrm{d}k = (\pi/Dt)^{1/2} \exp(-x^2/4Dt). \tag{6.27}$$

The integral in this expression is essentially the *Fourier transform* of a *Gaussian function*. Its evaluation, considered in Appendix 2, is obtained from eqn (A2.18) with suitable changes of variable. The form of solution needed for physical applications is obtained from eqn (6.27) upon multiplication by a constant and a shift in the x-coordinate by a constant amount x_0 to give

$$n(x, t) = \frac{N}{(4\pi Dt)^{1/2}} \exp\{-(x - x_0)^2/4Dt\}, \tag{6.28}$$

where N and x_0 are arbitrary. It is not too difficult to check by differentiation that this is indeed a solution of eqn (6.12).

The spatial dependence of the solution (6.28) at a given time t is a Gaussian function of x centred on coordinate x_0. The function has its maximum value at $x = x_0$, and the distribution is symmetrical around x_0. Some properties of the Gaussian function and its integrals are given in Appendix 2. Thus with the help of eqn (A2.8), integration of eqn (6.28) over the entire length of the x-axis gives

$$\int_{-\infty}^{\infty} n(x, t)\,dx = N. \tag{6.29}$$

Thus N represents the constant total number of particles in a volume of infinite extent parallel to the x-axis and unit cross-sectional area perpendicular to the x-axis. The spread of the distribution can be gauged by the mean-square particle displacement from the mean position x_0

$$\langle (x - x_0)^2 \rangle = \frac{1}{N}\int_{-\infty}^{\infty} (x - x_0)^2 n(x, t)\,dx = 2Dt, \tag{6.30}$$

where eqn (A2.12) has been used to evaluate the integral. The root-mean-square particle displacement therefore increases with the square root of the elapsed time; its zero value at $t = 0$ shows that all N particles are initially at coordinate x_0. Figure 6.5 shows the forms of the solution (6.28) for the particle concentration after 1, 10 and 100 s. The system tends towards a uniform distribution after an infinite time t.

The initial form of the density distribution (6.28) with all N particles concentrated at coordinate x_0 is known as a *delta function*. Since the solution is valid for any x_0, and since a sum of solutions of a linear differential equation is itself a solution, we can treat any arbitrary initial distribution of particles by building it up by appropriate summation of eqn (6.28). Consider an initial distribution in which there is a uniform concentration n_0 of particles for $x_0 < 0$ and zero concentration for $x_0 > 0$, the step shown in fig. 6.2. The required solution is obtained from eqn (6.28) by replacing N by n_0 and integrating x_0 along the negative x-axis

$$n(x, t) = \frac{n_0}{(4\pi Dt)^{1/2}} \int_{-\infty}^{0} \exp\{-(x - x_0)^2/4Dt\}\,dx_0. \tag{6.31}$$

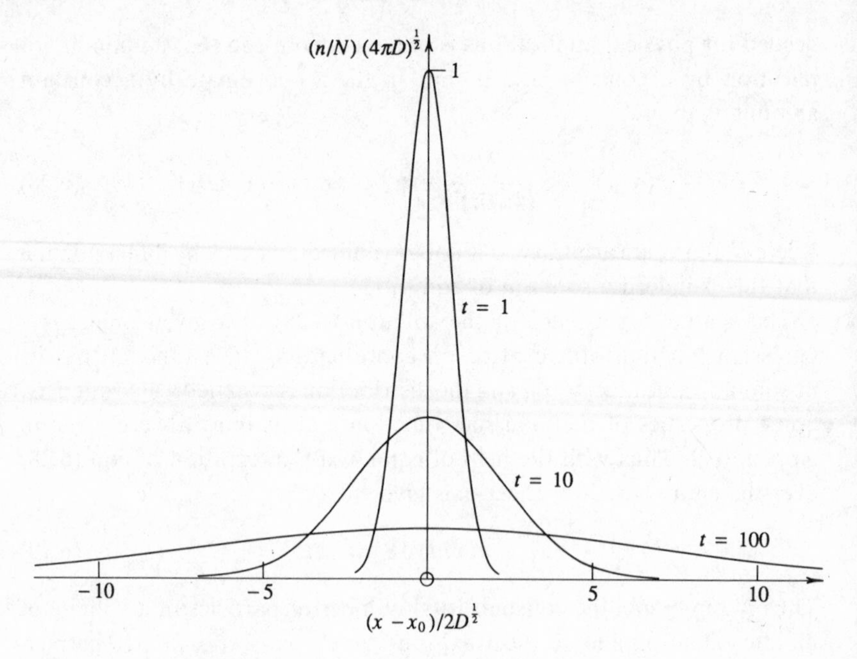

Fig. 6.5 Variations of particle concentration with position at three different times for an initial state in which N particles are located at the point $x = x_0$.

A change of variable to

$$w = (x - x_0)/(4Dt)^{1/2} \tag{6.32}$$

gives

$$n(x, t) = (n_0/\pi^{1/2}) \int_{x/(4Dt)^{1/2}}^{\infty} \mathrm{e}^{-w^2}\,\mathrm{d}w = \tfrac{1}{2} n_0 \operatorname{erfc}\left(\frac{x}{(4Dt)^{1/2}}\right), \tag{6.33}$$

where the *complementary error function* is defined as

$$\operatorname{erfc} s = (2/\pi^{1/2}) \int_{s}^{\infty} \mathrm{e}^{-w^2}\,\mathrm{d}w. \tag{6.34}$$

Some of the main properties of error functions are outlined in Appendix 2. Extensive tabulations of the functions exist (Abramowitz & Stegun, 1965), making it easy to obtain numerical values for the solution (6.33). The continuous curve in fig. 6.2, showing the density distribution that develops from an initial step, is plotted from the solution (6.33).

We now consider some applications of diffusion theory to physical systems.

6.5 Brownian motion

It has been shown in § 5.14 that particles of mass m moving vertically under gravity at temperature T form a suspension with density distribution

$$n(z) = n(0) \exp(-mgz/k_B T) \tag{6.35}$$

and average height

$$\langle z \rangle = k_B T/mg. \tag{6.36}$$

The mechanism by which suspended particles are maintained at finite elevations is their collision with atoms or molecules of the supporting fluid. Individual collisions have unobservably small effects, but the 10^{21} or so collisions that typically occur each second produce an overall motion of the suspended particles that can be observed through a microscope. The motion was first seen in pollen grains suspended in water by Brown in 1827, and its interpretation was established by Einstein in 1905. The *Brownian motion* of an individual particle has a random character, but the averaged motion of a population of similar particles obeys the diffusion equation. Indeed the derivation of the diffusion equation in §§ 6.2 and 6.3 can be reworked in terms of particle jumps caused by collisions with molecules of the supporting fluid.

The number density distribution (6.35) can be written more generally in the form

$$n(z) = n(0) \exp(-Fz/k_B T) \tag{6.37}$$

for particles subjected to a force F in the $-z$-direction. The force need not be gravitational; for example, in the case of charged particles it could be electrical. The corresponding particle flux in the direction of positive z resulting from diffusion is given by Fick's law (6.6) as

$$J = -D \frac{\partial n(z)}{\partial z} = \frac{DF}{k_B T} n(z). \tag{6.38}$$

For steady-state conditions, this flow must exactly balance the flow

$$n(z)\, v \tag{6.39}$$

in the negative z direction, where v is the velocity produced by the applied force F. Thus equating (6.38) and (6.39), we obtain the *Einstein relation*

$$v = DF/k_B T. \tag{6.40}$$

This result applies to any situation where 'backwards' diffusion balances the flow produced by an applied force field.

In the special case of spherical particles of radius r in a fluid of viscosity

η where the force and velocity are related by Stokes' law (5.134), F and v can be removed from eqn (6.40) to give

$$D = k_B T / 6\pi r \eta. \tag{6.41}$$

This is the *Stokes–Einstein expression* for the diffusion constant. Although its derivative above has considered the effects of an applied force on the particle distribution, the expression (6.41) for D is valid for the diffusion of spherical particles in the absence of any applied force, where the motion is governed by the diffusion equation (6.12) and its various solutions discussed in the previous section.

Experimental studies of Brownian motion are most simply made for diffusion of particles in a horizontal x-direction where the effects of the gravitational force can be ignored. Following the discussion of § 6.1, the averaged motions of individual particles in the random diffusion process are independent of the overall particle distribution $n(x, t)$. The predictions of diffusion theory can thus be applied to the averages of series of measurements made on single particles in a uniform distribution. Figure 6.6 shows observations of the positions of three particles at 30 s intervals made by Perrin (1909) using an optical microscope. With the Stokes–Einstein diffusion coefficient given by eqn (6.41), the mean-square particle displacement predicted by eqn (6.30) is

$$\langle (x - x_0)^2 \rangle = 2Dt = k_B T t / 3\pi r \eta, \tag{6.42}$$

an expression originally derived by Einstein. The equation involves only quantities that can be measured and it can thus be used to check theories of Brownian motion and diffusion. For the particles of approximate radius 0.2 μm immersed in water at room temperature, as in the experiments of fig. 6.6, the root-mean-square displacement after 30 s predicted by eqn (6.42) is

$$\langle (x - x_0)^2 \rangle^{1/2} = 8\,\mu\text{m}. \tag{6.43}$$

Experiments on Brownian motion give very good agreement with diffusion theory.

6.6 Carburizing of steel

The surface of steel can be hardened by diffusion of carbon into the metal from a carbon-rich atmosphere. Suppose that the initial concentration of carbon in the steel has the uniform value n_0 and that at time $t = 0$ its surface is put in contact with a gas mixture containing CH_4 or CO with a carbon concentration n_s. It is assumed that the carbon in the gas is

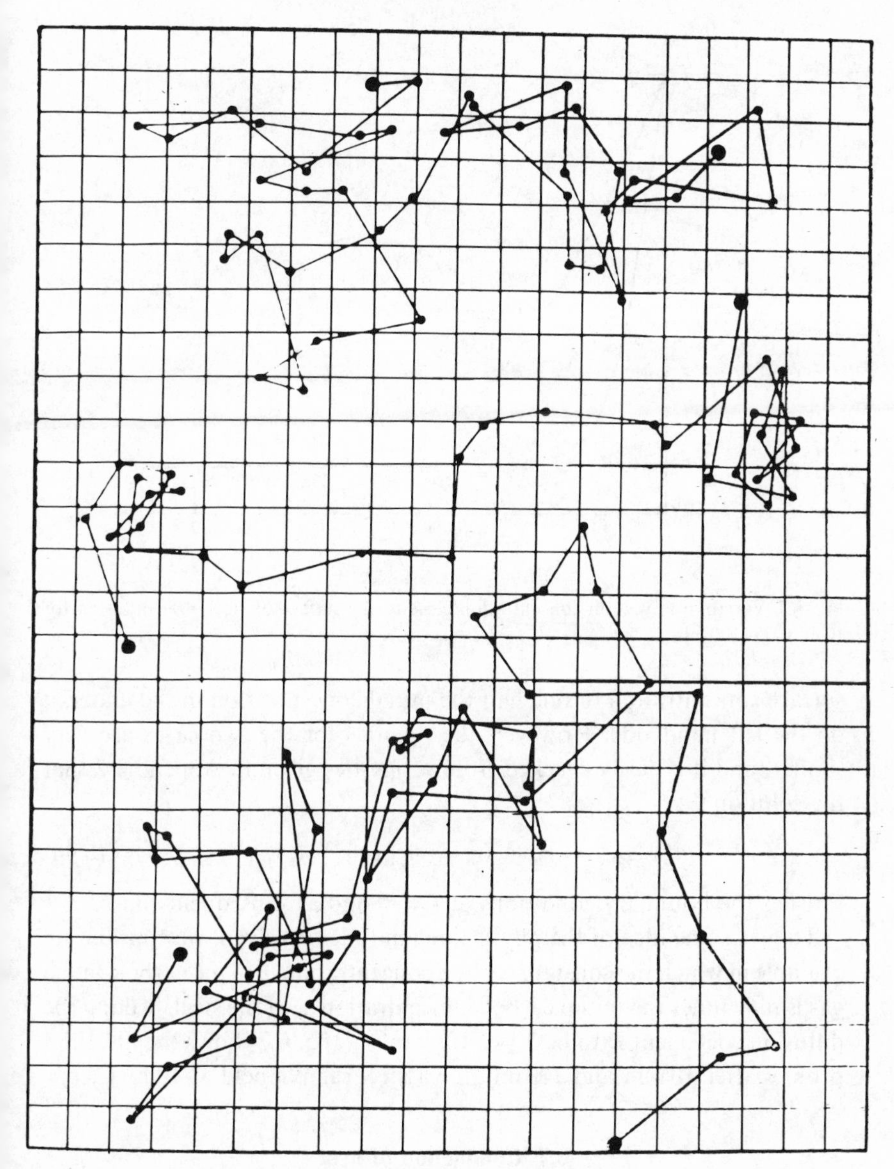

Fig. 6.6 Positions of gamboge particles in water as observed at 30 s intervals (from J. Perrin (1909). *Annales de Chimie et de Physique*, **18**, 1).

replenished so as to maintain its concentration at the surface of the steel at the constant value n_s. The conditions for this system are thus slightly different from those illustrated in fig. 6.2 where the diffusion of atoms from one material to the other produces a symmetrical distribution with a

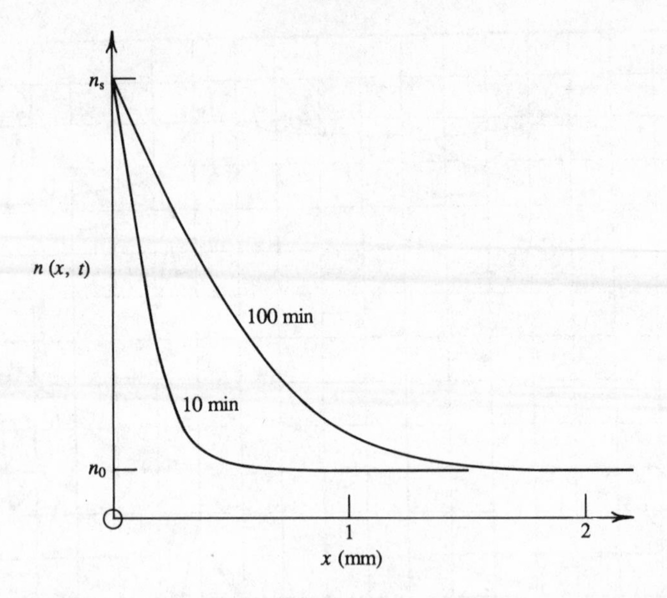

Fig. 6.7 Variation of carbon concentration close to the surface of steel after diffusion for 10 min and 100 min.

surface concentration of one half the initial concentration in the material on the left hand side. However, the theories for the two cases are very similar, and it is easily checked from properties given in Appendix 2 that the solution

$$n(x, t) = n_0 + (n_s - n_0) \operatorname{erfc} \{x/(4Dt)^{1/2}\} \qquad (6.44)$$

satisfies the boundary conditions at $t = 0$ and at subsequent times.

To give some idea of the physical magnitudes involved, we consider an example in which the constant carbon concentration n_s in the carbon-laden gas is nine times the initial carbon concentration n_0 in the steel. Taking the diffusion coefficient D to be $3.1 \times 10^{-11}\ \mathrm{m^2\,s^{-1}}$, fig. 6.7 shows the diffusion profiles after 10 min and 100 min for a steel sample held at 1000° C.

6.7 Conduction of heat

The conduction of heat is also embraced by standard diffusion theory, as we show below. In this section we neglect the other mechanisms for heat transfer, namely convection and radiation. Convection results from the relative motion of different parts of a material whose temperatures are not the same, and this mechanism is restricted to fluids. Radiation utilizes energy transfer via electromagnetic waves. The discussion is restricted to the macroscopic effects of heat conduction, and we ignore the microscopic atomic and electronic processes that are responsible for the conduction mechanism.

It is found experimentally that the heat flux Q in one dimension is proportional to the temperature gradient,

$$Q = -K_c(\partial T/\partial x), \tag{6.45}$$

where K_c is the thermal conductivity of the material and the units of Q are W m^{-2}. The negative sign ensures the flow of heat from regions of high temperature to regions of low temperature. It is seen that eqn (6.45) has the same form as Fick's law (6.6) for diffusion of matter, and as in the latter case, it is possible to prove an equation of continuity. Consider again fig. 6.4 with the particle current J replaced by the heat flow Q. The rate of flow of heat energy into the section of thickness δx and unit area perpendicular to the x-axis is, analogous to eqn (6.9)

$$-\frac{\partial Q}{\partial x}\,\delta x. \tag{6.46}$$

By energy conservation, this must equal the rate of change of the heat energy stored within the section, given by

$$\varrho C\frac{\partial T}{\partial t}\,\delta x, \tag{6.47}$$

where ϱ and C are the density and heat capacity of the material. The equality of (6.46) and (6.47) gives the continuity equation

$$\varrho C\frac{\partial T}{\partial t} = -\frac{\partial Q}{\partial x}, \tag{6.48}$$

analogous to eqn (6.11) for particle diffusion.

The equation for conduction of heat is now obtained by substitution of eqn (6.45) into eqn (6.48) to give

$$\frac{\partial T}{\partial t} = D'\frac{\partial^2 T}{\partial x^2}, \tag{6.49}$$

where

$$D' = K_c/\varrho C \tag{6.50}$$

is the *thermal diffusivity* or temperature conductivity of the material. The heat conduction equation (6.49) is mathematically identical to the one-dimensional diffusion equation (6.12), and it has an analogous three-dimensional generalization in the form

$$\partial T/\partial t = D'\nabla^2 T \tag{6.51}$$

when the temperature varies with all three coordinates.

An additional feature that can occur with heat conduction is the presence within the region of interest of sources or sinks of heat. Thus although energy, like the total number of atoms in a closed system, is a conserved quantity, there is the possibility of energy conversion from one form to another. The dynamical equation for heat energy alone can thus be generalized to

$$\frac{\partial T}{\partial t} = D'\nabla^2 T + \frac{S}{\varrho C}, \tag{6.52}$$

where S, usually a function of position, is the rate of creation (or destruction, when S is negative) of heat energy density by the sources or sinks of heat.

The equations for heat conduction and their solutions were first extensively treated by Fourier in 1822. The method of Fourier series was originally developed for the solution of these equations. In this connection, we have seen in eqn (6.22) that there exist solutions for arbitrary k with spatial dependence $\cos kx$ and $\sin kx$. Thus Fourier series consisting of sums of these basic solutions for different values of k can be used to generate more general solutions of the diffusion or heat conduction equation as in eqn (6.27). It is possible in this way to obtain solutions that match the requirements of a variety of boundary conditions. We do not treat these more general problems here but illustrate the application of the heat conduction equation to some simple systems.

6.8 Heat conduction problems

(i) Infinite slab. Consider a slab of thickness l in the x-direction and infinite dimensions in the y- and z-directions. In the steady state, when the left-hand side of eqn (6.49) vanishes, the most general spatial dependence of T is a linear one, analogous to the expression (6.17) for steady-state particle diffusion. Let

$$T = a + bx, \tag{6.53}$$

where a and b are constants. Then if one side of the slab at $x = 0$ is maintained at temperature T_0 and the other side at $x = l$ is maintained at temperature T_l, a and b are determined by the conditions

$$T_0 = a \quad \text{and} \quad T_l = a + bl. \tag{6.54}$$

Insertion of the solutions for a and b into eqn (6.53) gives

$$T = T_0 + (T_l - T_0)(x/l). \tag{6.55}$$

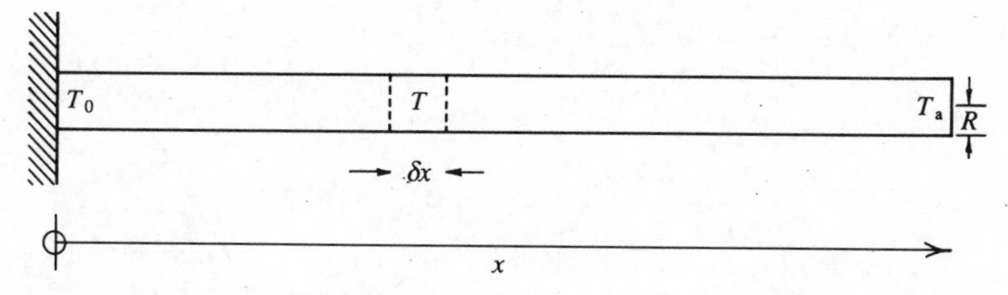

Fig. 6.8 Cross-section of circular rod viewed from the side.

(ii) Exposed uniform rod. This example illustrates the combination of radiative heat loss with heat conduction. Consider the rod of circular cross-section with radius R shown in fig. 6.8. One end of the rod, at $x = 0$, is maintained at temperature T_0 and the rod is allowed to lose heat by radiation into the surrounding air at temperature T_a. The rod is assumed to be sufficiently long that its far end is at the ambient temperature T_a, and sufficiently thin that the temperature is essentially uniform over a perpendicular cross-section.

The rate of radiative loss of heat energy from a surface at a temperature T slightly higher than its surroundings is proportional to the temperature difference and the surface area, the constant of proportionality being the *thermal emissivity* ξ. Thus the rate of change of energy per unit volume of bar for a short length δx is

$$S = -\xi(T - T_a)\frac{2\pi R\delta x}{\pi R^2 \delta x}. \tag{6.56}$$

In the steady state, the one-dimensional form of the heat conduction equation (6.52) with the thermal diffusivity inserted from eqn (6.50) gives

$$K_c \frac{\partial^2 T}{\partial x^2} = \frac{2\xi}{R}(T - T_a) \tag{6.57}$$

with solutions

$$T - T_a = A \exp\{\pm(2\xi/K_c R)^{1/2} x\}, \tag{6.58}$$

where A is a constant. The negative sign must be chosen to give $T = T_a$ at sufficiently large x, and A is fixed by the condition $T = T_0$ at $x = 0$, to give the final result

$$T = T_a + (T_0 - T_a)\exp\{-(2\xi/K_c R)^{1/2} x\}. \tag{6.59}$$

(iii) Periodic change of surface temperature. Consider a semi-infinite material in the half-space $x > 0$ whose surface in the $x = 0$ plane is subjected to a temperature differential from some mean value, whose

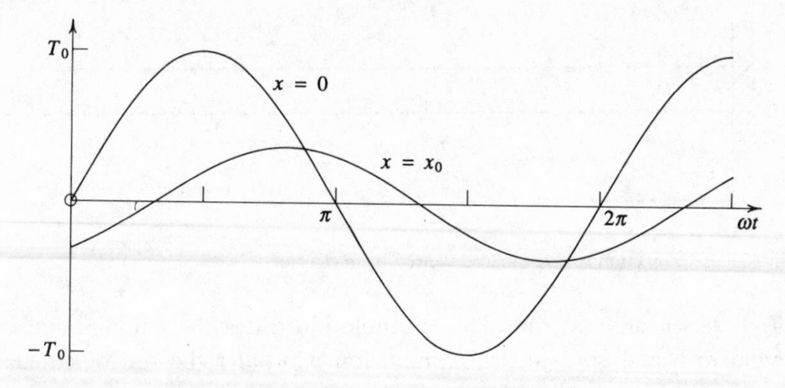

Fig. 6.9 Temperature oscillations at the $x = 0$ surface of a semi-infinite medium and at the characteristic depth, x_0.

magnitude

$$T(0, t) = T_0 \sin \omega t \tag{6.60}$$

oscillates with period $2\pi/\omega$. It would be expected that the temperature at $x > 0$ will also fluctuate with frequency ω. The appropriate solutions of the heat conduction equation (6.49) therefore have the oscillatory forms (6.26) previously obtained from the diffusion equation. The negative sign must be chosen in the exponential since the oscillations must diminish with increasing distance into the material. In addition, the sine form of solution must be chosen to match the boundary condition (6.60), and we obtain finally

$$T(x, t) = T_0 \exp\{-(\omega/2D')^{1/2}x\} \sin\{-(\omega/2D')^{1/2}x + \omega t\}. \tag{6.61}$$

Figure 6.9 shows the form of the temperature oscillations at the surface and at the characteristic depth

$$x_0 = (2D'/\omega)^{1/2} \tag{6.62}$$

at which the amplitude of the oscillations has fallen to $1/e$ of its magnitude at the surface. Note that the penetration depth decreases with increasing frequency of oscillation. The oscillations at coordinate x lag behind those at the surface by a time

$$\Delta t = x/(2D'\omega)^{1/2} \tag{6.63}$$

that also decreases with increasing frequency ω.

The above theory provides a crude model for the penetration of diurnal or annual temperature oscillations into the earth's surface.

6.9 Creep by diffusional flow: deformation resulting from the diffusion of point defects

The plastic flow of a crystalline solid resulting from the glide of dislocations on parallel sets of crystal planes was described in §§ 1.10 and 1.11. However, when solids deform in creep (§ 1.12), other mechanisms of flow may predominate. One of these, now usually called *diffusional flow* (but also known as *vacancy creep*) involves the stress-induced migration of vacancies, point defects found in crystals, introduced in § 1.11 and discussed further in § 6.1.

We describe diffusional flow here because it links topics discussed in both this chapter and the preceding one. It also emphasizes that the distinction made between solids and liquids on the basis of their viscosity and flow characteristics is somewhat arbitrary, as pointed out in §§ 1.6 and 5.12.

As mentioned in § 6.1, the processes of self diffusion become important in solids at temperatures exceeding roughly half the melting point expressed in degrees Kelvin. The equilibrium fraction of vacancies in a simple metal may then be high (typically 1/1000 at $0.8T_{\mathrm{M.P.}}$(K)), but the diffusion of vacancies within the interior of the crystal cannot alter its external shape. However, the migration of vacancies to or from the surfaces of a crystal (or discontinuities in a polycrystalline solid, e.g. grain boundaries, dislocations) does induce shape changes. The arrival or departure of a vacancy at a surface causes matter to be subtracted or added. This is the basis of vacancy creep – a slow diffusion-controlled change in shape at elevated temperatures in response to a constant applied stress.

The macroscopic elastic distortion produced by a deviatoric stress changes the symmetry of lattice sites, so that the energy barriers for the interchange of vacancies and atoms are not equally high in all directions. Thus when a single crystal, or a grain within a crystalline solid, is subjected to a non-hydrostatic stress, vacancies migrate to faces under compression and flow from faces in tension. Figure 6.10 shows the effects schematically for an idealized two-dimensional grain that is sheared by normal stresses. Atoms move along the paths indicated, so that their concentrations are non-equilibrium and differ at adjacent faces (vacancies migrate in the opposite directions). The driving force for the vacancy and atom fluxes is provided by the work done by the shear forces as the grain changes shape. The rate at which a grain of a solid undergoes creep can be calculated as follows:

Let υ be the atomic volume. Then the interatomic separation is approximately $\upsilon^{1/3}$ and the cross-sectional area of an atom is $\upsilon^{2/3}$. The force F

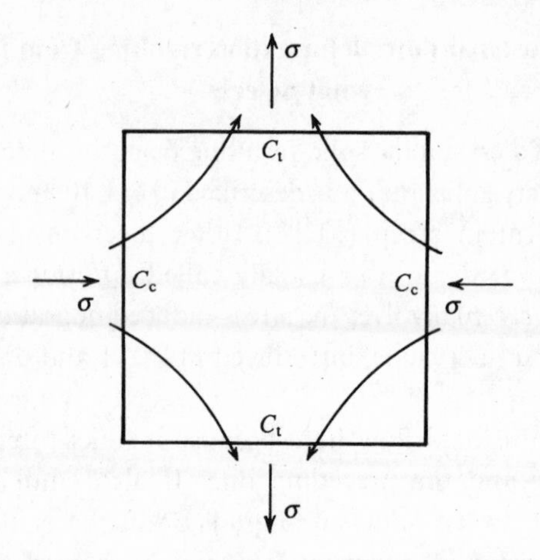

Fig. 6.10 The deformation of a single grain of a polycrystalline material by vacancy creep. The curved arrows show the directions of flow of *atoms* in response to the applied stresses, σ.

exerted on an atom in a surface in compression under a stress σ is

$$F = \sigma v^{2/3}, \tag{6.64}$$

so that the work done, W, when a single atom moves away from the surface by one interatomic distance, creating a vacancy, is

$$W = \sigma v^{2/3} \cdot v^{1/3} = \sigma v. \tag{6.65}$$

The formation energy of a vacancy at a compressed face (ignoring any difference between the formation energy of a vacancy at a surface and in the interior) is therefore

$$E = E_v - \sigma v, \tag{6.66}$$

where E_v is the energy required to create a vacancy in the unstressed grain. If the equilibrium concentration of vacancies in the unstressed grain is C_0, then from eqn (6.1)

$$C_0 = n/N = \exp(-E_v/k_B T). \tag{6.67}$$

Thus the concentration C_c of vacancies near a compressed face is

$$C_c = \exp\{(-E_v + \sigma v)/k_B T\}. \tag{6.68}$$

By the use of eqn (6.67) this becomes

$$C_c = C_0 \exp(\sigma v/k_B T) \simeq C_0(1 + \sigma v/k_B T). \tag{6.69}$$

Similarly at a face in tension, the concentration is

$$C_t \simeq C_0(1 - \sigma v/k_B T). \tag{6.70}$$

If the edge of the grain has length $2L$, so that a typical path length between adjacent faces is approximately equal to L, then the gradient of the number of vacancies diffusing across unit area per second is

$$\frac{\mathrm{d}n}{\mathrm{d}L} = -\frac{1}{\upsilon}\frac{\mathrm{d}C}{\mathrm{d}L} = \frac{C_t - C_c}{\upsilon L}$$
$$\simeq -2C_0\sigma/Lk_BT. \tag{6.71}$$

If the atomic self diffusion coefficient of the material is D, the vacancy diffusion coefficient D_v will be N/n times larger

$$\text{i.e. } D_v = ND/n = D/C_0. \tag{6.72}$$

Applying Fick's law (eqn (6.6)), we obtain the vacancy flux

$$J_v = -D_v\frac{\mathrm{d}n}{\mathrm{d}L}. \tag{6.73}$$

Substituting from eqns (6.71) and (6.72) yields

$$J_v = 2D\sigma/Lk_BT. \tag{6.74}$$

The crystal face moves distance $J_v\upsilon$ in unit time, so that the creep rate is

$$\dot{\varepsilon} = J_v\upsilon/L = 2D\upsilon\sigma/L^2k_BT. \tag{6.75}$$

This equation shows that vacancy creep is Newtonian viscous and hence the effective viscosity defined in accordance with eqn (5.120) is

$$\eta = \sigma/\dot{\varepsilon} = L^2k_BT/2\upsilon D. \tag{6.76}$$

Referring back to the general relationship between strain-rate and stress during creep (eqn (1.50)), we see that the exponent of stress is unity in this case of stress-induced diffusional flow. In the Newtonian definition of η (eqn (5.118)) the stress is a shear stress whereas our derivation above calculates a strain rate *in extension.* The equivalent viscosity for the normal stress σ used in our derivation is found to be

$$\eta = \sigma/3\dot{\varepsilon}, \tag{6.77}$$

which modifies eqn (6.76) to

$$\eta = L^2k_BT/6\upsilon D. \tag{6.78}$$

This result is known as the *Nabarro–Herring* equation. It may be used to estimate viscosities for materials under particular conditions and taking two examples we find that:
(a) if $L \simeq 10$ mm, $D \simeq 10^{-14}\,\mathrm{m^2\,s^{-1}}$ and $\upsilon \simeq 10^{-30}\,\mathrm{m^3}$, the viscosity $\eta \simeq 3 \times 10^{19}$ poise at 600° C. This is a high viscosity and the deformation

is extremely slow – the material exhibits solid properties, as we might expect.

(b) if $L \simeq 10\,\mu$m, and the other quantities are unchanged, then $\eta \simeq 1 \times 10^{13}$ poise – the behaviour of the solid becomes fluid-like. This is exactly what is observed with many very fine-grained materials, where high deformation rates can occur at moderate constant stress because the profusion of grain boundaries provides very many sources and sinks for vacancies.

When the above approach is applied to liquids, the path lengths for interchange are very small, of the order of the interatomic separation, and we take $L \simeq v^{1/3}$, which gives

$$\eta \simeq k_B T/6v^{1/3}D. \tag{6.79}$$

Substituting typical values for a liquid in this equation gives a viscosity $\simeq 10^{-2}$ poise at room temperature, which is in order of magnitude agreement with the experimental values given in Table 5.1. The form of eqn (6.79) agrees with the Stokes–Einstein relation, derived earlier (eqn (6.41)), namely

$$\eta = k_B T/6\pi r D. \tag{6.80}$$

Superficially, the last three equations appear to indicate that the viscosity increases with temperature. However, the diffusion coefficient varies with temperature according to

$$D = D_0 \exp(-E_b/k_B T), \tag{6.81}$$

where E_b is the activation energy to make an atom jump to an adjacent lattice site (see fig. 6.1). Combining eqn (6.81) with eqn (6.80) predicts that the viscosity of a solid decreases as the temperature is raised, as is generally observed.

The implications of the foregoing simple theory for stress-induced diffusional flow are far reaching. The close clearances at the tips of the hot, metal turbine blades in jet engines used to cause severe problems, with catastrophic failures when creep caused the blades to elongate. In recent years such blades (which are made by casting the metal alloy into moulds and then machining the resulting castings to high tolerances) have been forced to solidify from one end, so that all the grains of the alloy stretch from the root to the tip of the blade. Thus there are no internal surfaces (grain boundaries) transverse to the blade length, which is then unable to increase significantly by vacancy creep. Figure 6.11 illustrates the developments in the technology of overcoming the damaging effects of diffusional flow of turbine blades.

Fig. 6.11 The development of directionally-solidified impeller blades in jet turbine engines, which overcome the problems caused by diffusional flow. Left: the blade has a polycrystalline structure formed in the absence of directional solidification. Centre: blades are now made by pouring molten metal into preheated moulds which are then chilled from their bottom ends, so that crystals start to form there and then grow upwards in long columns. Diffusional flow cannot alter the length of the blades since there are no grain boundaries transverse to the length. Right: a blade in the form of a single crystal, resulting from very carefully controlled crystallization (courtesy of Dr. Maurice Gell, Materials Engineering, Pratt & Whitney, East Hartford, Connecticut, U.S.A. 06108).

6.10 Suggestions for further reading

Brownian motion

Einstein, A. (1956). *Brownian Movement*. New York: Dover Publications.

Lavenda, B.H. (1985). Brownian Motion. *Scientific American*, **252** (February), 56.

Diffusion

Clyne, T.W. (1985). *Diffusion Phenomena*: Illustrative programs for the Acorn micro-computer, model BBC B. Engineering Materials Software Series on disc. London: The Institute of Metals.

Crank, J. (1975). *The Mathematics of Diffusion*, 2nd edn. Oxford: Oxford University Press.

Jost, W. (1960). *Diffusion in Solids, Liquids and Gases*. New York: Academic Press.

Shewmon, P.G. (1963). *Diffusion in Solids*. New York: McGraw-Hill.

Walton, A.J. (1983). *Three Phases of Matter*, 2nd edn. Oxford: Clarendon Press.

Diffusional flow and creep

Herring, C. (1950). Diffusional Viscosity of a Polycrystalline Solid. *Journal of Applied Physics*, **21**, 437–45.

Poirier, J.P. (1985). *Creep of Crystals*. Cambridge: Cambridge University Press.

Heat conduction

Carslaw, H.S. & Jaeger, J.C. (1959). *Conduction of Heat in Solids*, 2nd edn. Oxford: Oxford University Press.

6.11 References

Abramowitz, M. & Stegun, I.A. (1965). *Handbook of Mathematical Functions*. New York: Dover.

Problems

1. The conversion of oxygen by an immersed electrode in a fuel cell is represented by the reaction

$$\tfrac{1}{2}O_2 + H_2O + 2e \rightarrow 2OH^-,$$

which should be assumed to be capable of maintaining a zero concentration of oxygen at the electrode surface. The flow of oxygen to the electrode is limited by diffusion through a thin static layer of electrolyte adjacent to the electrode. Calculate the rate of arrival of oxygen molecules at the electrode surface, if the diffusion layer is 6 μm thick, the oxygen concentration in the main volume of the electrolyte is 0.35 mole m^{-3} and the coefficient of oxygen diffusion in it is 2.4×10^{-19} m^2 s^{-1}.

2. Show that the rate of heat flow through a thick hollow sphere with inner and outer radii r_1 and r_2 is

$$4\pi K_c(T_1 - T_2)r_1r_2/(r_2 - r_1),$$

where K_c is the thermal conductivity, and T_1 (T_2) is the inner (outer) temperature, with $T_1 > T_2$.
Find an expression for the temperature half-way through the spherical shell.

3. Show that the rate of heat flow through a hollow cylinder of length L, and inner and outer radii r_1 and r_2 is

$$2\pi K_c L(T_1 - T_2)/\ln(r_2/r_1),$$

where K_c is the thermal conductivity, T_1 (T_2) is the inner (outer) temperature, with $T_1 > T_2$, and conduction through the ends of the cylinder is negligible. Show that the above expression reduces to that for heat conduction through an appropriate flat sheet when the cylinder wall is thin.
The passenger compartment of an executive jet airliner is essentially a tube 3 m in diameter and 20 m long. It is lined with a thermal insulating material 0.03 m thick for which K_c is 2.5×10^{-2} J m^{-1} K^{-1} s^{-1}. The ambient temperature in the compartment must be maintained at 20° C for comfort. Calculate the rate of heat input required if the external temperature is -30° C under operational conditions. (Neglect end effects.)

4. The air temperature in winter at the surface of a bare field oscillates according

to

$$T = 4 + 10 \sin \omega t,$$

where the temperature is expressed in °C and the period of oscillation is one day. If the thermal diffusivity of earth is $5 \times 10^{-7}\,\mathrm{m^2\,s^{-1}}$, find the maximum depth to which the earth freezes. If the air temperature is a minimum at 5 a.m., what is the time of day when this maximum depth of freezing occurs?

5. Prove from the form of the one-dimensional diffusion equation that the total number of particles

$$N = \int_{-\infty}^{\infty} n(x, t)\,\mathrm{d}x$$

is independent of the time t if the particle current at $x = \infty$ equals that at $x = -\infty$.

6. Consider a flow of particles parallel to the x-axis with a position-dependent velocity equal to ωx. Here ω is a constant and the velocity is therefore independent of time. Use the method of separation of variables to show that the particle concentration has solutions of the form

$$n(x, t) = Ax^{(B-\omega)/\omega} \exp(-Bt),$$

where A and B are arbitrary constants. What is the expression for the particle concentration at time t if the concentration had a uniform value n_0 at time $t = 0$?

7. The generalized equation

$$D \frac{\partial^2 n}{\partial x^2} = \frac{\partial n}{\partial t} - Cn$$

governs the particle concentration $n(x, t)$ in conditions of one-dimensional diffusion in a system where the diffusion coefficient is D and the particles are created at a rate Cn. Use the method of separation of variables to obtain a general form of solution in which the spatial contribution is oscillatory. Show that the concentration is independent of the time when the spatial oscillations have their wavelength equal to $2\pi(D/C)^{1/2}$.

8. Consider the diffusion equation (6.15) for a *two-dimensional* system in which the particle concentration $n(r, \theta, t)$ in polar coordinates is independent of θ. Find the value of the constant β such that the trial solution

$$n(r, \theta, t) = (\alpha/t) \exp(-\beta r^2/t)$$

satisfies the diffusion equation, using

$$\nabla^2 n = (1/r)\,\partial/\partial r(r \partial n/\partial r).$$

Find the value of α by normalizing the distribution for a total particle number N. The above distribution applies to particles initially at the origin. Show that the mean-square particle distance from the origin at time t is

$$\langle r^2 \rangle = 4Dt,$$

where D is the diffusion coefficient, and that the corresponding result for the x- or y-component of the particle displacement is

$$\langle x^2 \rangle = \langle y^2 \rangle = 2Dt$$

in agreement with eqn (6.30).

9. Consider the diffusion equation (6.15) for a *three-dimensional* system in which the particle concentration $n(r, \theta, \phi, t)$ in spherical polar coordinates is independent of θ and ϕ. Guess a suitable trial solution by analogy with the one-dimensional form (6.28) and the two-dimensional form given in the previous problem. Find the values of the constants in the trial solution, as in problem 8, using

$$\nabla^2 n = (1/r^2)\,\partial/\partial r(r^2\,\partial n/\partial r).$$

Hence show that the mean-square particle distances from the origin at time t are

$$\langle r^2 \rangle = 6Dt$$

and

$$\langle x^2 \rangle = \langle y^2 \rangle = \langle z^2 \rangle = 2Dt,$$

where D is the diffusion coefficient.

Appendices

1 Equations for the curvature of a surface

Consider the curve shown in fig. A1.1, with the tangents drawn at adjacent points (x, y) and $(x + \mathrm{d}x, y + \mathrm{d}y)$ on the curve separated by a distance $\mathrm{d}s$, so that

$$(\mathrm{d}s)^2 = (\mathrm{d}x)^2 + (\mathrm{d}y)^2 \tag{A1.1}$$

and

$$\mathrm{d}s/\mathrm{d}x = \{1 + (\mathrm{d}y/\mathrm{d}x)^2\}^{1/2}. \tag{A1.2}$$

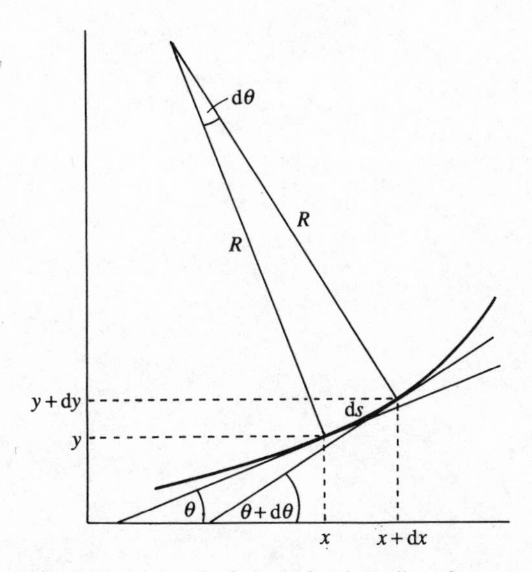

Fig. A1.1 The method of deriving the formula for the radius of curvature.

The slope of the first tangent is given by

$$\tan\theta = \mathrm{d}y/\mathrm{d}x \tag{A1.3}$$

and differentiation of this equation with respect to s gives

$$\sec^2\theta\,\mathrm{d}\theta/\mathrm{d}s = (\mathrm{d}^2y/\mathrm{d}x^2)(\mathrm{d}x/\mathrm{d}s). \tag{A1.4}$$

Normals to the tangents intersect at an angle $\mathrm{d}\theta$ at the centre of curvature. The radius of curvature R therefore satisfies

$$R\,\mathrm{d}\theta = \mathrm{d}s \quad \text{or} \quad 1/R = \mathrm{d}\theta/\mathrm{d}s. \tag{A1.5}$$

Thus, noting that

$$\sec^2\theta = 1 + \tan^2\theta = 1 + (\mathrm{d}y/\mathrm{d}x)^2, \tag{A1.6}$$

use of eqns (A1.2) and (A1.4) gives the curvature in the form

$$\frac{1}{R} = \frac{\mathrm{d}^2y/\mathrm{d}x^2}{\{1 + (\mathrm{d}y/\mathrm{d}x)^2\}^{3/2}}. \tag{A1.7}$$

An approximation valid for curves whose gradients are always small is

$$1/R \simeq \mathrm{d}^2y/\mathrm{d}x^2 \quad \text{for} \quad \mathrm{d}y/\mathrm{d}x \ll 1. \tag{A1.8}$$

2 Gaussian integrals

The basic Gaussian function has the form

$$G(x) = \exp(-\lambda x^2). \tag{A2.1}$$

Let

$$I_0 = \int_{-\infty}^{\infty} \exp(-\lambda x^2)\,\mathrm{d}x. \tag{A2.2}$$

It is difficult to perform this integral by standard methods but it can easily be evaluated by means of a trick. Since x is a dummy variable we can just as well write

$$I_0 = \int_{-\infty}^{\infty} \exp(-\lambda y^2)\,\mathrm{d}y, \tag{A2.3}$$

and the product of eqns (A2.2) and (A2.3) gives

$$I_0^2 = \int_{-\infty}^{\infty} \mathrm{d}x \int_{-\infty}^{\infty} \mathrm{d}y \exp\{-\lambda(x^2 + y^2)\}. \tag{A2.4}$$

The integration extends over the entire two-dimensional xy-plane and it is easier to perform in polar coordinates ϱ and θ where

$$x = \varrho \cos\theta, \quad y = \varrho \sin\theta \tag{A2.5}$$

and the element of area is replaced according to

$$\mathrm{d}x\,\mathrm{d}y \rightarrow \varrho\,\mathrm{d}\varrho\,\mathrm{d}\theta. \tag{A2.6}$$

Thus with care in fixing the limits of integration,

$$I_0^2 = \int_0^{\infty} \mathrm{d}\varrho \int_0^{2\pi} \mathrm{d}\theta\,\varrho \exp(-\lambda\varrho^2) = \pi \int_0^{\infty} \exp(-\lambda\varrho^2)\,\mathrm{d}(\varrho^2) = \frac{\pi}{\lambda}. \tag{A2.7}$$

The required integral is therefore

$$I_0 = (\pi/\lambda)^{1/2}, \tag{A2.8}$$

and it can be used to put the Gaussian function (A2.1) into a normalized form

$$g(x) = (\lambda/\pi)^{1/2} \exp(-\lambda x^2), \tag{A2.9}$$

which satisfies

$$\int_{-\infty}^{\infty} g(x)\,\mathrm{d}x = 1. \tag{A2.10}$$

This normalized function is plotted in fig. A2.1.

The basic integral can be used to evaluate more complicated integrals

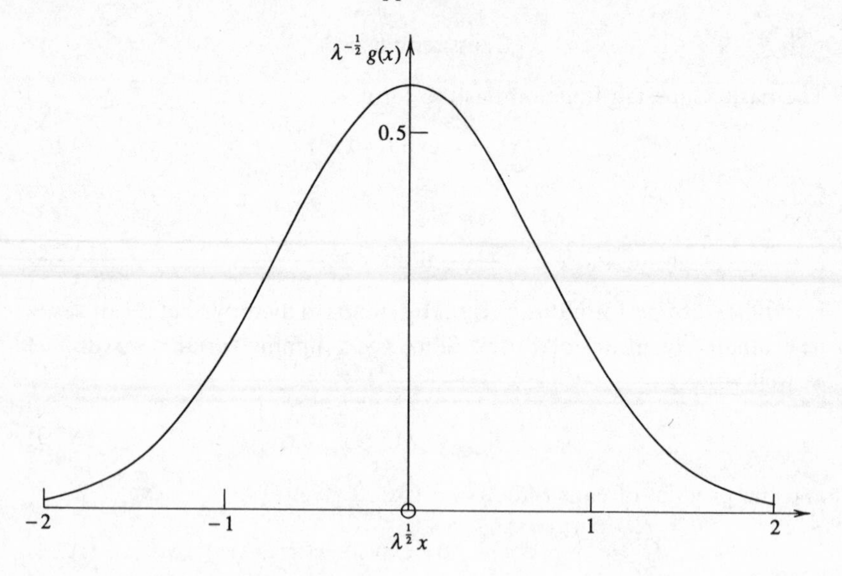

Fig. A2.1 Normalized Gaussian function.

such as

$$I_2 = \int_{-\infty}^{\infty} x^2 \exp(-\lambda x^2)\,\mathrm{d}x. \tag{A2.11}$$

This can be integrated by parts but it is much easier to use the differential property

$$I_2 = -\frac{\mathrm{d}}{\mathrm{d}\lambda}\int_{-\infty}^{\infty} \exp(-\lambda x^2)\,\mathrm{d}x = -\frac{\mathrm{d}}{\mathrm{d}\lambda}(\pi/\lambda)^{1/2} = \frac{\pi^{1/2}}{2\lambda^{3/2}}. \tag{A2.12}$$

Similarly

$$I_4 = \int_{-\infty}^{\infty} x^4 \exp(-\lambda x^2)\,\mathrm{d}x = -\frac{\mathrm{d}}{\mathrm{d}\lambda} I_2 = \frac{3\pi^{1/2}}{4\lambda^{5/2}} \tag{A2.13}$$

and so on.

All of the above integrals cover the range of the entire x-axis from $-\infty$ to ∞. Since the Gaussian function is symmetrical about $x = 0$, integrals over the ranges $-\infty$ to 0 or 0 to ∞ have one-half the values given above. More general ranges of integration cannot easily be evaluated since the trick of integration over the xy-plane in polar coordinates does not work when only a finite square area is to be covered. It is necessary in this case to resort to numerical tabulations. The *error function* is defined to be

$$\operatorname{erf} s = (2/\pi^{1/2}) \int_0^s \exp(-x^2)\,\mathrm{d}x = 2\int_0^s g(x)\,\mathrm{d}x \tag{A2.14}$$

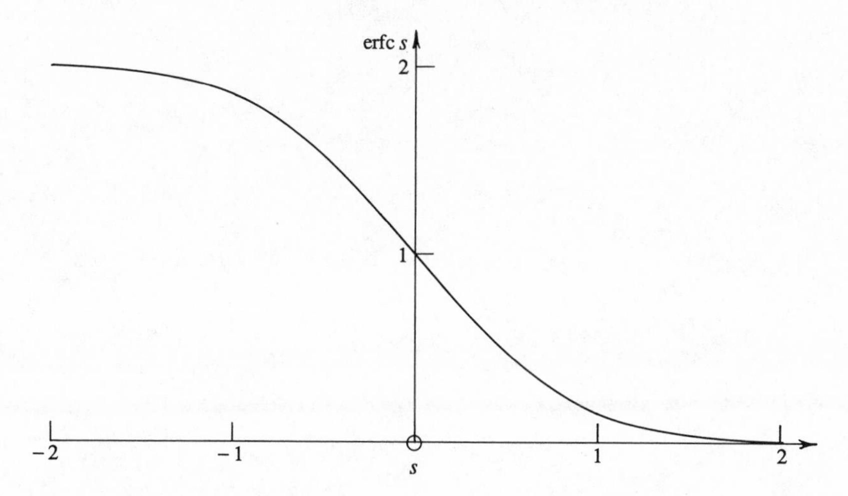

Fig. A2.2 The complementary error function.

and the *complementary error function* is

$$\text{erfc}\ s \;=\; (2/\pi^{1/2})\int_s^{\infty}\exp(-x^2)\,\mathrm{d}x \;=\; 2\int_s^{\infty} g(x)\,\mathrm{d}x, \quad \text{(A2.15)}$$

where $g(x)$ is the normalized Gaussian function (A2.9) with λ now set equal to unity. From eqn (A2.10) and the symmetry of $g(x)$

$$\text{erf}\ s + \text{erfc}\ s \;=\; 1; \quad \text{(A2.16)}$$

it is therefore only necessary to tabulate one of the functions and erf s is usually chosen. However, it is erfc s that is needed for the applications of Chapter 6, and its form is illustrated in fig. A2.2. Some special values available from the above elementary theory are

$$\begin{aligned} \text{erf}(-\infty) &= -1 \qquad & \text{erfc}(-\infty) &= 2 \\ \text{erf}\ 0 &= 0 & \text{erfc}\ 0 &= 1 \\ \text{erf}\ \infty &= 1 & \text{erfc}\ \infty &= 0. \end{aligned} \quad \text{(A2.17)}$$

The Fourier transform of the Gaussian function is

$$\int_{-\infty}^{\infty} G(x)\exp(\mathrm{i}qx)\,\mathrm{d}x \;=\; \int_{-\infty}^{\infty}\exp\{-\lambda(x-\mathrm{i}q/2\lambda)^2\}\exp(-q^2/4\lambda)\,\mathrm{d}x. \quad \text{(A2.18)}$$

The integral is the same as I_0 in eqn (A2.2) after a simple change of variable, and use of eqn (A2.8) thus gives

$$\begin{aligned} \int_{-\infty}^{\infty} G(x)\exp(\mathrm{i}qx)\,\mathrm{d}x &= \int_{-\infty}^{\infty}\exp\{-\lambda x^2 + \mathrm{i}qx\}\,\mathrm{d}x \\ &= (\pi/\lambda)^{1/2}\exp(-q^2/4\lambda). \end{aligned} \quad \text{(A2.19)}$$

Index